# AY's Neuroanatomy of C. elegans for Computation

Authors

**Theodore B. Achacoso, M.D.**

**William S. Yamamoto, M.D.**

Department of Computer Medicine
School of Medicine
The George Washington University
Washington, D.C.

Atlas, Edition 1
Software Database, Version 1.1

CRC Press
Taylor & Francis Group
Boca Raton London New York

CRC Press is an imprint of the
Taylor & Francis Group, an **informa** business

T0175057

CRC Press
Taylor & Francis Group
6000 Broken Sound Parkway NW, Suite 300
Boca Raton, FL 33487-2742

Reissued 2019 by CRC Press

A Library of Congress record exists under LC control number:

Publisher's Note
The publisher has gone to great lengths to ensure the quality of this reprint but points out that some imperfections in the original copies may be apparent.

Disclaimer
The publisher has made every effort to trace copyright holders and welcomes correspondence from those they have been unable to contact.

ISBN 13: 978-0-367-25025-6 (hbk)
ISBN 13: 978-0-367-25028-7 (pbk)
ISBN 13: 978-0-429-28562-2 (ebk)

Visit the Taylor & Francis Web site at http://www.taylorandfrancis.com and the
CRC Press Web site at http://www.crcpress.com

# PREFACE

Why would anyone want to tabulate neuroanatomical data on a nearly microscopic nematode, with virtually no associated neurophysiological observation, for computer use? More significantly, why would anyone want to tabulate illustration and textual information that already are so elegantly and meticulously catalogued? We brought this volume together as a companion to the software, so anyone dazzled by the power of computation could organize, rearrange, and simulate speculation upon data of a real nervous system. The venture forced upon us the realization that a network of about 300 interconnected objects is not amenable to pencil and paper speculation, nor accessible to manipulation without computer assistance ranging from simple redisplay to topological deformations and computation. In contrast to the study of artificial networks, one hopes that the data may encourage, through computation, the development of new methods for studying neural organization.

The size of the nematode nervous system leads to a database small enough for use by individuals on the small personal computer. Students, individual scientists, or philosophically oriented technologists can pursue an exploration of the morphological basis of the behavior of an organism. It is as if we had come upon the blueprint for an exotic Venusian computer chip, and propose to reverse engineer it. Or as neurologist and physiologist, to indulge our perhaps fatuous belief that *C. elegans* has a modicum of sentience --- a concept whose formal properties might be quantitatively described not by dissection but by study of anatomical gestalt projected upon the wall of a small cave.

After perusing some of the literature on artificial neural network simulation, it struck us as a refreshing change of tactics in the quest, to find an information processing system to which the property of sentience might be ascribed. Having no pineal, the nematode must display its repertoire of behavior and its sentience through the functioning of its nerve net --- a small net by standards of computer chips and nearly all other animals. Yet, it is the largest map of a complete system that has been fully described.

Several caveats accompany our work. First, the data are incomplete because in the choice between exhaustively collecting and collating data and speculating, we chose the latter. We are neither nematologist or microscopist, and no material derives from original observations on *Caenorhabditis elegans*. Except for those inconsistencies that lead to numerical or data format errors, no effort was made to correct or contribute to data available in library sources. Furthermore, no concordance was attempted where parallel data pertained to the same structure in different specimens. Because of this, most certainly any observations made from the data will require recomputation after each improvement on the quality of data on the file. Still, it is our belief that certain generalizations about design and pattern that we and others may make in consequence of computational manipulation have value. Numerical values we include were computed at various times in the completion of the data here collected. While perspectives may be sound, we are intensely aware that detailed recomputation is advisable with each anatomical updating of data.

Inaccuracies and deficiencies in this book may be immediately obvious to many specialists. To such errors, we are glad to make corrections, expressing gratitude for bringing them to our attention. We apologize for such errors and any misrepresentations that may have resulted from our inadvertence and ignorance. We invite, look forward to criticism and advice for improvement of this database.

Although the laboratory bench is still the altar of science in the quest for new knowledge, the end of the twentieth century finds human knowledge in such a state of rich poverty that some small alcove in the temple may be reasonably set aside for Scholastic contemplation for the organization of knowledge into understanding. It is in that alcove where we wish to plant this volume as a pennon for the heretics.

We gratefully acknowledge support and encouragement from the Department of Computer Medicine of The George Washington University, and our colleagues, and students in preparing this work.

Theodore B. Achacoso
William S. Yamamoto

Washington, D.C.
1991

# AUTHORS

**Theodore B. Achacoso, M.D.** (born 1961, Iba, Zambales, Philippines) received his Bachelor of Science degree in Biology from the University of the Philippines - Diliman at the age of 18, and his Doctor of Medicine degree from the University of the Philippines - Manila at the age of 22. After medical internship at the Philippine General Hospital, Dr. Achacoso trained and researched in neuroscience, particularly in computer-assisted neuroimaging, and in pharmacology, particularly in fertility regulation. He was Clinical Assistant Professor of Neurology and Science Research Specialist of Pharmacology at the College of Medicine of the University of the Philippines - Manila when he came to the United States as Visiting Assistant Professor of Medical Informatics at the Department of Computer Medicine of The George Washington University (Washington, D.C.) in 1988.

In the United States, Dr. Achacoso has taught, pursued research, and published in medical computation and imaging, in theoretical and computational neurobiology, and in biomathematical modeling and reverse engineering of the *Caenorhabditis elegans* nervous system. He introduced the terms *artificial ethology* and *computational neuroethology*, represented the synaptic and electrotonic connections of the *C. elegans* nervous system into matrix form, and investigated properties of this natural neural network. Dr. Achacoso is with the faculty of the Department of Computer Medicine of The George Washington University School of Medicine.

**William S. Yamamoto, M.D.** (born 1924, Cleveland, Ohio, U.S.A.) received his A.B. degree in Chemistry from Park College, Parkville, MO in 1945, and his Doctor of Medicine and Master of Science (Hon.) degrees from the University of Pennsylvania, Philadelphia in 1949 and 1971, respectively. Dr. Yamamoto is a respiratory physiologist, and has taught physiology, biomathematics, biomedical engineering, statistics, and related medical subjects at the University of Pennsylvania, UCLA, and The George Washington University. His research activities are in mathematical modeling of physiological systems, in medical computation, and in real and artificial nervous systems. Most of his publications are in the control of respiration in mammals, some are in health care industry computing, especially in communications, and recent ones include the identification of verbs in computer programs of physiological models.

Dr. Yamamoto is a professor at the Department of Computer Medicine (formerly the Department of Clinical Engineering) of The George Washington University School of Medicine, and was chairman of this department from 1971 to 1989. He is on the editorial board of *IEEE Transactions on Biomedical Engineering* and *Computers in Biomedical Research*, among others, and was on the editorial board of *Annual Review of Biophysics and Bioengineering*. Dr. Yamamoto was with the National Neural Circuitry Database Task Force, National Academy of Science, and is a member of the Biotechnology Resources Review Committee, National Institutes of Health.

# TABLE OF CONTENTS

# DEDICATION

Para kay Inay

-T.B.A.

To the gods and disciples of small science

-W.S.Y.

Chapter I

# Caenorhabditis elegans in Computational Research

> *The Third approached the animal,*
> *And happening to take*
> *The squirming trunk within his hands,*
> *Thus boldly up and spake:*
> *'I see,' quoth he, 'the Elephant*
> *Is very like a snake!'*
> -The Blind Men and the Elephant, J.G. Saxe

## A. A Brief History

"...we must move on to other problems of biology which are new, mysterious and exciting...the fields which we should now enter are development and the nervous system," reads Sydney Brenner's draft paper, after a long series of conversations with Francis Crick in late 1962. Thus, in a letter dated June 5, 1963 addressed to his chief, Max Perutz, head of the Medical Research Council's Laboratory of Molecular Biology in Cambridge, England, he wrote "...I would like to tame a small metazoan organism to study development directly." It was not until 1965, however, that Brenner chose the free-living nematode *Caenorhabditis elegans* over *Caenorhabditis briggsiae* which he originally proposed in 1963 to be a model animal for a concerted genetic, ultrastructural, and behavioral investigation of development and function in a simple nervous system.

The nervous system of insects was Brenner's next target after taming this nematode, but this was not to be. Nearly three decades from inception of "the worm project," *Caenorhabditis elegans* has become one of the most exhaustively studied model organisms in biology. The initial phase of "Sydney's madness" was spent in 4 to 5 years of genetic studies on the worm. By then, his "intellectual progeny" became afflicted. Their research in the 1970's produced an anatomy of *C. elegans* at the electron microscopic resolution, which provides a "wiring diagram" of the cell contacts of the animal, especially of its nervous system. Further research in the 1980's produced a documentation of complete cell lineages for both sexes, from zygote to adult, including locations and characteristics of all somatic cells in the adult. Biochemical and genetic studies abound in the 1990's, including a 10-year major effort to work out its full nucleotide sequence of 100 million bases.

This brief history of the use of *C. elegans* as an organism in formal biological research was summarized from references [7] and [12].

## B. The Organism

*Caenorhabditis elegans* (Phylum Nematoda, Class Secernentea or Phasmidia, Order Rhabditida, and Family Rhabditidae) is an one millimeter, transparent, free-living soil nematode found commonly in many parts of the world. A mature worm lives for about 17 days after reaching adulthood. *C. elegans* follows the typical nematode body plan of two concentric tubes, separated by a space called the pseudocoelom. The inner tube is in the intestine, the intestine and the gonad are in the pseudocoelomic space, and the outer tube consists of the cuticle, hypodermis, neurons, and muscles. In a live worm, shape is maintained by internal hydrostatic pressure.

An adult *C. elegans* is either a male, which exhibits mating behavior, or a hermaphrodite which lays eggs. The male tail is fan-shaped with 18 sensory rays, and at its base are two spicules which are inserted into the hermaphrodite vulva to aid in transfer of sperm during copulation (Figure 1.1). The hermaphrodite reproductive system consists of a bilobed gonad connecting to a common uterus that opens exteriorly through a ventrally protruding vulva.

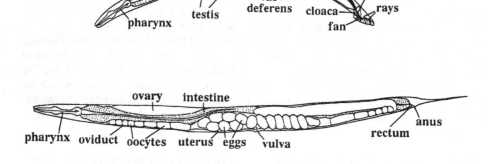

Figure 1.1 Line drawings of *C. elegans*. The male worm is illustrated at the top, and the hermaphrodite is shown at the bottom. Reprinted from reference [12], p. 2, with permission from reference [8].

*C. elegans* feeds on bacteria, which are crushed by the second lobe of a bilobed pharynx that pumps the food into the intestine (Figure 1.1). The central intestinal lumen connects to the anus near the tail. Probably also responsible for osmoregulation, the ventrally located excretory system consists of a pair of excretory canals, two side arms of a single, large, H-shaped cell which run the length of the animal and open exteriorly through a duct that connects to an anterior excretory pore located ventral to the terminal bulb of the pharynx.

On solid media, *C. elegans* crawls on one side. It propels itself forward or backward by undulatory movements. Locomotion is served by multi-sarcomere, obliquely striated body-wall muscle cells that are arranged lengthwise in four strips, two dorsal, and two ventral. Single sarcomere muscle cells are found in the alimentary system and sex muscles.

381 neurons with 92 glial/support cells and 302 neurons with 56 glial/support cells comprise the male and hermaphrodite nervous systems, respectively. The general structure of the *C. elegans* nervous system is made up of two units. 20 cells of a nerve

ring are contained in the pharynx, and is the central region of the neuropil in the animal. The other unit is composed of the rest of the neuropil, which are (1) a ventral cord, the main process bundle that emanates from the nerve ring, (2) a dorsal cord, axons of motor neurons that originate in the ventral cord and enter the dorsal cord via commissures, and (3) four sublateral processes that run anteriorly and posteriorly from the nerve ring (Figure 1.2). In this worm, neurons are grouped into the 9 identified nervous system ganglia.

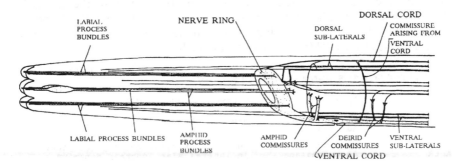

Figure 1.2 The two general units of the *C. elegans* nervous system. A nerve ring of 20 cells at the pharynx is the central region of the neuropil in the worm. The ventral cord and sublateral processes that emanate from the nerve ring, including the dorsal cord which arises from the ventral cord, comprise the other unit of the neuropil. Reprinted from reference [11], p. 17, with permission.

Aside from mating or egg-laying, feeding, and locomotion, *C. elegans* responds to mechanical, chemical, thermal, and osmolar stimuli by attraction or avoidance behavior. In a temperature gradient, it tends to remain at the temperature to which it was previously exposed. When food supply is short, immature *C. elegans* can transform into a "dauer" (or enduring) larva state that does not feed, resists desiccation, and can survive until 3 months. A dauer larva resumes normal development when food becomes available. For its anatomical and genetic simplicity, *C. elegans* displays a wide range of behavior. With the detail with which it has been described, it may be possible to correlate the behavioral repertoire with the known neuroanatomy, and to understand the development and wiring of the nervous system that underlies its behavior.

This section on anatomy and development of *C. elegans* was summarized from reference [12].

## C. A Model for Studying Nervous Systems by Computation

Since the details of anatomical parts and developmental processes of *C. elegans* provided in literature translate quite readily to mathematical representation, they can be manipulated computationally by following rules specific for the selected mathematical object into which the anatomical or developmental data have been represented. This makes *C. elegans* an interesting and useful computational model for the purpose of discerning generalizable patterns in its anatomy and development.

4

## C.1. *Data Collection and Refinement*

A segment of the complete parts list of the worm, as it appears in literature, is shown on Figure 1.3. The list accounts for all terminally differentiated somatic cells of the adult worm, both male and hermaphrodite. In this list is also a summary of each cell's lineage, coded in a meaningful alphabetical string, such that one can trace the way by which a cell is derived from a key blast cell (an embryonic founder cell or a blast cell present at hatching) by sequentially interpreting the letters of the string. The PARTLIST FILE SET (Chapter III, pp. 165-210) is an expanded and slightly modified version of this list, together with important comments about its consistency. The complete parts list is available electronically in the disks.

```
PARTS LIST FOR THE NEMATODE C. ELEGANS

AB              P₀ a              Embryonic founder cell

ADAL            AB plapaaaapp     Ring interneurons
ADAR            AB prapaaaapp

ADEL            AB plapaaaapa     Anterior deirids, sensory receptors
ADER            AB prapaaaapa     in lateral alae, contain dopamine

ADEshL          AB arppaaaa       Anterior deirid sheath cells
ADEshR          AB arpppaaa       "

ADEso           H2 aa        L&R  Anterior deirid socket
```

Figure 1.3 A segment of the *C. elegans* parts list. These first few lines appear exactly as in literature. Left column = part name, middle column = cell lineage string, and right column = part description. Reprinted from reference [12], p. 417, with permission.

There is mostly complete and detailed information on the embryology of *C. elegans*, especially on its cell generations and lineages. By referring to cell lineage charts, the cell lineage alphabetical strings in Figure 1.3 can easily be understood. A portion of the *C. elegans* cell lineage tree, showing the divisions of embryonic key blast cell labelled "D" to its terminus as body muscle is shown in Figure 1.4. The consistency of the cell lineage strings in the parts list (PARTLIS1.PRN in Chapter III, pp. 171-190) were cross-checked with these cell lineage trees provided in literature.

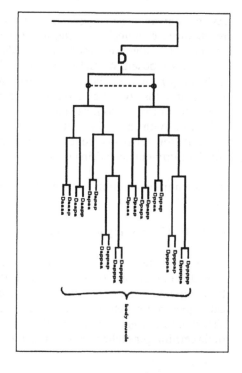

Figure 1.4    Divisions of key blast cell "D" in *C. elegans*. Data presented in Chapter III were cross-checked with illustrations like this. Reprinted from reference [12], p. 470, with permission from reference [9].

A wiring diagram of the generic connections among the identified neuron classes of the *C. elegans* nervous system (excluding the pharyngeal neurons) is available. Pre- and postsynaptic connections and gap junctions among the classes have been identified and labelled. Figure 1.5 shows a sample of the diagrams provided in literature. Beside the diagram is a table of the connections being represented. The NEURCLAS FILE SET (Chapter III, pp. 55-78) presents these data in tabular form, with added information about the cell classes. They are also presented in the same form in the disks. The tables presented in this book and in the disks provided may differ from actual manual counts of connections from the diagrams and tables provided in literature because redundant connections across diagrams, across tables (beside the diagrams), and across diagrams and tables have been removed to render the data unique and consistent.

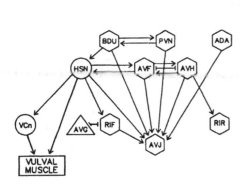

|     | PRE | POST | EJ |
|-----|-----|------|-----|
| ADA | AVB | | ADA |
|     | RIM | | AVD |
|     | SMD | | PVQ |
|     | RIP | | ASH |
|     | | | ADF |
| AVF | AVB | AIM | AVF |
|     | NMJ(VC) | | |
| AVG | AVB | PHA | |
|     | PVN | | |
| AVH | PVP | PHA | |
|     | SMB | | |
|     | ADF | | |
| AVJ | PVC | ADL | AVJ |
|     | AVB | PVR | RIS |
|     | RIS | PVC | PVC |
|     | AVE | | AVD |
|     | AVD | | |
| BDU | ADE | ALM | |
|     | | IISN | |
|     | | AVM | |
| IISN | AWB | PLM | IISN |
|     | BDU | | |
|     | AIZ | | |
| PVN | AVD | AVG | AVB |
|     | PQR | PDA | HSN |
|     | NMJ(VC) | | PVQ |
|     | AVA | | |
|     | AVL | | |
|     | PVC | | |
|     | VDn | | |
|     | PVT | | |
|     | DDn | | |
|     | PVW | | |
| RIF | AVB | AIA | |
|     | RIM | | |
|     | PVP | | |
|     | ALM | | |
| RIR | RIA | RIG | BAG |
|     | AIZ | ADF | |
|     | URX | DVA | |
| VCn | VCn | VCn | VCn |
|     | DDn | | |
|     | VDn | | |
|     | NMJ(VC) | | |

Figure 1.5 Sample of a diagram of connections among the neuron classes in *C. elegans*. Note the table indicating synaptic and/or gap junction connections between two classes beside the diagram. Reprinted from reference [12], p. 452, with permission.

An even more specific body of information is the detailed wiring diagram of each neuron of the nervous system of the *C. elegans* hermaphrodite (again, except the pharyngeal neurons). This information may be used by neural network scientists, computational scientists, mathematical biologists, behaviorists, and other researchers who seek new avenues to explore the morphologic basis of behavior. A schematic illustration of a neuron labelled in literature is reproduced in Figure 1.6. The schematic drawings were derived after the careful study of serial electron micrographs of the neurons and the connections among them. In Figure 1.7 are the important conventions used in the illustrations, which were used to translate the drawn connections into the lists provided in the disks as the LEGENDGO AND PREPOSGO FILE SETS for synaptic connections (also in Chapter III, pp. 95-164) and as the GAPJUNC FILE SET for gap junctions (also in Chapter III, pp. 79-94). Again, data were made consistent and unique whenever possible and from whatever information was available, e.g., from text, electron photomicrographs, tables, and additional diagrams provided in literature (Figure 1.8).

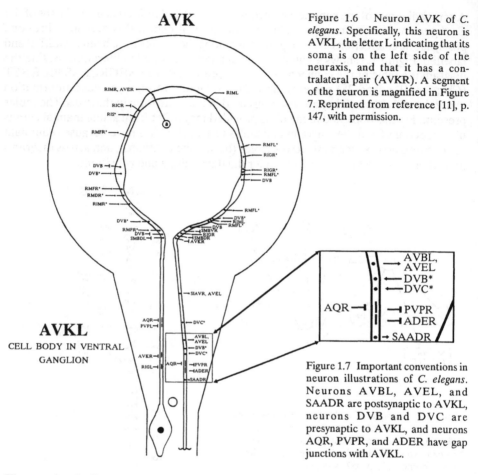

**AVK**

**AVKL**
CELL BODY IN VENTRAL
GANGLION

Figure 1.6 Neuron AVK of *C. elegans*. Specifically, this neuron is AVKL, the letter L indicating that its soma is on the left side of the neuraxis, and that it has a contralateral pair (AVKR). A segment of the neuron is magnified in Figure 7. Reprinted from reference [11], p. 147, with permission.

Figure 1.7 Important conventions in neuron illustrations of *C. elegans*. Neurons AVBL, AVEL, and SAADR are postsynaptic to AVKL, neurons DVB and DVC are presynaptic to AVKL, and neurons AQR, PVPR, and ADER have gap junctions with AVKL.

There exists in literature an abundance of text and diagrams on further detail on the anatomy of *C. elegans*. For example, the arrangement and movement of its muscles, the location and development of its nervous system ganglia, and the structure and formation of its nerve cords have already been studied. In the MOTORCON FILE SET (Chapter III, pp. 211-220, also in the disks), by synthesizing information available for both *C. elegans* and *Ascaris suum*, details of specific motoneuron-to-muscle cell connections in *C. elegans* are hypothesized and tabulated. Results of further research on the genetics of its cell lineage and morphology are likewise available in literature, together with more intricate details of its embryology. As previously mentioned, the ongoing studies are to sequence completely its genome.

In all data files available in this book, methods of data collection and refinement specific for each data file, complete with the references across which the data have been rendered consistent, are fully discussed for each data set in Chapter III.

In this subsection, specific data on the nervous system of *C. elegans* were taken from reference [11]. The rest of the data presented were derived from reference [12].

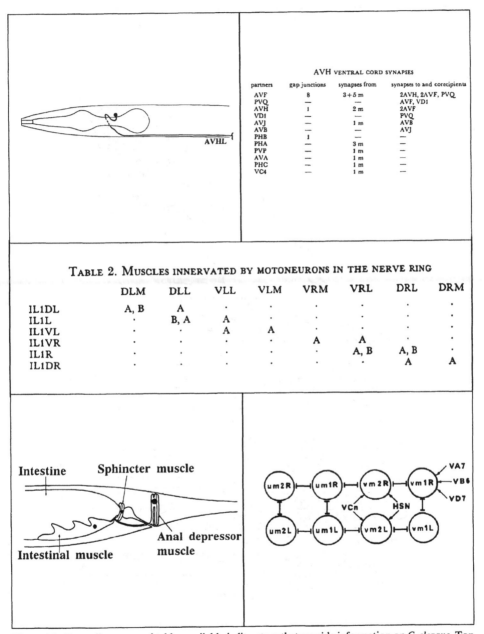

AVH VENTRAL CORD SYNAPSES

| partners | gap junctions | synapses from | synapses to and corecipients |
|---|---|---|---|
| AVF | 8 | 3 + 5 m | 2AVH, 2AVF, PVQ |
| PVQ | — | — | AVF, VD1 |
| AVH | 1 | 2 m | 2AVF |
| VD1 | — | — | PVQ |
| AVJ | — | 1 m | AVB |
| AVB | — | — | AVJ |
| PHB | 1 | — | — |
| PHA | — | 3 m | — |
| PVP | — | 1 m | — |
| AVA | — | 1 m | — |
| PHC | — | 1 m | — |
| VC4 | — | 1 m | — |

AVHL

TABLE 2. MUSCLES INNERVATED BY MOTONEURONS IN THE NERVE RING

| | DLM | DLL | VLL | VLM | VRM | VRL | DRL | DRM |
|---|---|---|---|---|---|---|---|---|
| IL1DL | A, B | A | . | . | . | . | . | . |
| IL1L | . | B, A | A | . | . | . | . | . |
| IL1VL | . | . | A | A | . | . | . | . |
| IL1VR | . | . | . | . | A | A | . | . |
| IL1R | . | . | . | . | . | A, B | A, B | . |
| IL1DR | . | . | . | . | . | . | A | A |

Intestine  Sphincter muscle

Intestinal muscle  Anal depressor muscle

Figure 1.8 Some diagrams and tables available in literature that provide information on *C. elegans*. Top left, the axon of neuron AVHL crosses to the contralateral side; top right, a table of the synapses and gap junctions of neuron AVH in the ventral cord; middle, the first few entries in a table of nerve ring motoneurons and the muscle cells they innervate, row labels are neuron names, column labels are DLM = dorsal left medial, VRL = ventral right lateral, etc., and A, B, C, D = sequence of muscle cells in each row, anterior to posterior; bottom left, the muscles of defecation; bottom right, gap junctions between uterine muscles (circles) are indicated by the I bars, and neurons that point by arrow to muscles innervate the muscles. (*continued next page*)

8

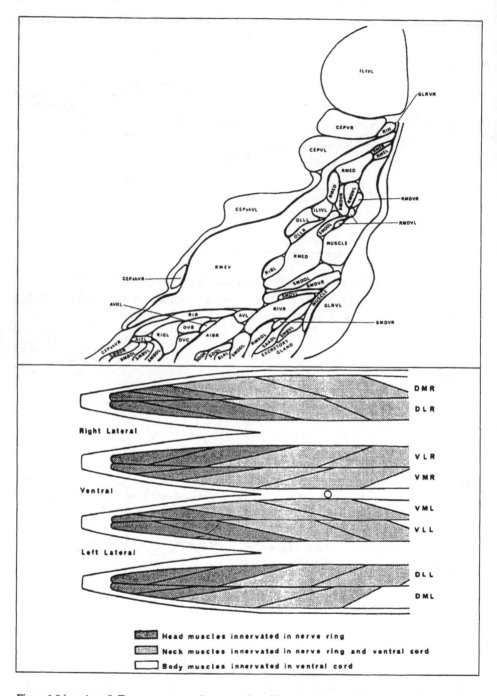

Figure 1.8 (*continued*) Top, reconstructed cross section of the ventral neuropile of the nerve ring; bottom, 'orange peel' projection of head muscles, muscle rows are labelled DMR = dorso-medial right, VLL = ventro-lateral left, etc. Data presented in Chapter III were rendered consistent and unique by referring to diagrams and tables like these. Reprinted from reference [11], pp. 23, 24, 43, 50, 142, with permission.

## C.2. *Data Translation and Manipulation*

How does one begin to study and analyze this wealth of information on a single organism? Now that the dissection has been done, how can the fragments conceptually be reintegrated so that the whole can be seen again, but with new understanding? What tools can be used to see the generalities of a whole? Viewing the structure and function of the *C. elegans* anatomical and developmental data in literature provides one with an almost intuitive choice of mathematical objects to use for biologically meaningful yet mathematically consistent investigations. Simultaneous with this choice is the recognition of which anatomical or developmental units can be discretized into the units of the mathematical object chosen. By following the rules of manipulation specific to the mathematical object, mathematical consistency can be achieved. By analyzing and relating the results to the anatomical or developmental units used, biological meaning may be derived. Examples in the succeeding paragraphs are initial excursions into this type of data translation, manipulation, and interpretation. In this chapter, they serve mostly to emphasize technique. They are discussed further in Chapter II, and are formally presented in reference [14]. Ultimately, if the entire organism is treated as one anatomical unit, the possibility exists that the whole organism may be integrated into one fabric of computation.

By treating the synaptic connection between a pair neuron classes as an element of a communication matrix, a neuron class connection matrix may be made, with presynaptic neuron classes as row labels, and postsynaptic neuron classes as column labels. The synaptic connection between any two neuron classes as the matrix element may be symbolized by an asterisk (Figure 1.9). By following rules specific to a matrix as a mathematical object, computations such as transitive closure by Boolean

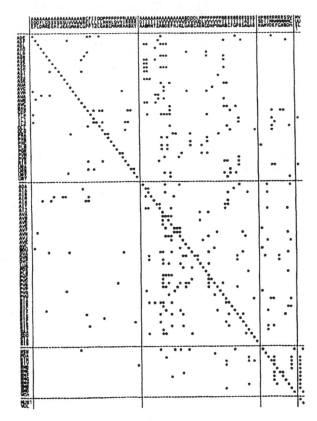

Figure 1.9 Matrix representation of the synaptic connections between neuron classes. Each asterisk (or matrix element) represents a synapse or synapses between a neuron class from the row (presynaptic class) and a neuron class from the column (postsynaptic class). Row and column labels correspond to neuron class names.

10

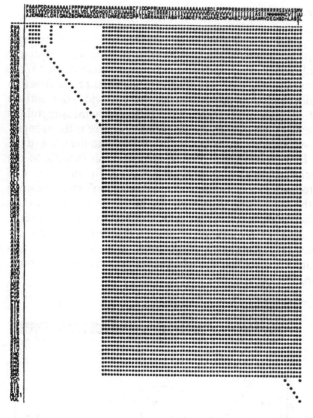

Figure 1.10 Transitive closure under Boolean multiplication of the matrix in Figure 1.9. To produce this matrix, further row-column exchanges were made to refine the resultant matrix after transitive closure. Through 12 synapses or less, every neuron might share in activity arising in any other neuron.

multiplication (Figure 1.10) can be done and analyzed. Computations leading to Figure 1.10 suggest the possibility that through 12 synapses or less, every neuron shares in activity arising in any other neuron.

In parallel, when a vertex adjacency matrix is used as a mathematical object in the investigation of individual neurons, presynaptic neurons become row labels, postsynaptic neurons become column labels, and a synapse or matrix element is represented by a dot or a pixel in a computer screen. If color is available, synaptic density (the number of times any two neurons are synaptically connected) may be coded in the color of each dot or pixel. This same representation may also be done with gap junctions (Figure 1.11). The vertex adjacency matrix, chosen in this case as the mathematical object by which to study anatomical

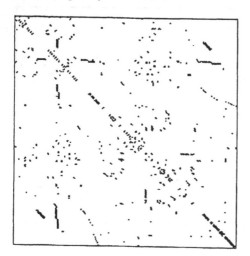

Figure 1.11 Matrix representation of the gap junctions between neurons. Each dot (or matrix element) represents one to several gap junctions between a neuron from the row and a neuron from the column. Although row and column labels corresponding to neuron names are no longer shown in Figures 1.11 to 1.16, neurons are practically in rostrocaudal sequence when sorted alphabetically, as in this matrix. Note that the matrix is symmetrical.

data, may again be manipulated through specific rules governing matrices. For example, by rearranging the order of rows and columns, certain overall characteristics of the nerve net can be made prominent (Figures 1.12 to 1.16). The sequence of neurons in the rows is the same as that in the columns for each topological equivalent. It is seen that random ordering of neurons in the rows and columns (Figure 1.12) corresponds to an unstructured nervous system. Anteroposterior sequence (Figure 1.13) and ganglionic order (Figure 1.14) of the neurons delineate the long and short tracts of the nervous system, and visually show cephalization, i.e., areas suspected of increased organization and function have more synaptic connections. This information is useful for any future attempts to fabricate electromechanically the synaptic connectivity of this worm, as components with short tracts can be grouped together, and long tracts can be grouped in a bus or back plane. The result of left-right arrangement of neurons in the matrix (Figure 1.15) suggest that lateralization exists, and may provide basis for the dextro- or sinistraversion of the worm during locomotion. Categorizing neurons as sensory (S), motor (M), and interneurons (I) in the matrix (Figure 1.16) shows dense connections between S to I, I to I, and I to M, and this may suggest the validation for the three-layer artificial neural network design.

The computer programs responsible for the display (DISPLAY.BAS, PLOT100.BAS) and rearrangement (MAKEDATA.BAS, MAKELIST.FOR) of the vertex adjacency matrix used in the examples above are source-listed and discussed in Chapter IV. Directions for running these programs from Disk 2 are also provided in the chapter. The DISPLAY.BAS program provides color displays of the matrix equivalents, as the synaptic and gap junction densities are encoded in the color. PLOT100.BAS provides a subroutine to print a hard copy of the display. MAKEDATA.BAS and MAKELIST.FOR create the input data for DISPLAY.BAS, PLOT100.BAS, and ELEGRAPH.BAS. ELEGRAPH.BAS is a program that generates a synaptic connectivity matrix with neurons in ganglionic order (like Figure 1.14), allows interactive manipulation of the matrix itself, and performs some elementary calculations on synaptic densities.

As in the case of neuron classes, transitive closure by Boolean multiplication can be performed on the neuronal synaptic connectivity matrix. The computer program that does this is shown as CONNECT.FOR in Chapter IV.

All computer programs introduced in this chapter are discussed fully in Chapter IV. Other mathematical objects used to investigate the *C. elegans* nervous system in detail are discussed in the next chapter.

### C.3. *Computational Anatomy*

Conventional anatomy proceeds by taking an organism apart (dissection) and putting its parts together again (integration). Dissection can be done at the macroscopic, microscopic, and even molecular levels. Although the integration of parts may be done physically after dissection, integration in conventional anatomy is usually done symbolically through text descriptions, pictures, drawings, diagrams, tables, charts, etc. that illustrate anatomy and anatomical relationships in an organism. Since the dissection of an organism can be done at different levels, integration can also be done at different

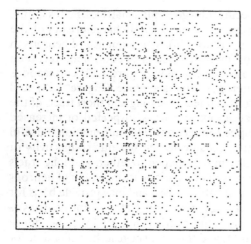

Figure 1.12 Random order. Each dot (or matrix element) represents a synapse or several synapses between a neuron from the row (presynaptic neuron) and a neuron from the column (postsynaptic neuron). Row and column labels (corresponding to neuron names) are in random order.

Figure 1.13 Anteroposterior order. Neurons represented by the row and column labels are arranged according to their anteroposterior appearance in the *C. elegans* neuropil. RUS = right upper submatrix, LLS = left lower submatrix, etc. LUS and RLS contain the short anterior to anterior (head) and posterior to posterior (tail) synaptic connections, respectively. RUS and LLS contain the long posterior to anterior (ascending) and anterior to posterior (descending) synaptic connections, repetively.

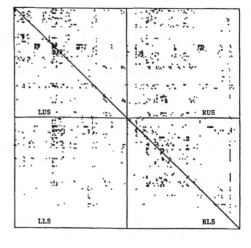

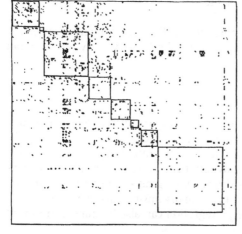

Figure 1.14 Ganglionic order. Neurons represented by the row and column labels are grouped into their respective ganglia. The neurons are sorted alphabetically within each ganglion, and the ganglia are sequenced anteroposteriorly. Intraganglionic synaptic connections are contained in each box.

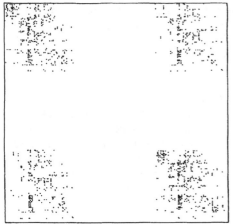 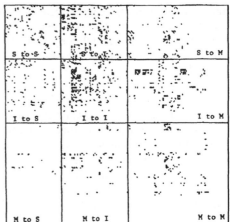

Figure 1.15 Order by anatomic laterality. Neurons represented by the row and column labels are ordered from left to right, as lateralized in the *C. elegans* neuropil. Homolateral synapses are located in the LUS and RLS while commissural connections are in the RUS (left to right connections) and LLS (right to left connections).

Figure 1.16 Order by neuron type: sensory, interneuron, motor. Neurons represented by the row and column labels are grouped as either sensory neurons (S), interneurons (I), or motor neurons (M). The neurons retain their anteroposterior order within each group. S to I, I to I, and I to M synaptic connections are dense.

levels, e.g., at the cellular, tissue, organ, and system levels. Integration puts together the pieces from dissection into an image that relates all the parts coherently, thereby leading to an understanding of the organism's anatomy.

Computational anatomy deals only with the integrative portion of conventional anatomy, i.e., providing a way of putting the parts together again. Dissection of *C. elegans* has already been done, and its anatomy has been integrated extensively in literature using practical and conventional representations. As a tool for integration, however, computation does not only integrate anatomy. Computation integrates anatomy with function. Thus, when pieces of anatomy taken from dissection are put together by computation, the picture produced through the computational process becomes not only descriptive of part, but also of probable function.

The previous subsection (C.2) describes the processes involved in computational anatomy, and in particular, the examples illustrate computational neuroanatomy. Upon presentation of suitably detailed anatomical or developmental data, computational anatomy initially involves, almost simultaneously, (a) the identification of the discrete anatomical parts and developmental processes that can be translated into mathematical terms, and (b) the choice of mathematical objects into which anatomical and developmental data can be mapped and manipulated. Once the initial decisions are made, data are collected and rendered consistent for manipulation. Data are translated and formatted into symbols used by the mathematical object chosen, and then manipulated in ways that are not only mathematically consistent, but are also physiologically and biologically meaningful. In subsection C.2, the synapses and the neurons were the discrete anatomical units mathematically mapped into and manipulated in the chosen mathematical object, a communication matrix. The choice of mathematical object and the manipulation of data, however, were made within the framework of specific physiologic and general biologic principles of neurons and nervous systems. Only within the context of these principles were the meaning and interpretation of the results derived.

Aside from prudence in interpreting the results of initial computations on the *C. elegans* anatomy, limitations in accuracy of the data being manipulated were always taken into consideration. Some exploratory calculations in computational neuroanatomy run into situations where connections that had to be present for logical consistency were absent in the *C. elegans* nervous system connectivity data presented in literature. This emphasizes the need for further improved data, and one can only hope that results from computational anatomy and neuroanatomy would be meaningful enough to inspire anatomists to return to the bench and work towards this goal.

Computational anatomy seeks to present in one image derived from computation (as in the examples in C.2 above), or in one process of computation (see Chapter II), a meaningful perspective or realization on the anatomy, physiology, or development of an organism which otherwise cannot be perceived or realized by simply viewing data presented in conventional formats. Rather than study the *C. elegans* nervous system meticulously as a physically dissectable or biochemically constructed entity, the methods of computational neuroanatomy attempt and aspire to integrate by computation relevant fragments of these valuable studies in order to produce another view of its whole. The new perspective gained may provide further understanding and insight into the relationship between morphology and function. By elucidating this relationship, the methods of computational anatomy can be used as tools in exploring the morphologic basis of behavior, an avenue of research in ethology. Conceivably, a possible bridge between computational anatomy and ethology that would allow some verification of the results of computational anatomy is artificial ethology [39]. A probable function of parts may emerge or may be discerned during the process of computation in computational anatomy. When this function is equated with behavior, it may be possible to verify or reproduce this behavior in an artifice. Artificial ethology has been defined to consist of (1) the replication and/or production of behavior by an artificial medium, (2) observation of the artifice producing the behavior, and, possibly (3) the modification of the behavior expressed by the artifice and/or of the artifice itself for some useful purpose. By this definition, artificial ethology has been proposed to encompass artificial intelligence (AI, replication of "intelligent" behavior in a machine), neurocomputation (replication of neural behavior in a machine), and computational neuroethology [39]. Rather than proceed from function (AI) or mathematical theory (neurocomputation) to arrive at a working design for computer software and hardware, computational neuroethology proceeds in the opposite direction, i.e., it accepts a natural, working nervous system design and seeks the function or theory behind the design. The processes in AI and neurocomputation may or may not conform to nature since man can impose his will on machine behavior, and it is the machine that is particularly malleable. Computational neuroethology, on the other hand, starts with a nervous system design that is already successfully functioning in nature (e.g., the nervous system of *C. elegans*), and seeks the design principles that make it function. Computational neuroethology may serve as the specific bridge between computational neuroanatomy and neuroethology, since computational neuroethology involves (1) the translation of a natural nervous system into synthetic form, based on currently known facts about its anatomy and physiology, (2) the actual creation or technological synthesis (or "hardwiring," in case of electromechanical translation) of such a biologically based synthetic neural system, and (3) the observation, analysis, improvement, and application of the resultant behavior of such an artificial or synthetic nervous system. In fact, in order to translate a natural nervous system accurately into synthetic form, computational neuroanatomy is the first step in computational neuroethology. These are the major motivations for computational anatomy and neuroanatomy.

Chapter II

Computations on the Nervous System:  Some Results

*The Fourth reached out an eager hand,*
*And felt about the knee.*
*'What most this wondrous beast is like*
*Is mighty plain,' quoth he;*
*'Tis clear enough the elephant*
*Is very like a tree!'*
-The Blind Men and the Elephant, J.G. Saxe

## A. Motivation, Tools, and Early Observations

In the principal source of neuroanatomic data tabulated in this atlas, J.G. White and his associates, Southgate, Thomson, and Brenner write, "The availability of the complete connectivity data for a nervous system generates an almost irresistible desire to speculate extensively on the functions of such a structure. We will, however, try to resist this temptation and leave such speculation for future work, when we hope that they can be backed up by corroborative experimental data. We will, therefore, try to confine our comments to the general features of the connectivity, some of which may not be obvious from the connectivity diagrams, and to the functional aspects of those parts of the circuitry for which there is some relevant experimental data" [11, p. 52]. In this chapter, we shall succumb to the temptation. When information collected becomes very detailed, however, the graphic means for conveying the information to others reaches a complexity that mirrors the complexity of the anatomical object. Even speculation unsupported by experiment becomes too difficult to pursue without special tools.

One tool is the theory of graphs. A graph is a simple technique for representing components connected to each other by directed or undirected arcs. Two nodes which are joined by an arc are adjacent. The adjacency arcs, in turn, may be represented by a matrix in which a row and its corresponding column are given the same name. The entry may indicate either presence or absence of a connection. A connection, when present, may also be weighted in some way to indicate the magnitude of a flow going over that path. Such a matrix is called an adjacency matrix, and becomes accessible to a number of theorems and algorithms useful in deriving generalizations about the network.

A few results that follow are some of the simplest conjectures that one may derive about the neuroanatomy. They are compact characterizations which result from the application of simple graph theory to detailed morphology winnowed from numerous tissue sections. This approach at once evades the problem of the image being as complex as the object. Generally, the results are not expressed in language familiar to anatomists, and seem only to stimulate casual curiosity among physiologists. Although ideationally sound, they are not backed by experiment. Instead, they institute a regimen akin to design, an interesting bridge between analysis and synthesis. Thus, one formalizes the process of conjecture about the "life in a structure." In this endeavor, the digital computer's powers for keeping in control a large mass of data are indispensable.

As stated in the first chapter of this volume, the first process in the computational investigation of the *C. elegans* nervous system is the transcription of the anatomical data into a form usable by simple computers. The form utilized is that of the simple spreadsheet, which is readily identified as a form of relational database. In this form, data can be checked immediately for inconsistencies, such as erroneous and duplicate entries, and necessary symmetries (e.g., if Neuron A sends to Neuron B, then the table should also show that Neuron B receives from Neuron A). One also finds that the nervous system cannot be reported fully without including afferent and efferent connections to the exterior or to non-neural tissue. These have been tabulated as four nodes (discussed below, see Figure 2.4). A most elementary generalization is enforced by the graph theoretic approach: "Nervous systems are networks suspended between an internal and external world. The organism is incomplete without its internal and external environments." In some cases where there were no neuron-to-neuron connections tabulated, connecting a neuron once to the appropriate internal or external environmental node was at the least necessary to establish logical connectivity. This is evident in the twenty pharyngeal neurons where we were unable to obtain detailed synaptic maps.

Two categories of activity must be embodied in the anatomy of the network. The problem is not unlike attempting to study a telephone network, and the culture from which the messages arise. In the telephone system, we can identify those messages that pertain only to the net, commonly known as control lines or signals (e.g., dial tone, ring, and busy signal), and those that deal with Teddy talking to his mother, or with a particular 911 call. Without the world of messages, however, the telephone network is irrelevant and incomprehensible. Similarly, one tries to separate the study of neuroanatomy from those aspects that deal with the integrity of the system, and those that relate to the life functions of an organism in its environment.

## B. Numerical Characterization

A nerve cell connects to another multiply (*multiplicity of connection*), and to multiple other nerve cells (*degree of connection*). Figure 2.1 shows distributions of the highest multiplicity per neuron arranged in descending order by the number of efferent (Figure 2.1a) and afferent (Figure 2.1b) connections. The net is not uniformly weighted. Messages from some neurons seem to be more "important" than others, and thus, some neurons may be thought to "heed" some inputs more than others. In contrast, Figure 2.2 shows the distribution of connections as an equally weighted, p-1 graph. Afferent (indegree) connections show a high count of 45 in one neuron, and 27 cells show counts of 5 per cell, while each of 21 neurons have but one afferent connection. The mean density is 7.2 with a standard deviation of 7.2. Similarly, efferent (outdegree) connection densities range from 39 (on 2 cells) to 1 (on 14 cells). Equal weighting extracts message importance from the traffic routing pattern. Now, some neurons appear to be "busier hubs" than others. The network is not evenly dense. At every level of efferent connectivity, there is a wide range of afferent connection density, thus producing convergent and divergent hubs.

Presumably, in lower organisms like *C. elegans*, the routing is primarily a consequence of epigenetic mechanisms and non-trophic factors [46]. A crude form of epigenesis can

**Figure 2.1a. Distribution of Highest Efferent Multiplicity/Neuron**

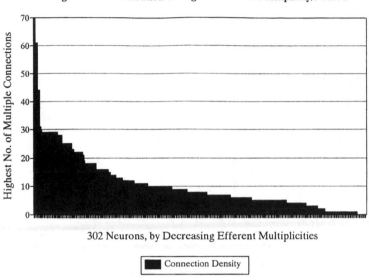

302 Neurons, by Decreasing Efferent Multiplicities

**Figure 2.1b. Distribution of Highest Afferent Multiplicity/Neuron**

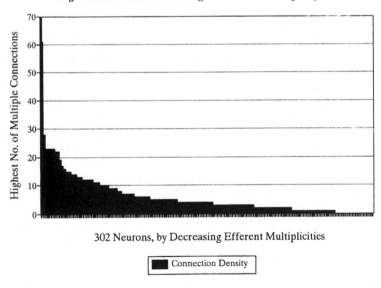

302 Neurons, by Decreasing Efferent Multiplicities

Figure 2.1. Connection densities of the highest multiplicity made by a neuron in an efferent connection (Figure 2.1a) and received by a neuron in an afferent connection (Figure 2.1b). The sequence of the 302 neurons in Figure 2.1a is different from Figure 2.1b.

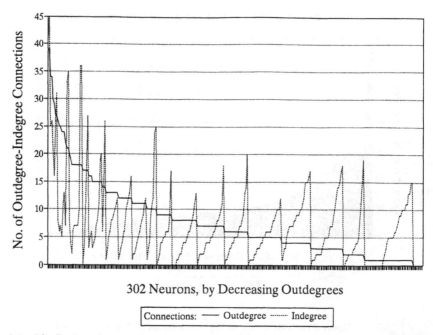

302 Neurons, by Decreasing Outdegrees

| Connections: —— Outdegree ········ Indegree |

Figure 2.2. Distribution of connections as an equally weighted, p-1 graph. When the multiplicity of connections is reduced to 1, the traffic routing pattern appears. Some neurons seem to be "busier hubs" than others, and since the network is not evenly dense, there are convergent and divergent hubs.

be simulated on digital computers using Monte Carlo methods. For example, by setting the range of a rectangularly distributed random variable to 302, and choosing about 3% as the possible connectivity, a large number of network matrices may be constructed. Using a larger computer [49], several hundred networks are easily generated and compared with the *C. elegans* distribution, with respect to common statistical parameters. Table 2.1 shows a sample of such simulation for groups of 100 networks, each the result of varying the parent distribution over some common types. For comparability, the connection density was kept at 3% (slightly higher than that of *C. elegans*, which is 2.62% as computed from the data), and the other characteristics of the distributions were adjusted to select a model to be grossly near the nematode. These parameters include setting a magnitude for at least the first and second moments, and in the case of infinite tail distributions, truncating the distribution at two or three standard deviations or representing the cell locations on an infinite line congruen modulo 302 (which is similar to plotting the distribution on a circular base line) Generally, the origin of the distribution for efferent and afferent connections was taken to be the same. In circular computation, however, the origins may be allowed to slip ; set number of cells. Unequal origins permit adjustment for correlation between ef ferent and afferent connections. It is promptly learned that in terms of the first o second moments, the nematode connection densities represent networks whose firs and second parameters are many decimal orders of magnitude away from any networ that may likely occur in a sample. If *C. elegans* typifies a neural net, and is the result c

Table 2.1. Networks of 302 Neurons Simulated By Monte Carlo Methods Using Different Probability Distributions. Each sample consists of 100 networks.

| Distribution | Mean In | Mean Out | SD In | SD Out | Correl | Skew In | Skew Out |
|---|---|---|---|---|---|---|---|
| Rectangular | 8.86 | 8.86 | 2.93 | 2.94 | -0.008 | 7.91 | 8.74 |
| Bernoulli | 9.06 | 9.06 | 2.96 | 3.01 | -0.007 | 7.95 | 8.61 |
| Poisson | 9.01 | 9.01 | 2.95 | 2.95 | -0.006 | 8.57 | 8.45 |
| Gaussian sd = 75 | 9.03 | 9.03 | 5.23 | 4.29 | 0.68 | 52.53 | 55.26 |
| Log Normal | 8.88 | 8.88 | 7.81 | 7.79 | 0.87 | 329 | 334 |
| Exponential mean = 115 shift = 0 | 8.79 | 8.79 | 6.80 | 6.76 | 0.82 | 254 | 249 |
| Exponential mean = 115 circular shift = 15 | 8.86 | 8.86 | 6.83 | 6.82 | 0.51 | 312 | 305 |
| C. elegans* | 7.20 | 7.73 | 7.20 | 6.98 | 0.52 | 554 | 792 |
| ** | (8.03) | (8.61) | (7.60) | (7.37) | (0.52) | - | - |

\* Data for *C. elegans* include 282 neurons, omitting the 20 pharyngeal neurons and "cells" WE, WI, WM, and WN (see Figure 2.4 below)
\*\* Numbers in parentheses are scaled to 302
Note: Skew refers to the third central moment

a "random" point process for connection, a model of greater complexity would appear necessary. Perhaps, embryogenesis or cell lineage models leading to realistic neural networks may be developed by probabilistic automata [28].

In chapter one we indicated briefly attempts to answer the query: "How is the whole fabric of this simple nervous system laid out?". We return to this issue here. When the graph is laid out as its matrix, with presynaptic cells along the rows of the matrix and correspondingly with postsynaptic cells along the columns, the graphically printed matrix evokes visual patterns which convey important aggregate impressions. These impressions derive from the general principle used to determine the ordering of the cells along the rows. Thus, randomizing the order shows that the connections are sparse but even (Figure 1.12); rostrocaudal order (Figure 1.13) separates long from short paths, reminiscent of the evolution of tracts; ganglionic order (Figure 1.14), that the statistical pattern relates to the anatomy of convergence and divergence; right and left (Figure 1.15), that there is fundamental asymmetry in a symmetrical organism; and by functional type (Figure 1.16), that the system is based upon sequential signal paths. In each case the pictorial display of the matrix provides a Gestalt of the network.

If the cells are grouped into ganglia, with cells in each ganglion ordered alphabetically, and the ganglia are ordered from rostral to caudal (Figure 1.14), a calico patterning appears in which dense rows seem to associate with dense columns. Such a pattern seems to confirm the idea that ganglia exist as hubs or central exchanges, and are the locations for collections of physically "busy" neurons. Mechanical processes in embryogenesis and the needs for organ masses may be the counterpoised forces leading to separation between ganglia. The calico pattern relates to the statistical characterization of synaptic density when ganglia are examined. In Figure 2.3, the synaptic density distributions for each ganglion are shown. It appears that ganglia contain two kinds of neurons: afferent collectors and efferent senders. In between are neurons with relatively few connections that may perform the "logic," reinforcing the suggestion that the collection and distribution of particular signals are indeed the functions of some other cells. Perhaps, this argues for the emergence of centralization of decision function in ganglia. Viewed as logical elements, cells which have many inputs, in consequence of the feature of threshold, have great logical capability. But this logic is directed, it would seem, at establishing the equivalent "meaning" of many subsets of inputs. It can be argued that the features of convergence and threshold require that efferent messages be determined by the factorial expression $nCr$, for $n$ things taken $r$ at a time. The combinatorial grows rapidly with $n$. On the other hand, cells with few inputs and many outputs are logically weak, but serve as "broadcasting" neurons. For example, fixed motor connections lead to behavioral patterns like startle reactions, which have been accessible to study by neuroethologists in many species [46]. Ganglia are seen then to be sites for logical units that switch on broadcast units. Thus, ganglia in *C. elegans* appear not to be simple exchanges, as one might be prone to generalize from the structure, say, of sympathetic ganglia in mammals. There are, however, ganglia to suggest that such designs exist in *C. elegans*. Of these, the retrovesicular and the anterior ganglia may be the examples.

Reordering the neurons in the matrix, so that all left-sided neurons come first in ganglionic order, and right sided neurons, last, with the unpaired neurons in the middle in roughly anteroposterior order (Figure 1.15), displays right-left symmetry of the connections. Commissural connections are emphasized in the right upper submatrix (RUS, left to right connections) and in the left lower submatrix (LLS, right to left connections). Homolateral connections are in the left upper and right lower submatrices. As in Figure 1.13, short tracts are near the main diagonal, and long tracts are seen farther away from it. Right and left neuronal asymmetry contrasts with the dorsoventral symmetry of the musculature and innervation. By virtue of surface forces in the film of water on the particles of soil on which they live, these nematodes live in a planar environment. Their locomotion is constrained by their dorsoventral muscular symmetry. This asymmetry of the sensory environment may be impressed upon the other sensory and logical processes. One thinks of the flounder, which, having exchanged its right and left for belly and back, respectively, has also moved both eyes to what was once the left side of the body (most of the time). Perhaps, the flounder adapts to the flattening of an otherwise three dimensional world as it matures to live on the ocean bottom. The study of genetic mapping may lend insight into this issue, the evolution of handedness. Does handedness arise from neural asymmetry determined by genetic or embryogenic circumstance?

In Figure 1.16, the neurons are reordered again according to presumed function: sensory (S), interneuron (I), or motor (M). Some ambiguity is inevitable beyond the fact that the data are incomplete, but this display shows the anatomy in a nine submatrix form, classified by function. The submatrix above the center (S to I) are sensory

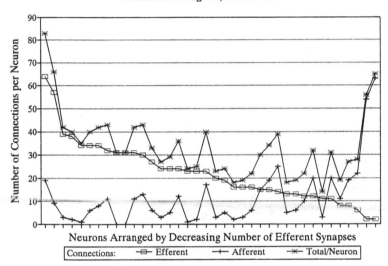

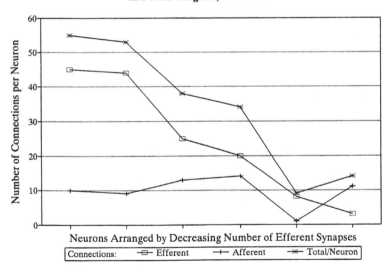

Figure 2.3. Synaptic density distributions for each of the nine ganglia. Neurons in each ganglion are arranged by decreasing number of efferent connections. Afferent connections are arranged by increasing number.

24

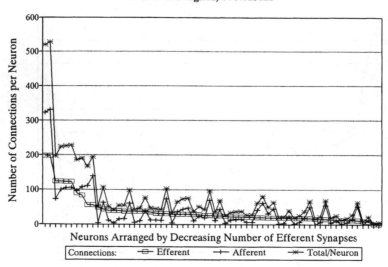

**Density of Synaptic Connections
in Lateral Ganglion, 64 Neurons**

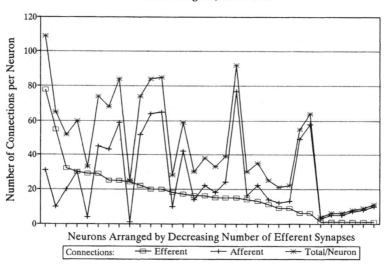

**Density of Synaptic Connections
in Ventral Ganglion, 32 Neurons**

Figure 2.3. (*continued*)

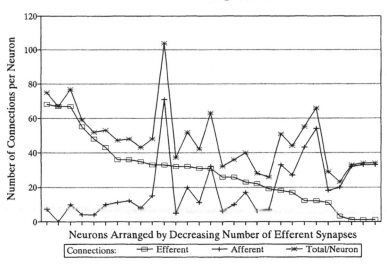

**Density of Synaptic Connections
in Retrovesicular Ganglion, 29 Neurons**

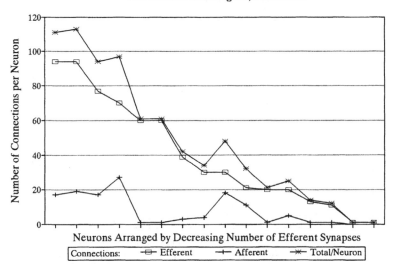

**Density of Synaptic Connections
in Posterolateral Ganglion, 16 Neurons**

Figure 2.3. (*continued*) By strict ganglionic classification, only six neurons belong to the Posterolateral Ganglion. As explained in Chapter III, p. 102, 10 unclassified neurons were included arbitrarily in this ganglion.

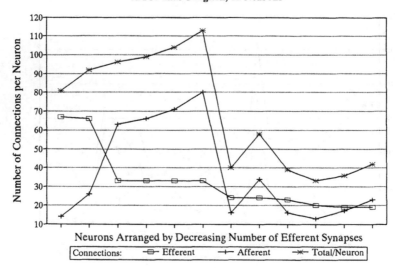

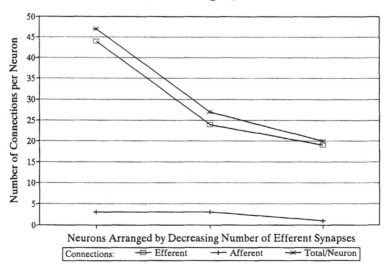

Figure 2.3. (*continued*)

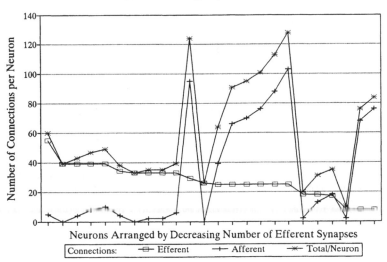

Figure 2.3. (*continued*)

connections, including interoceptive ones. The central submatrix (I to I) becomes the CNS of the nematode, showing connections only between interneurons. The submatrix to the right of the center (I to M) shows connections to motor neurons, and includes those to internal organs and neuroendocrine cells. The other submatrices are sparse, and provide some insight on the importance of proprioception and sensory feedback control at this phylogenetic level. The sequential nature of signal flow (S to I, I to I, and I to M, where synaptic connections are dense) conforms with our notion of computation order. It is also like the three-layered structure favored by investigators in artificial intelligence, trying to simulate types of behavior using neural networks with "perceptron-like" elements. In *C. elegans*, a few cells are found to have both motor and sensory functions, and these cells cross all three layers. With the exception of axon-reflex arcs, this phenomenon is not seen in the mammalian nervous system. The left column of submatrices deal with sensory feedback: from the sensory cell level, from the interneuron mass, and motor proprioception. As may be expected from the phylogenetic level of *C. elegans*, all three are sparse.

In Figure 2.4, the three-layer structure suggests that a fruitful linkage between connection diagrams and electronic networks may be made through a representation of the nematode as a four-terminal network by function. WE and WI are, respectively, sources for external or internal input signals, and may represent sense organs or free endings. WM and WN represent output destinations such as somatic musculature (WM), and internal organs such as anus, uterus, and pharynx or endocrine organs (WN). In this view, the worm itself, S-I-M, is seen imbedded between an external and internal environment. The further study of the network from this perspective will now be pursued.

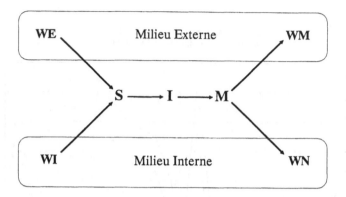

Figure 2.4. Representation of *C. elegans* nervous system as a four-terminal network by function. WE, WI = external and internal input signals, respectively; WM, WN = external (somatic musculature) and internal (endocrine or internal organs) output signals, respectively; S = sensory neurons; I = inter-neurons; M = motor neurons.

## C. Anatomical Computations

The preceding examples of computational manipulation of the data serve to extract by file manipulation, anatomical generalizations implicit in simple display. In addition to this, properties for networks of the same morphology can also be computed. To do this, some dynamic properties must be assumed for the elemental neuron.

An initial, intrinsically interesting inquiry on the data is the question: "Is the nervous system in a single organism in one or several pieces?". In the simplest of metazoan organisms, the triggering of one subsystem (e.g., a nematocyst) appears not to require global involvement of the animal. On the other hand, from Aurelia to Man, we intuitively accept the idea that a nervous system is all connected as one piece. There are known anomalies, however, like Siamese twins where intraorganism communication involves mechanical or audio-visual communication through an environment outside the network. On this hypothesis, computation may be used to provide evidence for correctness of connectivity in the data, and test the belief that each individual, indeed, has one nervous system.

If there are ten neurons, each with an efferent axon, then a complete interconnection would require 90 links, assuming that a neuron is not externally linked to itself ($n*(n-1)$). On the other hand, if there are less than 9 connections, it is not possible that all the neurons are connected. For the minimum of 9, only one arrangement would assure that the neurons communicated in a complete fashion. There are 302 neurons in the *C. elegans* hermaphrodite, many with branched axons, totalling some 6334 connections. Of the 90,902 possible connections, only 6334 odd arcs are found. It seems quite possible then that the *C. elegans* nervous system is, indeed, divided into several disconnected pieces. The number 6334, however, does exceed the minimum of 301 undirected arcs necessary to make a "one piece" network that is deliberately constructed like a ring. A generic answer is sought for all arrangements that would decide connectedness in relation to the number of connections. Is there some general rule to which the nematode nervous system conforms?

To begin answering this question, the multiplicity of a connection is reduced to one, and the axonal link is considered bidirectional. For the resultant system, Hastings [23a-23c], May [27], and others [23d] provide a theorem for a lower bound on the number of connections necessary to assure connectedness of a system. The graph of a matrix $M$ is almost surely connected if the fraction of connections $C$, among $n$ neurons, exceeds the natural logarithm of $n$, i.e., $nC$ $(1+\varepsilon)(\log n)$. Connectance, C, is the number of cells in the adjacency matrix that have entries divided by $n^2$ (the total number of possible entries). A network constructed with $C$ less than $(\log n)/n$ is probably not connected in one piece. For *C. elegans* with 302 neurons, this theorem requires that 1.89% of the matrix entries be non-zero. Since the data in this book cover only connections for 282 neurons of the anatomy, the effective matrix size becomes 282x282, and its recorded 2081 connections comprise 2.62% of the total possible connections. This is greater than the theoretical value of 2.00%.

A network is connected when it is possible to trace a path from any node (or neuron) to any other node in the net. Intuitively, if this is not possible, the network must be in at least two pieces, as in the highway networks of England and the United States. With this idea, a computation may be performed by expressing the neuroanatomy as the matrix of an undirected graph, and with unit synaptic density for each connection. Successive Boolean polynomials can then be computed, $(1+M)^k$ until $(1+M)^{k+1} = (1+M)^k$ or up to $k = N\text{-}1$ where $N$ is the number of neurons in the net. Calculating this for a system with 306 nodes (corresponding to the neurons of the *C. elegans* hermaphrodite and the 4 environmental termini), the polynomials equalize above $k = 4$, even though of the 93,636 possible arcs possible in a 306 node graph, only 2449 arcs (2.50%) are present. Transitive closure of the network as an undirected Boolean graph establishes that the nervous system is physically connected in one piece, and that no neuron is more than four sequential neurons away from any other. This is partly an artifact, if one considers that neurons for which we have no internal connection data were included (e.g., the pharyngeal neurons which were recorded as being connected to the pharynx, an internal environment or WN site). If the pharyngeal neurons are excluded and only 282 of the neurons which remain in the data are used in the computation, however, transitive closure computation still yield 282 neurons in one piece.

Since the undirected graph closes, the stricter calculation on the neuron matrix as a directed graph can be performed. This will preserve the unidirectional character of chemical synapses. When this is done for 282 neurons (without the 20 pharyngeal neurons), it is found that closure occurs in six synaptic junctions for 280 neurons. As expected from the data entered, two neurons, CANL and CANR, are isolated. What this means is that even if an impulse cannot retrace its axon, every neuron can reach every other neuron in 6 or fewer synapses. Furthermore, it implies that all sensory inputs are liable to feedback from both motor and interneurons, and that all motor neurons are in feedback control circuits. All the ganglia are, therefore, interconnected, and the nervous system is all in one piece. Manipulation of the matrix of path length = 1 leads mathematically to statements which have the character of "gross" neuroanatomy. Similar manipulation of the connectivity graph and its matrix should allow determination of properties like bandwidth and cutpoints, which help define functional subsystems. These, in turn, would require "traditional" neurophysiological function studies through ablation and stimulation experiments. The paradoxical statement regarding the role of CANL and CANR neurons may be resolved by such studies.

Another use of the matrix of coefficients is to investigate whether the organism, as portrayed, is implicitly consistent with an elementary requirement of existence: stability in time. In this case, the matrix represents the flow of signals between nodes of the graph. Each entry in the adjacency matrix is taken as an excitatory coupling of the neuron. The result can be viewed when values are added to the principal diagonal of the matrix of a system of equations. Conditions on equation systems that represent neuronal networks resemble representations for ecological systems, wherein animal species are considered to exist in prey-predator-environment food chains. The dissimilarity between models is substantial, but the parallels are plausible. Neither negative animals nor negatively excited neurons propagate. At the neuronal level, nonlinear, multiplicative terms are absent. They are, however, present in the cross product of populations and in the terms which describe the probability of prey-predator encounter. In consequence, single neuron equation systems are of linear form, but neural models representing collections of neurons as nuclear masses (populations of similar neurons) restore the necessity for algebraic products. Since neurons adapt but show no significant self-excitation, there is no equivalent of birth rates. All units have negative self-coefficients. The algebra of neural activity seems limited to the one operation, addition, and it has no inverse, although both positive and negative signals occur. In this sense, the network operates over a groupoid. Time lags are a common feature because of synaptic delays and transmission lags. Even in the face of such difficulties, the stability problem will be addressed using some parallel arguments from mathematical ecology. Assume that neuroanatomical connection is the matrix for a system of ordinary differential equations (or linear difference equations) with constant coefficients. As networks become larger and more richly connected, they tend to become unstable unless the strength of interactions is weakened. May's theorem [23a, 23b] proposes an upper bound for stability of networks that relate these variables. For a system of $n$ components and connectivity $C$, with a strength of connection $\alpha^2$, the product $\alpha^2 (nC)$ should be less than one, i.e., $\alpha^2 (nC) < 1$. Variations of this analysis are discussed by May [27]. As indicated above, C. elegans has a connectivity of .0262 > $(log\ n)/n$, sufficient to assure connectedness. Since $nC = 7.38$, assigning a value of .367 or less to $\alpha$ would meet the conditions of the theorem .

Under the assumptions made, point stability may be estimated by computing the eigenvalues of the adjacency matrix. The required result is that no eigenvalue of the system must have a real part greater than zero. Setting all values on the principal diagonal to -1, and the force of connection to be 1 for excitation and -1 for inhibition, and using synaptic multiplicities of 1, computation yields over 100 eigenvalues with positive real parts. The system under this assignment is unstable. To seek a stable configuration, the numerical entries may be adjusted by considering the physical meaning and a theorem of Gerschgorin: Every eigenvalue $E$ of an $nxn$ matrix $A$ lies in at least one of the circular discs with center $a_{ii}$ and with radius as the sum of all off-diagonal entries on each row ($\sum_{j \neq i} |a_{ij}|$) [32, p. 71]. Since all entries in the *C. elegans* adjacency matrix are real, the discs have centers on the real axis. Physical interpretation of the constants correspond to adjusting both the first order damping coefficients on the principal diagonal and the strength of synaptic coupling in the off-diagonals. As can be seen in Table 2.2, stability can be achieved by decreasing the magnitude of the entry on the principal diagonal, or by decreasing the strength of coupling, or both. The postsynaptic potentials always decay, and the coefficients are always negative and do not bear a necessary relationship in magnitude to the force of synaptic connection. However, this analysis suggests that some magnitude relationships not in our data must exist for stability.

Some attention needs to be directed at the magnitudes of the off-diagonal elements and their signs . In many of the simulations, positive and negative entries (excitation and inhibition) were made equally probable, although Gerschgorin's theorem should be indifferent to these assignments. The assignment of sign opens another neurophysiological issue, viz., the distribution of inhibitory connections. Before proceeding to this issue, the calculations may be summarized: The *C. elegans* neuronal net is sparse, but is above the boundary to assure connectedness, and is below the bound to assure stability within a range of modest adjustments of membrane adaptation constants.

## D. Inhibition and Stability

*C. elegans* seems to be too small to attract the study of its nervous system by electrophysiological methods. Inhibition, where reported, has been by identification of neurotransmitters at specific nerve endings. Our available sources of anatomical data are undoubtedly incomplete, and further biochemical anatomy is needed. 34/302 about 11.2% of the neurons are identified as having GABA, serotonin, or other transmitters known to be inhibitory in other species. As one may expect from the Gerschgorin theorem, the distribution of inhibitory neurons should have less effect on stability than selection of magnitudes. The eigenvalues, however, are far from insensitive. Some implications for the role of sign in neuronal function are found in the factor $\alpha^2$, relating to the strength of a single connection in the stability theorem. For systems constrained by a conservation principle, like solute diffusion in multicompartment models, the row sums of the connection matrix should be zero, and all the eigenvalues should be negative. In the stability theorem for systems of interest, the product of the number of cells and connectivity is generally greater than one. The connection strength ($\alpha^2$), of necessity, is less than one, and is equated to a normalized variance for the distribution of magnitudes in the connection strengths. In the simplest instance, computations for synaptic strengths (Table 2.2) may be considered to be taken from distributions with mean zero and equiprobable excitation of $+1$ or inhibition of $-1$. The multiplicity of a single connection was restricted to one. The $\pm 1$ distribution has mean zero and variance $= 1$. By assigning positive magnitudes to excitation and negative ones to inhibition, any other discrete distribution may be chosen. For example, with an excitatory magnitude inversely proportional to the proportion of excitatory neurons among all neurons, and similarly, with inhibitory neurons in inverse proportion to their numbers, one obtains

Table 2.2. Effect of Changing Strength of Connection and Adaptation Constants on Point Stability of a Network.

| Run No.[1] | Diagonal Entry[2] | Sum of All Entries[3] | Average Row Sum[4] | Prob. Excit.[5] | Mean Row Sum[6] | Val(+) Excit.[7] | Val(-) Inhib.[8] | No. of (+) Eigenval.[9] |
|---|---|---|---|---|---|---|---|---|
| Set A | | | | | | | | |
| 1 | 0 | -42 | 2.65 | 0.5 | -0.139 | 1 | -1 | 125 |
| 2 | -1 | -344 | 1.65 | 0.5 | -0.139 | 1 | -1 | 49 |
| 3 | -2 | -646 | 0.654 | 0.5 | -0.139 | 1 | -1 | 9 |
| 4 | -2.7 | -857.4 | -0.0453 | 0.5 | -0.139 | 1 | -1 | 0 |
| 5 | -3 | -948 | -0.345 | 0.5 | -0.139 | 1 | -1 | 0 |
| 6 | -10 | -3062 | -7.345 | 0.5 | -0.139 | 1 | -1 | 0 |
| Set B | | | | | | | | |
| 7 | -1 | -1180 | 1.895 | 0.3 | -2.907 | 1 | -1 | 43 |
| 8 | -1 | -920 | 1.701 | 0.35 | -1.931 | 1 | -1 | 42 |
| 9 | -1 | -746 | 1.589 | 0.4 | -1.47 | 1 | -1 | 43 |
| 10 | -1 | -542 | 1.614 | 0.45 | -0.795 | 1 | -1 | 47 |
| 11 | -1 | -344 | 1.65 | 0.5 | -0.139 | 1 | -1 | 49 |
| 12 | -1 | -304 | 1.531 | 0.51 | -0.00662 | 1 | -1 | 46 |
| 13 | -1 | 290 | 1.799 | 0.55 | 0.96 | 1 | -1 | 50 |
| 14 | -1 | 208 | 2.225 | 0.7 | 0.689 | 1 | -1 | 45 |
| 15 | -1 | 642 | 3.943 | 0.8 | 2.119 | 1 | -1 | 45 |
| 16 | -1 | 1028 | 5.925 | 0.88 | 3.403 | 1 | -1 | 44 |
| 17 | -1 | 1132 | 6.349 | 0.9 | 3.748 | 1 | -1 | 46 |
| 18 | -1 | 1320 | 7.079 | 0.95 | 4.37 | 1 | -1 | 46 |
| 19 | -1 | 1358 | 7.101 | 0.96 | 4.497 | 1 | -1 | 44 |
| Set C | | | | | | | | |
| 20 | -1 | -13.9 | -0.124 | 0.5 | -0.046 | 0.33 | -0.33 | 0 |
| 21 | -1 | -15.5 | -0.0177 | 0.5 | 0.0513 | 0.37 | -0.37 | 0 |
| 22 | -1 | -16 | 0.0088 | 0.5 | -0.053 | 0.38 | -0.38 | 1 |
| 23 | -1 | -16.8 | 0.0619 | 0.5 | -0.0556 | 0.4 | -0.4 | 1 |
| 24 | -1 | -29.8 | 0.885 | 0.5 | -0.0987 | 0.71 | -0.71 | 26 |
| Set D | | | | | | | | |
| 25 | 0 | 1391.9 | 8.253 | 0.8874 | 4.609 | 1 | -1 | 160 |
| 26 | -1 | 1089.9 | 7.253 | 0.8874 | 4.609 | 1 | -1 | 45 |
| 27 | -2 | 787.9 | 6.253 | 0.8874 | 4.609 | 1 | -1 | 15 |
| 28 | -3 | 485.9 | 5.253 | 0.8874 | 4.609 | 1 | -1 | 9 |
| 29 | -3.75 | 259.4 | 4.503 | 0.8874 | 4.609 | 1 | -1 | 4 |
| 30 | -4 | 183.9 | 4.254 | 0.8874 | 4.609 | 1 | -1 | 2 |
| 31 | -5 | -118.1 | 3.253 | 0.8874 | 4.609 | 1 | -1 | 1 |
| 32 | -8.3 | -1114.6 | -0.0465 | 0.8874 | 4.609 | 1 | -1 | 0 |
| 33 | -9 | -1326.1 | -0.7465 | 0.8874 | 4.609 | 1 | -1 | 0 |

*(continued next page)*

Table 2.2.  Effect of Changing Strength of Connection and Adaptation Constants on Point Stability of a Network. (*continuation*)

| Run No.[1] | Diagonal Entry[2] | Sum of All Entries[3] | Average Row Sum[4] | Prob. Excit.[5] | Mean Row Sum[6] | Val( +) Excit.[7] | Val(-) Inhib.[8] | No. of ( +) Eigenval.[9] |
|---|---|---|---|---|---|---|---|---|
| Set E | | | | | | | | |
| 34 | -1 | -127.8 | -0.02295 | 0.8874 | -0.423 | 0.12 | -0.88 | 0 |
| 35 | -1 | -5.4 | 0.568 | 0.3 | -0.0179 | 0.7 | -0.3 | 9 |

| | |
|---|---|
| 1 | Experimental run number |
| 2 | Adaptation constant |
| 3 | Sum of all strengths of connection |
| 4 | Column entries in a given row were summed for all rows, and mean of the sums was computed. Numbers in this column thus represent average afferent connectivity. |
| 5 | Probability that a synapse is excitatory |
| 6 | Column entries in a given row, except the diagonal, were summed for all rows, and mean of the sums was computed. Numbers in this column thus represent the average of off-diagonal entries. |
| 7 | Strength of the excitatory connection |
| 8 | Strength of the inhibitory connection |
| 9 | Number of eigenvalues with positive real parts, computed as a measure of network stability |
| Set A | Equiprobable sign of synapses, but increasing adaptation; 0 to -10 as diagonal entries; note relation to average row sum |
| Set B | Varying excitation probability between 0.3 and 0.96; there is little effect as expected from Gerschgorin's theorem |
| Set C | Constant probability; symmetrical changes in synaptic coupling strength |
| Set D | Prob. Excit. is the observed fraction of excitatory neurons in *C. elegans* data; increasing adaptation |
| Set E | From *C. elegans* data, with synaptic strength inverse to observed fraction of excitatory neurons; decreasing adaptation |

a Bernoulli distribution of mean zero. Although the rationale for choosing distributions needs defense, stability is achieved for a given neural net by adjusting the principal diagonal. In summary, stability for a matrix of connectivity which allows for excitation and inhibition is achieved by adjusting the magnitudes of the parameters of a binomial distribution that describe the sign and intensity of inhibition and the proportion of such cells. It is gratifying to observe that point stability for the matrix representation can be achieved using parameters which seem to have plausible physical meaning. Inhibitory entries, indeed, tend to stabilize the system.

Stabilizing effect of inhibition was observable by randomly changing the signs of the excitatory entries to negative in the adjacency matrix. Random distribution of synaptic connections, however, does not conform to the general pattern in nervous systems capsulized in the Dale-Feldberg rule: a neuron does not act as both an excitatory and an inhibitory neuron. A neuron's processes are all excitatory or all inhibitory. In the system matrix, the rule leads to the matrix acquiring a striped appearance. If each row represents the coefficients for afferent connections to a neuron equation represented by the row, then in each column are the efferent connections from a neuron and has either all entries as negative numbers or zero, or as all positive numbers or zero. Computer simulation shows that stable networks occur as readily when randomizing the matrices by column or when randomizing by element. The phenomenon is difficult to demonstrate when applied to randomly connected matrices of dimension less than 10. Connectivity and inhibition interact significantly when the matrix is striped.

We have not considered the problems posed by the existence of several neurotransmitters of the excitatory and inhibitory types. Each chemically identified transmitter may partition the nerve net functionally, without obviating the analysis here described [23d]. This makes the analysis more complex. For the present purpose, the conclusion is that stability may be achieved by adjusting coupling constants in an manner consistent with neuroanatomical and neurophysiological ideas. Computer modeling on the *C. elegans* data allows the combination of ideas on complexity, stability, excitation, inhibition, specific neural sign, and synaptic strength, with morphological data. Computer simulation in the context of stability of the *C. elegans* network opens to modeling some broad questions on the workings of the nervous system. While not yet forging a bridge between structure and function, *C. elegans* data seem to open a way toward a complete simulation of the nervous system of one organism.

After these results, some previously unremarked questions acquire importance. For example, "How prevalent are inhibitory cells in any nervous system?" Anyone who has driven a microelectrode through the mass of a mammalian brain to observe spontaneous activity, finds more long regions of silence in the electrode track rather than a cacophony of nerve impulses. Yet, the very absence of this cacophony of nerve impulses is what permits localization and neuronography. Does the mammalian brain do much of its computing in the silences?

Another perspective concerns synaptic multiplicity, which refers to the number of terminals (boutons) or physical synapses between a given pair of neurons. As the statistics of the network and the figures show, on average there are about three (6345/2334) endings for each connection. Among neurophysiologists and most conventional neural system modelers, the generally agreed upon property of neurons as computing elements, is spatial summation. The neuron is viewed as a logical majority organ [37]. To propagate an impulse, a neuron must receive the additive conjunction of $K$ inputs. Physically, this appears to be a consequence of the ratio between postsynaptic potentials and threshold. The reduction of such structures to cascades of logical "and" gates seems to form the basis for a widespread analogy between binary digital computer design and workings of the nervous system. The analogy has opened the way to much fruitful research. The technique of equivalencing of this logic to binary systems conceals a feature of single neurons in terms of computing potential. The combination of "fixed wiring" with threshold indicates a flexibility in computation for single neurons that depends upon the network in which it is embedded. In *C. elegans*, if Neuron A connects to Neuron B threefold, and the ratio of threshold to postsynaptic potential is three, then the connection A-to-B is monosynaptic. If each connection has multiplicity = 1, the logical versatility of the neuron with, say, 7 termini and a threshold of 3 is $7C3$ = $7!/(4!3!)$ = 35. Thirty five patterns of inputs on the neuron are logically equivalent. With varying multiplicity, a single neuron may have a considerable repertoire of input combinations to produce a single output. One might suppose that a more apt model for the nervous system is one that computes on synonyms. Choice of the coupling constant, therefore, lead to a way to estimate the logical power of single neurons.

Since the stability-connectivity constraints are quite general, calculation about hypothetical nervous systems may be scaled up to many neurons, even up to $10^{10}$ (the commonly stated estimate of neurons in the mammalian brain, [42, 43]). Such reasoning demonstrates that the number of unique connections per cell, $u$, rises together with the logarithm of the number of neurons $n$ in a system. Without changing the conceptualization of a neuron as an arithmetic majority organ, the logical versatility of each neuron increases as $uCp$. If the ratio of threshold to the number of boutons necessary to transmit, $p$, depends upon membrane properties, $p$ may remain relatively constant, or increase at a rate slower than $log\, n$. The factor $u!$ in $uCp$ makes the number of logically equivalent inputs rapidly more numerous for a single neuron. An hypothetical organism $X$, through scaling up under connectivity constraints, may attain a convergence of 20 while allowing bouton size to decrease threefold. Then, in organism $X$, $20C9$ (the combination of 20 taken 9 at a time) is the logical versatility of the larger net. The number is so large as to be implausible, but in any case one sees that the *C. elegans* system and system $X$ are qualitatively different. Scaling up a common design, in this case, increases the logical power of a single element, to appear as an emergent property of large nets......perhaps not unlike men in societies.

### E. Other Speculation

Graph theory provides other tools for anatomically unconventional perspectives, of which we now cite a example. White and his co-authors [11] remark on the observation of the graph theoretic triple operation [20, p. 147]: "One of the striking features of the connectivity diagram is the high incidence of triangular connections linking three classes." The connection reference is one in which the path A→B is paralleled by A→C→B. Considering also that three neurons can only be connected as a cycle of three, e.g., as a path in line or a fork and line, dissection in accordance with graph theory may lead to a portrayal by multi-unit patterns. Are there other representative substructures forming repetitive patterns? Starting from the adjacency matrix in Boolean form with the principal diagonal set to zeroes, a Boolean multiplication of the matrix by itself will show all paths of length 2. Further multiplications elaborate paths of greater length. In the present data on 282 neurons, there are 2,081 arcs of length one out of a possible 79,524. Squaring the matrix produces 23,695 paths of length 2. Of these, 4,419 are A→C→B arcs in a triple operation. It is, perhaps, not surprising that anatomists see such connections often.

Consider paths of length 2 which close upon themselves: Neuron A connects to Neuron B and Neuron B connects to Neuron A. There are 396 of these, involving 165 unique neurons. This number is over half the total number of neurons in the nematode. Drawn diagrammatically, such a structure may be portrayed in a form reminiscent of a daisy petal. Because 396 daisy petals with distinct ends requires 792 cells, while only 165 neurons are involved, some petals must share a common node. The graphical structure thus produced is formally called a roseace, but we shall refer to them as daisies. We tabulate them in Table 2.3.

396 corresponds to the number of symmetrical entries on the Boolean matrix. In a matrix of actual multiplicities, the symmetrical entries are usually unequal. Since this calculation excludes external connections, the graph is drawn from 1,784 synaptic connections. Figure 2.5 illustrates some patterns formed by two neuron loops. Interestingly, symmetrical connections preserve right left asymmetry in large measure.

Table 2.3. Enumeration of Symmetrically Connected Neurons by Unique Centers

| No. of Unique Neurons as Centers (Daisies, D) | No. of Neurons Connected to Centers (Petals, P) | Total No. Cells Connected per Class of Daisy (DxP) |
|---|---|---|
| 4 | 8 | 32 |
| 1 | 7 | 7 |
| 7 | 6 | 42 |
| 9 | 5 | 45 |
| 13 | 4 | 52 |
| 19 | 3 | 57 |
| 49 | 2 | 98 |
| 63 | 1 | 63 |
| Totals    165 | 30 | 396 |

Unpaired interneurons seem as much concerned with rostrocaudal connectivity as right-left communication.

The daisies are not connected evenly. Some neurons are busier than others, and this seems to account for the "calico" pattern in the dot matrix representation of the nervous system (Figure 1.14). This is also readily seen by printing only the symmetrical connections of the p-1 graph of the whole nervous system. The number of petals per daisy range from 8 to 1. There are just four neurons with eight petals. They are RIAL, RIAR, PVCL, PVCR. Symmetrical left-right neuron pairs RIAL and RIAR are located rostrally in the lateral ganglion, and symmetrical left-right neuron pairs PVCL and PVCR are located caudally in the lumbar ganglion. The two rostral, 8-petal daisies are not directly connected to the caudal ones, but all four daisy centers are interneurons. The two pairs of daisies also differ in configuration. The rostral pair (centers: RIAL and RIAR) share seven out of eight petal ends, and are connected at tips like the fingertips of a pair of Durer praying hands. In contrast, the caudal pair (centers: PVCL and PVCR) share no petal tips, but are connected center to center like two wheels (of 7 petals each) joined by an axle (the common 8th petal). Considering that these are topological descriptions for the signal flow of a computation, puzzlement increases. By analogy to human conversations, the rostral pair of daisies are two separate chairmen who receive reports from one and the same committee, and the caudal pair of daisies are two separate committees reporting to their respective chairmen, who, in turn, talk to each other. Although committee members do not talk to each other in this representation, this is not necessarily true. Altogether, the petals are analogous to pairwise conversations between persons in a large gathering.

Only one daisy of seven petals is found in the data, and its center is HSNR. This motor neuron, which is reported to be involved primarily with control of the vulval egg-laying muscles, is paired with the motor neuron HSNL. HSNL, in turn, connects to only three petals. As shown in Figure 2.5, HSNR seems to have an almost Pythagorean complement of associated daisies at its petal ends. Indeed, its petal ends become centers for 1

(PVQL), 2 (RIFR), 3 (HSNL), 4 (BDUR), 5 (none), and 6 (AVJL) petalled daisies. To the extent that the connection matrix might be considered as the matrix of a linear system of equations, the matrix of daisies with petals which are all excitatory or all inhibitory is Hermitian, and its eigenvalues are real and its eigenvectors are linearly independent. The question of what such first order loops perform in the context of signal processing will be revisited below, but we shall proceed next with the consideration of other paths of length 2.

Comparing the matrix of degree 1 with degree 2, there are 4,419 triple operations with unique, ordered pairs of neurons at their bases. Many baseline pairs are paralleled by paths of length 2 through different third neurons. In addition, triple operations originating in the external sensory environment (WE) number 205. Throughout this simple organism is a high degree of sensory feedback. At the same time in the interneuronal net, triple operation indicates a high degree of feed forward. On the average, each triple operation is paralleled 2.8 times. Only 33 of the 282 neurons used in this computation are found not to originate a triple operation, but each of 23 others originate a triple operation only once. The distribution of feed forward paths is highly uneven. The most multiply connected are PVCR (141), PVCL (130), PVNR (103), PVNL (105), AVAR (119), and AVAL (121), which have over 100. Only 9 cells, however, have between 50 and 100 triple operations, namely, AVG (99), AVDR (71), FLPR (66), AQR (64), DVA (64), PVPR (55), AVER (53), AVJL (51), and AVJR (50). The latter set of neurons seems to have no pattern with regard to laterality, in contrast to the former set. In the most densely connected triple operations, the neurons are paired and are nearly the same ones that dominate daisies distribution. An anatomist, indeed, might see nearly three times as many triple operations as reciprocal connections, since about 15% of all connection are in triple operations.

If the diagonal cells of the $M^2$ matrix is set to zero, and the product $MxM^2$ is formed, it can be deduced that there are 158 neurons connected in 698 triangular cycles. As in the case of roseaces, the triangular neuron patterns share arcs. As in the case of symmetrical pairs, composite figures consisting of 3-cycles (deltas) show plate-like mosaics. These are tabulated in Table 2.4.

Table 2.4. Partial List of 3 Neurons in Cyclical Arrangement Listed by Number of Cycles Originating from Unique Neurons. Only clusters of 10 or more are shown. There are 24 neurons that participate in only one 3-cycle, and 144 that participate in none.

| No. of Centers | N |
| --- | --- |
| 1 | 53 |
| 1 | 44 |
| 2 | 38 |
| 1 | 37 |
| 1 | 36 |
| 1 | 31 |
| 1 | 28 |
| 1 | 24 |
| 1 | 23 |
| 2 | 22 |
| 1 | 21 |
| 1 | 20 |
| 2 | 19 |
| 1 | 18 |
| 2 | 17 |
| 3 | 16 |
| 6 | 15 |
| 2 | 14 |
| 3 | 13 |
| 6 | 12 |
| 2 | 11 |
| 6 | 10 |
| 8 | 9 |
| 1 | 8 |
| 10 | 7 |
| 8 | 6 |
| 8 | 5 |
| 13 | 4 |
| 15 | 3 |
| 24 | 2 |
| 24 | 1 |

38

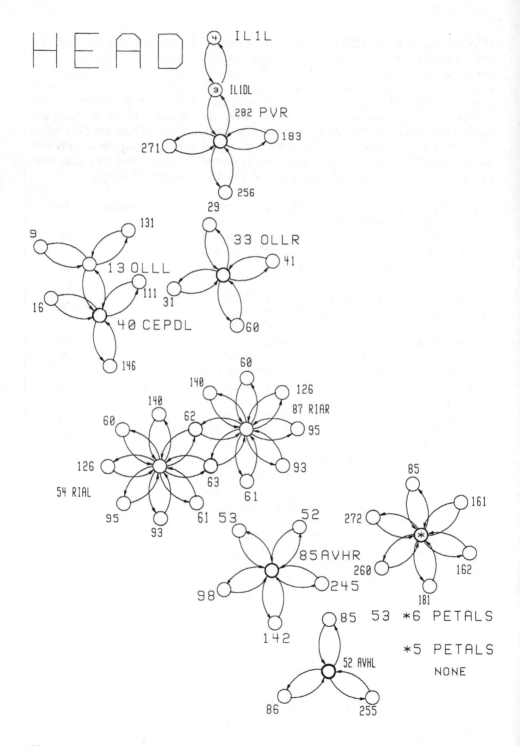

Figure 2.5. Samples of two-neuron cycles

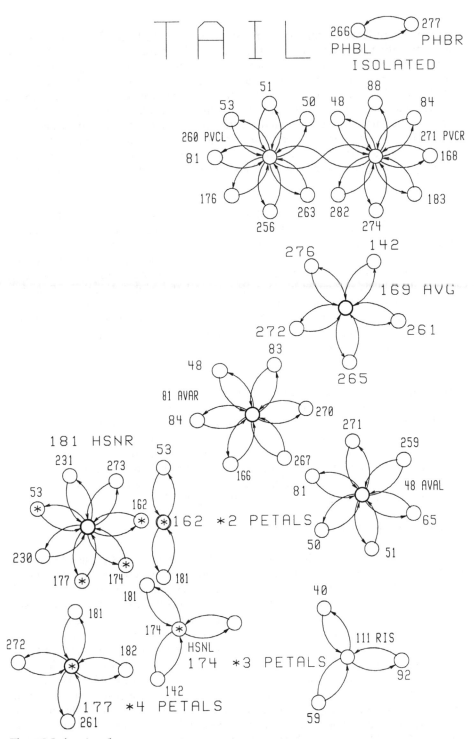

Figure 2.5. (*continued*)

Unlike cycles of length 2, the range of complexity is much larger from 53 triangular plates emanating from a single center, PVCR, to 24 neurons which participate as a corner only once. None of the latter stand disconnected from other triangles. The other most complicated centers in rostrocaudal order are: AVJL (38), AVJR (22), AVAL (36), AVAR (38), which are interneurons in the lateral ganglion; HSNL (19), HSNR (37), which are the egg laying neurons in subventral site; and PVCL (44), PVCR (53), which are the interneurons in the lumbar ganglion. The most caudal busy neuron is PVR (17), a sensory neuron of the lumbar ganglion. These nine neurons are right-left interconnected and rostrocaudally connected as parts of a common 3-cycle system, extending the length of the nematode. Some other configurations are shown in Figure 2.6.

It should be apparent that the neuroanatomy may be examined as configurations of cycles of length 4, 5, and greater, even though transitive closure occurs in 6 steps. Such representations would be of interest to one pursuing the significance of reverberating circuits [25, 30, 31, 33, 34, 38]. The analysis suggests that simple configurations form higher order patterns within the prevailing transmission of signals. Path synthesis of the synaptic data leads to figures like the daisies and the triple operations. Stated as magnitude time functions, the two-neuron loop produces not surprisingly a time-characterized magnitude time function different than the input function to either neuron, and three neurons connected as a two-petal daisy differs quite significantly from three similar neurons connected in a cycle. Similar study of loops of increasing size, were considered by MacGregor [25].

In summary, the logic of computing transitive closure to see if the nervous system is, indeed, a unit in communication as well as a unit by physical connection, leads to the consideration of powers of the adjacency matrix. The intervening powers of the matrix, explored briefly, shows that the neuroanatomy may be reconstructed as a plication of simpler patterns. These seem related to signal processing significance, even though one may be uncertain about the proper way to represent signal processing in a single cell. The richness of the connectivity in the anatomical data promotes a need for methods from graph theory to computational neuroethology [17].

## F. Suggestions on the Signal Processing Representation

Let us briefly discuss formulations for representing the nervous system as a network of similar cells. Given a structure, it is possible to conjecture its function by according the elements with dynamic physical properties. Like Harvey, who assigned hydraulic properties to the anatomy of the circulation, we can today use information processing concepts to animate neurons. The qualitative idea that neurons are arranged in specific patterns of connections for functional purpose is as old as neuroanatomy. The idea that particular patterns have special functions certainly dated to an early part of the century before the availability of techniques for recording neuronal activity. One thinks particularly of R. Lorente de No, or of electroencephalographers like Grey Walter [47], in proposing dynamic models for neural activity. The quest for reverberating circuits for continuing activity in a system with no obvious pacemakers probably originates with them. Quantitative representations also have a long history. As suggested by Householder [45] and many thereafter, the network anatomy is readily recast as difference equations. Such equations feature time lags on the physiological basis of the absolute refractory period, synaptic delay, and transmission lags. These properties were observed early and in all nervous systems that propagate activity by impulse signals. Representation by difference equations places the focus upon the time dependent properties of neural systems. The triple operation and the daisy petal can be recast as difference equations. Thus, taking $A$ as the output cell, $B$ as the input cell, $X(t)$ as input signal, $C$ as the bypassing cell, $p$ as an adaptation constant, and $q, r, s,$ and $u$ as synaptic coupling coefficients, one might write (viewed from the axon hillock at a given instant):

$$A(t) = pA(t\text{-}T) + qB(t\text{-}T) + rC(t\text{-}T)$$

$$B(t) = pB(t\text{-}T) + X(t\text{-}T)$$

$$C(t) = pC(t\text{-}T) + sB(t\text{-}T).$$

The output $A(t)$ is a weighted sum of an input $X(t)$ and its past. It may be viewed also as an hybrid digital filter, or an autoregressive moving average process, or any of several other models that emphasize time delays in the system [18].

Similarly the daisy petal is :

$$A(t) = p\,A(t\text{-}T) + q\,X(t) + uB(t\text{-}T)$$

$$B(t) = p\,B(t\text{-}T) + r\,Y(t) + sA(t\text{-}T).$$

This pair suggests that each neuron in the structure computes recursively a weighted time average of its own activity, summed with that of its mate over a theoretically infinite past. The autoregressive form requires representation of adaptation. Even at the complexity of one neuron, a form of memory finds expression. Because of time lags, nervous systems are intrinsically autoregressive moving average networks in time, and less like logical state machines. Logical operation depends upon the branching of networks to achieve basic behavior. In each case, the anatomy defines equation form.

We have seen fit, in part as whimsy, to excerpt from the poem on blind men examining an elephant, to them an unknown but certainly coherent object. Mathematical dissection also tolerates many views of how *C. elegans* may work. Almost certainly, each of these views will be partly in the right. Thus, it cannot be our purpose to select the "correct" or best mathematical representation for the signal processing behavior of neurons. Nor can we claim it within our experience to review this very large field of investigational study. From the first widely acknowledged symbolic and formal model of neurons attributed to Drs. Warren McCulloch and Walter Pitts [36], there have been many representations, some of almost overwhelming elegance and complexity. It is a problem to which it seems nearly every serious biomathematician has turned his hand at one time or another. Many substantial reviews of this work are available. In recent groping among them, we came upon two sets of representative works. "Neural Computing Architectures" edited by Igor Aleksander, 1989 [15], references "Parallel Distributed Processing" by D. E. Rumelhart and J. L. McClelland, Volumes 1 and 2, MIT Press, 1986. These books catalog many of the approaches that emphasize computation

42

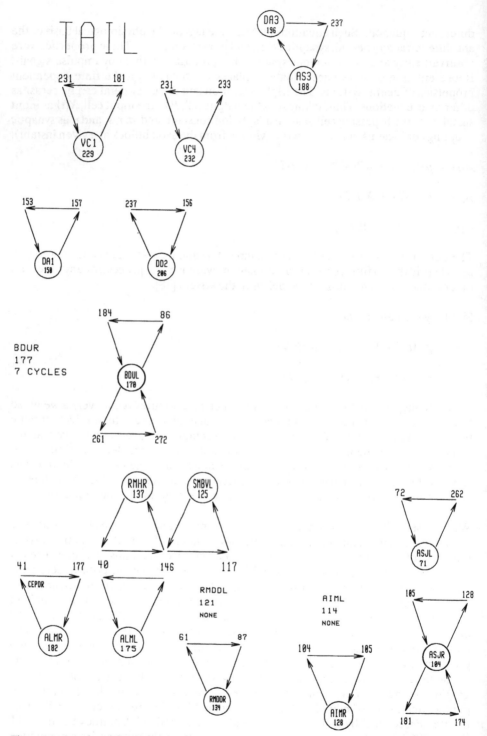

Figure 2.6. Samples of three-neuron cycles

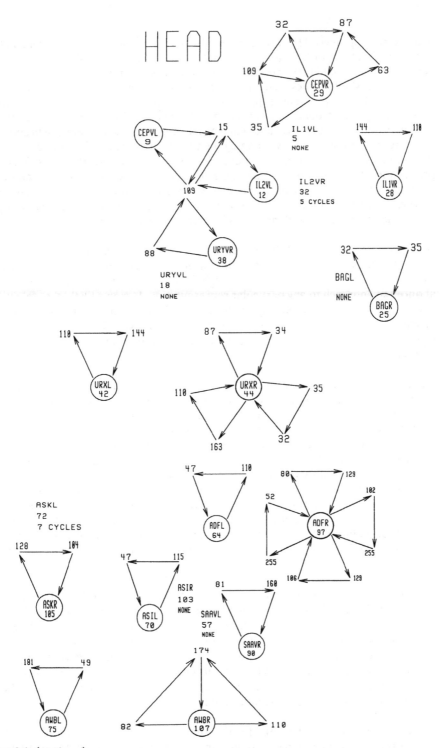

Figure 2.6. (*continued*)

and engineering views. The second book is by R.J. MacGregor [25], "Neural and Brain Modeling." This book is also computer-oriented, but emphasizes biological verisimilitude. Should these not suffice, any library or technical bookstore can provide a twenty-foot shelf of titles. Scientific journals include many papers which are specifically devoted to the problem of appropriate representation. Suffice it to say that for the present purposes, models of neurons may be divided along the somewhat blurred boundary line between logical and signal processing models, and between probabilistic and deterministic systems. Overlapping the first dichotomy, another line divides finite state from continuous variable representations. Most of the work seems directed at production of behavior by engineering a network in software, or less commonly, in hardware. The *C. elegans* problem focuses on a question: "Given an anatomically correct network of unit neurons that behave according to some limited set of operations, what does the network do that is interpretable as 'wormness'?"

A perspective over some of the models is provided by Uwe an der Heiden in a study entitled Structures of Excitation and Inhibition [16]. It is a comfortable framework in which to seek how *C. elegans* anatomy might be reified for simulation without binding at once the approach to any particular representation. It seems likely that selection of one neuronal representation may produce a significantly different behavior from that produced by selecting another. It is also possible, however, that over some subset of representations, animation of the nematode network may be in the large, insensitive to model choice.

Heiden's perspective is cast in the form of integral equations, and focuses on the statement of separable functions: the production of impulse trains from slow (generator) potentials, and the conversion of impulse strings into slow membrane potentials. Features of neural function, such as refractory period, conduction time, postsynaptic potentials, and threshold, find identity as formal entities in the system. Impulse generation and the action of the postsynaptic membrane are treated as a transform pair, which converts a function into a transform domain and then recovers the function. The pattern mirrors familiar engineering uses of convolution transforms like the LaPlace, Fourier, Hilbert, Hartley, Walsh and others [19, 21]. Operations performed on the transformed functions have rigorous representation in the domain of the original function. The generalization by An der Heiden is shown to encompass at least the Ratliff-Hartline equations, J.D. Cowan's differential equation model, and the McCulloch-Pitts formulation.

Looking through the same telescope from the other end is the work of Marmarelis and Marmarelis [26]. This work addresses the interpretation of experimental data from nonlinear control and neural systems. The result of their methodology is the determination of one or more kernels to a convolution representation of the system under study. The justification for system description is developed in progressively more practical mathematics starting with the Volterra representation, and passing via Wiener-Lee orthogonalization to the Lee-Schetzen method of using correlation functions. In both convolution and correlation, the independent variable is time.

A neurologically coherent exemplification of neuronal activity which fits the convolution perspective is that by Cowan, as further developed by Stein et al. These models are

membrane, $(a+b)T/\pi$. If the two pulse trains $Z(a)$ and $Z(b)$ are added, even if some pulses overlap, the total number of pulses is preserved in any sum over time. Adding the functions before filtering produces $(a+b)T/\pi$. The procedure is linear in the average, i.e., the average magnitude of the sum function is the sum of the average magnitudes of the functions. Any linear low-pass filter would also generate the linear magnitude relationship. If time is in any way quantized, the applicable mathematical tools for the single neuron would seem to be that of finite time series, i.e., autoregressive moving average representations [18]. One may conclude that, except for scaling factors, a low-pass filter recovers the average sum. Although, in general, low-pass filters do not recover the signal from frequency modulated form [29], in this case, impulse interval modulation followed by linear filter is a transformation pair that seems to satisfy Heiden's generalization. Angle modulation as we have defined it, is a transformation of a continuous function in the time domain into a discrete function in the time domain. As interpreters of anatomy, we find this description awkward but hewing more closely to observation, and thus assisting in the quest for a functional neuroanatomy.

Realistic inclusion of several non-linear features such as saturation, threshold, and refractory period is necessary. None of these appears to modify the characteristic that only one binary operation, addition, is necessary. It has no inverse. The significance of the impulse is limited to positive (excitation), negative (inhibition), and absence 0. A simple arithmetic of markers may be constructed on this basis, and it is non-Boolean. Because of the inverse relation between signal magnitude and time interval, sums are not strictly arithmetic. For example, two magnitudes with a common divisor have augmented influence greater than their sum. As in other approaches, the neuron performs the operations on signals but the network anatomy forms the equations.

The inverse transform is commonly attributed to a specialized part of postsynaptic membrane. We have described it as a low-pass filter, but both physical and theoretical considerations recommend the existence of a low frequency cutoff, particularly in the interneuron level. Within this bandwidth, time intervals are divided by addend and augend impulses. Impulses from a single connection cannot be spaced closer than a minimum interspike interval, and commonly do not sum. Spatial summation of inputs, the principal form of calculation, is limited by a summation time determined by the lower "skirt" of the filter. These considerations permit the view of the computation at the neuron as a majority organ [36, 37], and that the impulse size denotes a logical (Boolean) magnitude in a sequence of logical assertions. At the same time, temporal spacing of impulses from a single source marks off interval magnitudes in a discretized time domain. A large range of time intervals (milliseconds to minutes) may be marked off by impulses. Convergence of two signals, at a neuron with threshold greater than zero, might produce after-summation and inversion (a string that marks off the beat frequency of the original, but looks like a logical "and"). If close enough spaced, the signals would reinforce, and if far apart, neither would be remarked. Logical coincidence would report some form of harmony. Variations in signal timing dependent on signal strength thus become the "logical" operation of a fixed anatomical connection, i.e., a relationship between spacings rather than between assertions.

Many nervous systems, particularly those like *C. elegans*, have gap junctions which appear to behave like electrical synapses. These can be shown to be equivalent to a "transform" pair different from the chemical synapse. Further, electrical synapses are evidently more dependent upon the space constant established by the physical location on the soma or dendrite. While this is certainly a consideration also for chemical synapses, the inclusion of these phenomena is omitted from the present discussion.

The logic of selecting which inputs are to interact is the function of the genetically determined anatomy and any subsequent trophic changes. The computation and its "significance" occurs at the synaptic site, where time magnitudes are recognized and combined into other time magnitudes representing recognition of some harmony or dissonance, rather than truth or falsehood in the inputs in a rigorously logical sense. Within the neuronal network, variables are measured in physical time, and no driving clock is needed. George Bishop, in a provocative essay of four decades ago, asserted in the heyday of the Boolean logic model of the nervous system "that the chief and most characteristic functions of nervous and other excitable tissues are performed by means of graded responses" [41]. In this recollection, we state this somewhat differently: "The CNS computes in its silences, by marking off all magnitudes by the passage of time in an otherwise silent sensorium." In a world where the only variable is time, oddly enough, no clock is necessary, and harmony may drive the nervous system at its most elementary levels.

### G. Refractory Period and Threshold

Using the pulse frequency model, a few other physiological features need consideration. A very small signal, $f(t)$, may not produce a pulse for a long time since $f(t)t$ must at some time evaluate to $\pi$. Similarly, since any physical system is eventually limited in bandwidth by power or inertia considerations, large signals in the system must be limited by a minimum interspike interval. Neurons characteristically show time periods of unresponsiveness in their electrical activity, such as the absolute refractory period, synaptic transmission delay, and conduction time. The presence of an absolute refractory period sets the highest frequency of response pulses. For a refractory period of duration $R$, the highest angular velocity (magnitude) effective $X$, is $XR = \pi$. The corresponding frequency is $1/2R$. For $R \approx 1$ msec, $f \approx 500$ hertz. It is important to remember that the frequency domain is physically mapped into the time domain. The fastest impulse potentials observed in neurons in common reference are, perhaps, in the Renshaw cells of the mammalian spinal motor system, and have been reported as high as 1700 per second.

To estimate a lowest frequency, one must consider adaptation. This differs according to neuron type. Some sensory receptors show very low adaptation and are functionally low-pass. Others, like most interneurons, adapt quickly, perhaps, as briefly as the postsynaptic potentials. If the adaptation time is taken as the time constant for the low frequency cutoff, one gets an estimate of the slowest effective signal. In a low-pass system, however small the input signal angular velocity (magnitude $x$), if the threshold is $S > 0$, eventually, after a time interval, $xT > S$, and the spacing $xT = n\pi$. Alternatively, threshold may be used to determine a lowest frequency $f = 1/2T = S/2\pi$. Threshold appears to be an independent variable of the system, not necessarily related to $R$. In a bandpass system, both $S$ and kernel participate through convolution with the input signal to determine the lowest effective frequency for impulse transmission. This characteristic, however, pertains to temporal summation, and does not necessarily affect the significance of delimited long interspike intervals. The bandwidth of a neuron may thus be estimated to lie between an upper bound $f = 1/2R$ ($500$ hz) and a lower bound which may be estimated in terms of convolution kernel and threshold. The lower bound, however, does not appear to limit the range of signal magnitudes, but rather, the amount of error tolerated in establishing logical coincidences.

The kernel and threshold establish a weighting sequence for summation of sequentially arriving impulses, which, on the time scale, may be considered to be quantized into refractory periods. A relationship between temporal summation and lowest end of the bandwidth may be conjectured, using the difference equation $X(nT) = c*X(n-1)T + Y_k(nT)$, where $n \geq 0$, and $Y(nT)$ is taken to have the value 1 only for $n(mod(K)) = 0$ and zero otherwise, and $c < 1$. The limiting solution for this equation is $x(nT) = (1-c^{nK})/(1-c^K)$. For some $n$, $X(nT)$ must exceed the threshold $S$ to propagate. The adaptation constant $c$ and the threshold $S$ determine the maximum time window in which temporal summation can occur. Impulses arriving more widely spaced than $nK$ do not interact. Thus, the convolution kernel determines the lower edge of the neuronal transmission bandwidth. The indicated bandwidth, of course, is different from the range of time intervals delimited by impulses, which, according to the hypothesized mapping, may lead to representation of magnitudes smaller than that provided by the lower end of the summation bandwidth. Threshold, interacting with the convolution kernel, makes variable the number of convergent signals necessary for a neural network to propagate signals by spatial summation [44].

If an organism has neurohumoral mechanisms to adjust threshold, it has mechanisms for controlling the "observation time" for neurons, as well as the number of logical conditions that need to be fulfilled for transmission. Both the angular velocity domain and the "frequency" domain denoted above by $\omega$ or $f$ are physically in the time domain. The angle modulation thus described converts a time varying signal into a time varying interpulse interval between fixed-sized pulses. The linear operator, $h(t)$, through convolution, converts the pulse sequence into a time varying magnitude. The time constant of a neuronal membrane depends upon the physical properties of the membrane and its total size. The latter is a function, at least, of the cell size. For uncoupled cells, the literature reports time constants ranging from 1 to 30 milliseconds, e.g., 29 ms for a neuron in Heliostoma, 4 ms in the spinal motor neuron of the cats, and 2.5 ms as the physical time constant of the mammalian unit membrane. The time constant defines the rate of rise of potential in response to a step function, and may be related to the bandwidth of the neuron as we have shown. The principal purpose of the preceding discussion is to motivate the use of the detailed anatomy of C. elegans as a basis for attempting to deduce how a network represents and manipulates it inputs. It is certain that the nervous system animates existence of the worm. The goal, which is to reconstruct through a formal, active representation of the neuron and through a formal representation of its anatomy, the worm, admits to a hubris that Gods dislike. What this approach seeks is the modeling of function from structural information. To this end, we have belabored computable descriptions for the neuron, action essential to computed neuroethology. Still, one has to admit that the idea has its limits. If you are going to model the worm, there are many things that anatomy simply does not provide, even with much physiological consideration. Yet it may be the only way we can proceed today.

Neural network research, on the other hand, for the most part, and for most of its successes to date as measured in terms of successful imitations of specified behavior, use artificial configurations of artificial neurons and external processes such as learning sets, annealing, training, and probabilistic application of trophic processes like Hebb's rule, and study nervous systems to develop structure from specified behavior. This essentially is artificial intelligence.

Lord E.D. Adrian in his Waynflete lectures of 1947 [40] quotes from a visualization of the signals passing in a perceiving nervous system as moving points of light in a huge dark image of neuronal channels in the brain, not unlike perhaps the lights of motor vehicles traversing a network of roads on a darkened hill. A view which has not the contemplative charm of Lord Adrian, but recognizes differently the transactions of impulses, would compare them to the conversations at a large political cocktail party where many simultaneous conversations occurring pairwise among threes and small circles assemble. The neurophysiologist, like an observant news reporter, must divine whether the purpose of the gathering is to establish a sense of the party, or a plank in a platform, or a decision to march upon city hall......through a field of daisies......

Chapter III

Data in Machinable Form

## A. README.1ST

### A.1. *Purpose*

The following computer readable database has been prepared from published documents for the purpose of making small computer manipulations of the neuroanatomy of the nematode, *Caenorhabditis elegans*. Practical accuracy of the data is limited by the accuracy of the data referenced, and of the manual entry of data into the computer via keyboard. Inconsistencies in literature, when observed, have been resolved arbitrarily by the editors. Large groups of entries, particularly those in the table of neuromuscular connections (MOTORCON file set), are based on plausible conjecture by extension from precise data on representative somatic segments.

These data are made available for machine reproduction and may be duplicated according to the Distribution Agreement below. Please communicate any corrections or new specific information to the editors.

### A.2. *Media and Contents*

Two (2) 5.25 inch, 360 Kb floppy diskettes accompany this book.

Disk 1 is the data diskette, and contains seven (7) files:

(1) NEURCLAS.PRN, (2) GAPJUNC.PRN, (3) LEGENDGO.PRN,
(4) PREPOSGO.PRN, (5) PARTLIS1.PRN, (6) PARTLIS2.PRN, and
(7) MOTORCON.PRN.

Disk 2 is the program diskette, and contains the sample computer programs discussed in Chapter IV.

### A.3. *Directions*

Files in this database are grouped into six (6) file sets:

1) The NEURCLAS file set contains the synaptic connections of the NEURon CLASses of the C. elegans hermaphrodite.

2) The GAPJUNC file set contains the GAP JUNCtions of the individual neurons of the C. elegans hermaphrodite.

3) The LEGENDGO file set contains the LEGENDs used to arrange the PREPOSGO file set in Ganglionic Order. This file set is a prerequisite in understanding the PREPOSGO file set.

4) The PREPOSGO file set contains the PREsynaptic and POStsynaptic connections of the individual neurons of the C. elegans hermaphrodite. These neurons are arranged in Ganglionic Order. A prerequisite for understanding this file set is the LEGENDGO file set.

5) The PARTLIST file set lists all the cells (neural and non-neural) that comprise C. elegans as an organism, both male and hermaphrodite, including the cell lineage and a brief description of each cell.

6) The MOTORCON file set contains the MOTOR (muscle/effector) cell CONnections of the motoneurons of C. elegans.

Files in each set have names that terminate in either the .TXT or the .PRN extension. Files named with the .TXT extension (TeXT files) are printed in this book, and files named with the .PRN extension (PRiNt files) are printed in this book and stored in machinable form on Disk 1. Text files, which contain the description, derivation, and comments about corresponding data in the print files, are not included on the diskettes. Except for the PARTLIST file set where a single text file describes two print files, each file set is composed of a text file-print file pair.

Files on Disk 1 (which are print files with the .PRN file name extension) can be imported into electronic spreadsheets like Lotus™ 123™ (version 2 or higher) and Quattro Pro™ (release 1 or higher), and can be manipulated and saved either as electronic worksheets or also as print files. As electronic worksheets, they can also be imported into database software like Paradox™. The authors cannot be responsible for any revisions or modifications of any of the electronic spreadsheet programs above which would make the files with the .PRN extension in Disk 1 incompatible with manufacturer-provided software or its descendants. No electronic spreadsheet software or documentation is provided with this database.

**A.4.** *Nomenclature and Conventions*

Neuron names consist of 3 upper case letters (e.g., AQR), the last letter replaceable by a number of up to 2 digits. Geometric symmetry descriptors added to the first three letters are the letters D or V (dorsal or ventral) and L or R (left or right). The first three characters of the neuron name identify the neuron class. For example, neurons OLQDL, OLQDR, OLQVL, AND OLQVR belong to neuron class OLQ, and conversely, neuron class IL1 has neurons IL1DL, IL1DR, IL1L, IL1R, IL1VL, AND IL1VR as members. For those with no symmetrical relations, the third character of the neuron name is a numeral which represents the anterior to posterior location of a neuron relative to the members of its class, e.g., VA12, and this numeral is replaced by the lower case n, e.g., VAn, to refer to the neuron class [11].

**A.5.** *Distribution Agreement*

**A.5.a. Copyright**

"AY's Neuroanatomy of *C. elegans* for Computation" software database, including its documentation, is copyrighted. You may copy or otherwise reproduce the software or any part of it, as expressly permitted in this Distribution Agreement. To prevent any inadvertent file alteration, the editors encourage that you duplicate the two original master diskettes for back-up purposes prior to any file manipulation, provided that you reproduce all copyright notices and other proprietary legends on such copies.

**A.5.b. Restrictions on Use and Transfer**

Copies of the original software (a) may not be edited and distributed freely, and (b) may not be sold independently for profit or monetary gain. In the interest of keeping all editions of the software database consistent with the editors of this neuroanatomic atlas, provision (a) above disallows you from distributing any edited version, wholly or in part, of this software database. At the same time, the editors strongly encourage you to send any corrections, additions, or suggestions to them, so that future versions of the software database can reflect these revisions. This practice will assure that future distribution will be a current and a standard machine readable data form for *C. elegans*. Provision (b) implies that this software database itself was compiled from original sources and edited by the editors mainly in the interest of scientific study. While back-up and/or edited copies of this software may be used personally, any research or other publication resulting from its use should cite it and original literature as a bibliographic reference (see Terms and Conditions of Use below).

**A.5.c. Software Medium and Warranty**

The software will be available only in 5.25 inch, 360 Kb floppy diskettes. Defective master diskettes will be replaced upon return of the original.

**A.5.d. Limitations of Liability**

The editors make no warranties, express or implied, with respect to the software database or its fitness for any purpose. The software is distributed only on an "as is" basis, mainly for the purpose of stimulating further scientific activity on this field. The entire risk as to its quality, use, and performance is with the user. In no event will the compiler-editors be liable for direct, indirect, incidental, or consequential damages resulting from any defect or errors in the database, even if you have been advised of the possibility of such damage.

**A.5.e. Terms and Conditions of Use**

Aside from following the provisions above, any publications, activities, or products resulting from or stimulated by this database, should acknowledge the use of, further use of, or reference to this database.

--- End of File ---

## B. NEURCLAS FILE SET

### B.1. *NEURCLAS.TXT*

#### B.1.a. Description

The NEURCLAS data file has seven (7) columns.

1) The **Entry Number** column provides a sequential numbering system for each of the neuron class entries.

2) The **Class Name** column contains the neuron classes arranged alphabetically.

3) Neuron classes in the Class Name column are presynaptic to neuron classes in the **Pre** column.

4) Neuron classes in the Class Name column are postsynaptic to neuron classes in the **Post** column.

5) The **Gap Junc** column contains the neuron classes which pair with the neurons in the Class Name column to form gap junctions.

6) The **Chart ID** column contains the derivation of the data, and is explained below.

7) The **Type** column was added to classify the functional type of each neuron class. The conventions are: S = Sensory neuron type, I = Interneuron type, M = Motoneuron type, and B (Body) = Non-neural cell.

#### B.1.b. Derivation

Data were mainly derived from:

Wood, W.B. The Nematode *Caenorhabditis elegans* (Monograph 17), edited by W. Wood and the Community of C. elegans Researchers. Cold Spring Harbor: Cold Spring Harbor Laboratory, 1988.

As stated, the Chart ID column contains the derivation of the data. The convention for each entry is one number, followed by one letter, e.g., 3F. The numbers used are 1 to 6, and represent the pages from the reference (above) from which the data were taken:

1 - page 450

2 - page 451

3 - page 452

4 - page 453

5 - page 454

6 - page 455.

56

The letters used are F and T, and refer either to the (F)igure or to the (T)able which are laid out side by side on each of the cited pages of the reference above (see Figure 1.5, p. 5). Thus 3F means that data were taken from the Figure illustrated on page 452, and 5T means that data were taken from the Table on page 454.

In Figure 1.5, p. 5, classes of sensory neurons (S) are represented by triangles, inter-neurons (I) by hexagons, and motor neurons (M) by circles [12]. Together with non-neural cells (B), this type of representation is the basis for entries in the Type column.

## B.1.c. Comments

The format of the data referenced in literature is shown in Figure 1.5, p. 5.

Neurons within a class have the same, or very similar, patterns of connectivity when compared to members of other classes [11]. Although the total of 118 neuron classes may overestimate the number of types of neurons that are intrinsically different, there is usually striking morphological or ultrastructural differences between classes [12].

The Type column provides the simplest functional grouping for the neuron classes. Realistically, however, these functions may overlap within one class [11].

The NEURCLAS data file integrates the pre- and postsynaptic neuron class connections derived from both figures and tables in the reference. To verify a connection from the reference, use the Chart ID as a guide. Redundant connections of classes among figures, among tables, and among figures and tables have been removed.

General nomenclature was discussed in subsection A.4. Conventions specific to this .PRN file include (1) the "/" format for a neuron group and several neuron classes, i.e., IL2L/R for neuron group IL2L and IL2R, AVD/E for neuron classes AVD and AVE, VD/DD for VDn and DDn, VA/DA for VAn and DAn, and VB/DB for VBn and DBn, (2) the lowercase "sh" appended to CEP, i.e., CEPsh which stands for cephalic sheath cells (non-neurons, see PARTLIS1.PRN, p. 171-190), (3) muscle, written as MUSn, where the lowercase n is either 1, 3, or 5 depending on the page of the figure referenced (same meaning as the numbers in the Chart ID column, and as seen in Figure 1.5, p. 5, MUS3 is vulval muscle), (4) neuromuscular junction (NMJ) found in the ventral cord (VC) or dorsal cord (DC), (5) general anatomical parts stated in words, i.e., PHARYNX and Anal, taken directly from the reference, (6) motor neuron classes with the lower case n (e.g., ASn) taken from a table, and without the lower case n (e.g. AS) taken from a figure, and (7) an entry indicating uncertainty, OLQ?, copied from a table.

--- End of File ---

**B.2.** *NEURCLAS.PRN*

| Entry Number | Class Name | Pre | Post | Gap Junc | Chart ID | Type |
|---|---|---|---|---|---|---|
| 1 | ADA | SMD | | PVQ | 3T | I |
| 2 | ADA | RIP | | ASH | 3T | I |
| 3 | ADA | RIM | | AVD | 3T | I |
| 4 | ADA | AVJ | | | 3F | I |
| 5 | ADA | AVB | | ADA | 3T | I |
| 6 | ADA | | | ADF | 3T | I |
| 7 | ADE | RMD | AVM | | 2T | S |
| 8 | ADE | RIG | | | 2F | S |
| 9 | ADE | OLL | | | 2F | S |
| 10 | ADE | FLP | FLP | | 2F | S |
| 11 | ADE | CEP | | | 2F | S |
| 12 | ADE | AVA | BDU | | 2T | S |
| 13 | ADE | | IL2L/R | AVK | 2T | S |
| 14 | ADF | SMB | | ADA | 4T | S |
| 15 | ADF | RIR | | RIH | 4T | S |
| 16 | ADF | RIA | | ADF | 4T | S |
| 17 | ADF | AUA | AWB | | 4F | S |
| 18 | ADF | AIZ | | | 4F | S |
| 19 | ADF | | | AIA | 4T | S |
| 20 | ADL | AVD | | ADL | 4T | S |
| 21 | ADL | AVB | | OLQ | 4T | S |
| 22 | ADL | AVA | | RMG | 4T | S |
| 23 | ADL | AIB | | | 4F | S |
| 24 | ADL | AIA | | | 4F | S |
| 25 | AFD | AIY | AWA | AIB | 4F | S |
| 26 | AFD | | | AFD | 4T | S |
| 27 | AFD | | AIN | | 4F | S |
| 28 | AIA | RIF | | ADF | 4T | I |
| 29 | AIA | AWC | | AIA | 4T | I |
| 30 | AIA | AIB | ASH | ASI | 4F | I |
| 31 | AIA | | AIZ | | 4F | I |
| 32 | AIA | | AWC | | 4F | I |
| 33 | AIA | | PVQ | | 4F | I |
| 34 | AIA | | ADL | AWA | 4F | I |
| 35 | AIA | | ASE | | 4F | I |
| 36 | AIA | | ASG | | 4F | I |
| 37 | AIA | | ASI | | 4F | I |
| 38 | AIA | | ASK | | 4F | I |

| Entry Number | Class Name | Pre | Post | Gap Junc | Chart ID | Type |
|---|---|---|---|---|---|---|
| 39 | AIA | | AIM | | 4F | I |
| 40 | AIB | SAAD | RIB | RIS | 4T | I |
| 41 | AIB | RIM | DVB | DVC | 4T | I |
| 42 | AIB | RIB | RIM | RIG | 4T | I |
| 43 | AIB | AVB | DVC | DVB | 4T | I |
| 44 | AIB | | ASH | | 4F | I |
| 45 | AIB | | AWC | | 4F | I |
| 46 | AIB | | FLP | RIV | 4T | I |
| 47 | AIB | | AIZ | | 4F | I |
| 48 | AIB | | AIA | | 4F | I |
| 49 | AIB | | ASG | AFD | 4F | I |
| 50 | AIB | | ADL | | 4F | I |
| 51 | AIB | | ASI | | 4F | I |
| 52 | AIB | | ASK | | 4F | I |
| 53 | AIB | | ASE | | 4F | I |
| 54 | AIM | CEPsh | | SIBD | 4T | I |
| 55 | AIM | AVF | | AIM | 4T | I |
| 56 | AIM | ASK | | | 4F | I |
| 57 | AIM | ASJ | ASK | | 4F | I |
| 58 | AIM | ASG | | | 4F | I |
| 59 | AIM | AIA | | | 4F | I |
| 60 | AIN | RIB | | | 4T | I |
| 61 | AIN | CEPsh | | | 4T | I |
| 62 | AIN | BAG | | AIN | 4T | I |
| 63 | AIN | ASE | | ASG | 4F | I |
| 64 | AIN | AFD | | AUA | 4F | I |
| 65 | AIY | RIB | | RIM | 4T | I |
| 66 | AIY | RIA | | AIY | 4T | I |
| 67 | AIY | AIZ | AWC | | 4F | I |
| 68 | AIY | | AWA | | 4F | I |
| 69 | AIY | | ASE | | 4F | I |
| 70 | AIY | | AFD | | 4F | I |
| 71 | AIZ | SMB | RIR | ASG | 4T | I |
| 72 | AIZ | RIM | RIH | | 4T | I |
| 73 | AIZ | RIA | AWA | AIZ | 4T | I |
| 74 | AIZ | AVE | AWB | | 4T | I |
| 75 | AIZ | AIB | AWA | ASH | 4F | I |
| 76 | AIZ | AIA | AIY | | 4F | I |
| 77 | AIZ | | HSN | | 4T | I |
| 78 | AIZ | | AWB | | 4F | I |

| Entry Number | Class Name | Pre | Post | Gap Junc | Chart ID | Type |
|---|---|---|---|---|---|---|
| 79 | AIZ | | ADF | | 4F | I |
| 80 | ALM | PVC | | AVM | 2F | S |
| 81 | ALM | CEP | | | 2T | S |
| 82 | ALM | BDU | | PVR | 2T | S |
| 83 | AQR | RIA | DVA | AVK | 2T | S |
| 84 | AQR | PVC | | PVP | 2F | S |
| 85 | AQR | BAG | | | 2F | S |
| 86 | AQR | AVD | | RIG | 2T | S |
| 87 | AQR | AVB | | RIG | 2F | S |
| 88 | AQR | AVA | | | 2F | S |
| 89 | AS | VD/DD | AVB | AVA | 5F | M |
| 90 | AS | MUS5 | AVD/E | | 5F | M |
| 91 | AS | | AVA | | 5F | M |
| 92 | ASE | RIA | | | 4T | S |
| 93 | ASE | AWC | AWA | | 4F | S |
| 94 | ASE | AIY | ASI | | 4F | S |
| 95 | ASE | AIB | | | 4F | S |
| 96 | ASE | AIA | AIN | | 4F | S |
| 97 | ASG | AIB | | | 4F | S |
| 98 | ASG | AIA | AIM | AIN | 4F | S |
| 99 | ASG | | | AIZ | 4T | S |
| 100 | ASH | RIA | | ASH | 4T | S |
| 101 | ASH | AVD | | RIC | 4T | S |
| 102 | ASH | AVB | | RMG | 4T | S |
| 103 | ASH | AVA | | ADA | 4T | S |
| 104 | ASH | AIB | | | 4F | S |
| 105 | ASH | AIA | | AIZ | 4F | S |
| 106 | ASI | AWC | | | 4F | S |
| 107 | ASI | ASE | | AIA | 4F | S |
| 108 | ASI | AIB | | | 4F | S |
| 109 | ASI | AIA | | | 4F | S |
| 110 | ASI | | | ASI | 4T | S |
| 111 | ASJ | PVQ | AIM | | 4F | S |
| 112 | ASK | AIM | | | 4F | S |
| 113 | ASK | AIB | AIM | | 4F | S |
| 114 | ASK | AIA | PVQ | PVQ | 4F | S |
| 115 | ASK | | | RMG | 4T | S |
| 116 | ASK | | ASJ | ASK | 4T | S |
| 117 | AUA | RIB | URX | AUA | 4T | S |
| 118 | AUA | RIA | | URX | 4T | S |

60

| Entry Number | Class Name | Pre | Post | Gap Junc | Chart ID | Type |
|---|---|---|---|---|---|---|
| 119 | AUA | AVE | | | 4T | S |
| 120 | AUA | AVA | | | 4T | S |
| 121 | AUA | | ADF | AWB | 4F | S |
| 122 | AUA | | | AIN | 4F | S |
| 123 | AVA | VA/DA | DVA | VA/DA | 5F | I |
| 124 | AVA | VAn | FLP | LUA | 6T | I |
| 125 | AVA | VAn | ADE | AVA | 2T | I |
| 126 | AVA | SAB | SAA | AVA | 6T | I |
| 127 | AVA | SAB | AVD/E | SAB | 5F | I |
| 128 | AVA | SAB | AUA | VAn | 2T | I |
| 129 | AVA | PVC | LUA | | 6F | I |
| 130 | AVA | PVC | RIC | PVC | 2F | I |
| 131 | AVA | PVC | PVC | PVC | 5F | I |
| 132 | AVA | LUA | AVD | PVC | 6F | I |
| 133 | AVA | DAn | RIC | RIM | 6T | I |
| 134 | AVA | DAn | SAA | RIM | 2T | I |
| 135 | AVA | ASn | AUA | URY | 6T | I |
| 136 | AVA | ASn | DVC | URY | 2T | I |
| 137 | AVA | AS | AVB | AS | 5F | I |
| 138 | AVA | | ASH | | 5T | I |
| 139 | AVA | | LUA | | 2T | I |
| 140 | AVA | | AVD | | 2F | I |
| 141 | AVA | | DVA | | 6F | I |
| 142 | AVA | | SDQ | | 2F | I |
| 143 | AVA | | PVC | | 6F | I |
| 144 | AVA | | AVB | | 6T | I |
| 145 | AVA | | PLM | | 6F | I |
| 146 | AVA | | PVC | | 2F | I |
| 147 | AVA | | LUA | | 5T | I |
| 148 | AVA | | SDQ | ASn | 6T | I |
| 149 | AVA | | AVE | | 2F | I |
| 150 | AVA | | PLM | | 2F | I |
| 151 | AVA | | RIB | | 5T | I |
| 152 | AVA | | PLM | | 5T | I |
| 153 | AVA | | ADE | | 6T | I |
| 154 | AVA | | DVA | | 2T | I |
| 155 | AVA | | AVE | | 6T | I |
| 156 | AVA | | DVC | | 6F | I |
| 157 | AVA | | ADL | SAB | 6T | I |
| 158 | AVA | | PQR | | 2T | I |

| Entry Number | Class Name | Pre | Post | Gap Junc | Chart ID | Type |
|---|---|---|---|---|---|---|
| 159 | AVA | | AQR | DAn | 6T | I |
| 160 | AVA | | SDQ | | 5T | I |
| 161 | AVA | | ADE | | 5T | I |
| 162 | AVA | | PHB | | 2T | I |
| 163 | AVA | | PHB | | 5T | I |
| 164 | AVA | | ASH | VAn | 6T | I |
| 165 | AVA | | PQR | | 6F | I |
| 166 | AVA | | RIB | SAB | 2T | I |
| 167 | AVA | | PVP | | 6F | I |
| 168 | AVA | | PHB | | 6F | I |
| 169 | AVA | | SAA | RIM | 5T | I |
| 170 | AVA | | PQR | | 5T | I |
| 171 | AVA | | PVP | AVJ | 5T | I |
| 172 | AVA | | RIB | | 6T | I |
| 173 | AVA | | FLP | URY | 5T | I |
| 174 | AVA | | ADL | ASn | 2T | I |
| 175 | AVA | | AUA | | 5T | I |
| 176 | AVA | | PVD | | 6F | I |
| 177 | AVA | | RIC | AVA | 5T | I |
| 178 | AVA | | FLP | | 2F | I |
| 179 | AVA | | ASH | DAn | 2T | I |
| 180 | AVA | | PVP | | 2F | I |
| 181 | AVA | | DVC | SAB | 5T | I |
| 182 | AVA | | ADL | | 5T | I |
| 183 | AVA | | AVB | | 2F | I |
| 184 | AVA | | AQR | | 2F | I |
| 185 | AVA | | AQR | | 5T | I |
| 186 | AVB | HDC | RIM | RIB | 2T | I |
| 187 | AVB | HDC | RIF | AVB | 5T | I |
| 188 | AVB | AVE | AVM | | 2F | I |
| 189 | AVB | AVD/E | PVC | VB/DB | 5F | I |
| 190 | AVB | AVD | SDQ | SDQ | 2F | I |
| 191 | AVB | AVA | | RID | 5F | I |
| 192 | AVB | AVA | PVC | | 2F | I |
| 193 | AVB | ASn | RIF | AVB | 2T | I |
| 194 | AVB | AS | | DVA | 5F | I |
| 195 | AVB | | AIB | VBn | 2T | I |
| 196 | AVB | | AIB | | 5T | I |
| 197 | AVB | | PVN | | 5T | I |
| 198 | AVB | | ASH | SIBV | 2T | I |

| Entry Number | Class Name | Pre | Post | Gap Junc | Chart ID | Type |
|---|---|---|---|---|---|---|
| 199 | AVB | | ADA | RID | 2T | I |
| 200 | AVB | | AVF | | 2T | I |
| 201 | AVB | | ADL | | 5T | I |
| 202 | AVB | | PVP | | 2F | I |
| 203 | AVB | | AVF | | 5T | I |
| 204 | AVB | | AVG | | 2T | I |
| 205 | AVB | | URX | | 5T | I |
| 206 | AVB | | AVM | SIBV | 5T | I |
| 207 | AVB | | RIM | RIB | 5T | I |
| 208 | AVB | | PVN | | 2T | I |
| 209 | AVB | | URX | | 2T | I |
| 210 | AVB | | SDQ | | 5T | I |
| 211 | AVB | | AVF | DBn | 2T | I |
| 212 | AVB | | FLP | | 2F | I |
| 213 | AVB | | ADL | PVN | 2T | I |
| 214 | AVB | | PVR | PVN | 5T | I |
| 215 | AVB | | ASH | SDQ | 5T | I |
| 216 | AVB | | AVG | | 5T | I |
| 217 | AVB | | PVR | DVA | 2T | I |
| 218 | AVB | | AQR | | 5T | I |
| 219 | AVB | | PVP | | 5T | I |
| 220 | AVB | | FLP | | 5T | I |
| 221 | AVB | | ADA | | 5T | I |
| 222 | AVB | | AQR | | 2F | I |
| 223 | AVD | VAn | ASH | AVJ | 2T | I |
| 224 | AVD | VAn | ADL | FLP | 6T | I |
| 225 | AVD | SAB | AQR | FLP | 2T | I |
| 226 | AVD | SAB | ASH | ADA | 6T | I |
| 227 | AVD | LUA | PLM | | 6F | I |
| 228 | AVD | DAn | FLP | AVJ | 6T | I |
| 229 | AVD | DAn | ADL | ADA | 2T | I |
| 230 | AVD | AVA | PVC | AVM | 2F | I |
| 231 | AVD | AVA | LUA | | 6F | I |
| 232 | AVD | ASn | AQR | AVM | 6T | I |
| 233 | AVD | ASn | PQR | | 2T | I |
| 234 | AVD | | PVN | | 6F | I |
| 235 | AVD | | PVC | | 6F | I |
| 236 | AVD | | PQR | | 6F | I |
| 237 | AVD | | PVW | | 6F | I |
| 238 | AVD | | LUA | | 2T | I |

| Entry Number | Class Name | Pre | Post | Gap Junc | Chart ID | Type |
|---|---|---|---|---|---|---|
| 239 | AVD | | PHB | | 6F | I |
| 240 | AVD | | PVN | | 2T | I |
| 241 | AVD | | FLP | | 2F | I |
| 242 | AVD | | PLM | | 2F | I |
| 243 | AVD | | AVB | | 2F | I |
| 244 | AVD/E | VA/DA | DVA | | 5F | I |
| 245 | AVD/E | SAB | PVN | ADA | 5T | I |
| 246 | AVD/E | SAB | AVB | | 5F | I |
| 247 | AVD/E | AVA | PVC | | 5F | I |
| 248 | AVD/E | AS | | | 5F | I |
| 249 | AVD/E | | AIZ | | 5T | I |
| 250 | AVD/E | | RIS | | 5T | I |
| 251 | AVD/E | | PHB | | 5T | I |
| 252 | AVD/E | | AVJ | | 5T | I |
| 253 | AVD/E | | AQR | RIM | 5T | I |
| 254 | AVD/E | | PQR | RME | 5T | I |
| 255 | AVD/E | | OLL | | 5T | I |
| 256 | AVD/E | | CEP | | 5T | I |
| 257 | AVD/E | | RIB | | 5T | I |
| 258 | AVD/E | | ADL | AVJ | 5T | I |
| 259 | AVD/E | | FLP | AVM | 5T | I |
| 260 | AVD/E | | ASH | FLP | 5T | I |
| 261 | AVD/E | | BAG | | 5T | I |
| 262 | AVD/E | | RMG | | 5T | I |
| 263 | AVD/E | | URY | | 5T | I |
| 264 | AVD/E | | RIG | | 5T | I |
| 265 | AVD/E | | AVK | | 5T | I |
| 266 | AVD/E | | PLM | | 5T | I |
| 267 | AVD/E | | AUA | | 5T | I |
| 268 | AVD/E | | ALA | | 5T | I |
| 269 | AVD/E | | LUA | RMD | 5T | I |
| 270 | AVD/E | | URX | | 5T | I |
| 271 | AVE | VAn | URY | RME | 2T | I |
| 272 | AVE | SAB | FLP | RIM | 2T | I |
| 273 | AVE | DAn | URX | RMD | 2T | I |
| 274 | AVE | AVA | CEP | | 2F | I |
| 275 | AVE | ASn | DVA | AVE | 2T | I |
| 276 | AVE | | AVK | | 2T | I |
| 277 | AVE | | AVB | | 2F | I |
| 278 | AVE | | RMG | | 2T | I |

| Entry Number | Class Name | Pre | Post | Gap Junc | Chart ID | Type |
|---|---|---|---|---|---|---|
| 279 | AVE |  | BAG |  | 2F | I |
| 280 | AVE |  | RIG |  | 2F | I |
| 281 | AVE |  | RIS |  | 2T | I |
| 282 | AVE |  | AVJ |  | 2T | I |
| 283 | AVE |  | AIZ |  | 2T | I |
| 284 | AVE |  | OLL |  | 2F | I |
| 285 | AVE |  | AUA |  | 2T | I |
| 286 | AVE |  | RIB |  | 2T | I |
| 287 | AVE |  | PVC |  | 2F | I |
| 288 | AVE |  | ALA |  | 2T | I |
| 289 | AVF | NMJ(VC) |  |  | 3T | I |
| 290 | AVF | HSN |  |  | 3F | I |
| 291 | AVF | AVJ | HSN |  | 3F | I |
| 292 | AVF | AVH | AVH | AVH | 3F | I |
| 293 | AVF | AVB | AIM | AVF | 3T | I |
| 294 | AVG | PVN |  |  | 3T | S |
| 295 | AVG | AVB | PHA |  | 3T | S |
| 296 | AVG |  |  | RIF | 3F | S |
| 297 | AVH | SMB |  |  | 3T | I |
| 298 | AVH | RIR | AVF | AVF | 3F | I |
| 299 | AVH | PVP | PHA |  | 3T | I |
| 300 | AVH | AVJ |  |  | 3F | I |
| 301 | AVH | AVF |  |  | 3F | I |
| 302 | AVH | ADF |  |  | 3T | I |
| 303 | AVJ | RIS | PVC | PVC | 3T | I |
| 304 | AVJ | PVC | ADL | AVJ | 3T | I |
| 305 | AVJ | AVE |  | AVD | 3T | I |
| 306 | AVJ | AVD |  |  | 3T | I |
| 307 | AVJ | AVB | PVR | RIS | 3T | I |
| 308 | AVJ |  | PVN |  | 3F | I |
| 309 | AVJ |  | AVH |  | 3F | I |
| 310 | AVJ |  | BDU |  | 3F | I |
| 311 | AVJ |  | RIF |  | 3F | I |
| 312 | AVJ |  | HSN |  | 3F | I |
| 313 | AVJ |  | AVF |  | 3F | I |
| 314 | AVJ |  | ADA |  | 3F | I |
| 315 | AVK | SMD |  | AQR | 6T | I |
| 316 | AVK | RIM | RMF | AVK | 6T | I |
| 317 | AVK | PDE | PDE | DVB | 6F | I |
| 318 | AVK | HDC |  | RIC | 6T | I |

| Entry Number | Class Name | Pre | Post | Gap Junc | Chart ID | Type |
|---|---|---|---|---|---|---|
| 319 | AVK | AVE | RIG | SMB | 6T | I |
| 320 | AVK | | | RIG | 6T | I |
| 321 | AVK | | | ADE | 6T | I |
| 322 | AVK | | DVB | | 6F | I |
| 323 | AVK | | PVM | PVP | 6F | I |
| 324 | AVL | VD12 | | | 6T | I |
| 325 | AVL | SAB | | | 6T | I |
| 326 | AVL | NMJ(VC) | | | 6T | I |
| 327 | AVL | | PVP | DVB | 6F | I |
| 328 | AVL | | DVB | PVM | 6F | I |
| 329 | AVL | | | DVC | 6F | I |
| 330 | AVM | PVR | | | 2T | S |
| 331 | AVM | PVC | | AVD | 2F | S |
| 332 | AVM | BDU | | | 2T | S |
| 333 | AVM | AVB | | ALM | 2F | S |
| 334 | AVM | ADE | | | 2T | S |
| 335 | AWA | ASE | | | 4F | S |
| 336 | AWA | AIZ | | | 4F | S |
| 337 | AWA | AIY | | | 4F | S |
| 338 | AWA | AFD | | AIA | 4F | S |
| 339 | AWA | | | AWA | 4T | S |
| 340 | AWB | RIA | | AWB | 4T | S |
| 341 | AWB | AVB | | RMG | 4T | S |
| 342 | AWB | AIZ | | | 4F | S |
| 343 | AWB | ADF | | AUA | 4F | S |
| 344 | AWC | AIY | ASE | | 4F | S |
| 345 | AWC | AIB | | | 4F | S |
| 346 | AWC | AIA | ASI | | 4F | S |
| 347 | AWC | | AIA | | 4T | S |
| 348 | BAG | RIG | RIG | | 2F | S |
| 349 | BAG | RIB | | BAG | 2T | S |
| 350 | BAG | RIA | AIN | RIR | 2T | S |
| 351 | BAG | AVE | AQR | RIG | 2F | S |
| 352 | BDU | PVN | PVN | | 3F | I |
| 353 | BDU | HSN | | | 3F | I |
| 354 | BDU | AVJ | | | 3F | I |
| 355 | BDU | ADE | ALM | | 3T | I |
| 356 | BDU | | AVM | | 3T | I |
| 357 | BDU | | HSN | | 3T | I |
| 358 | CEP | URB | | | 2F | S |

| Entry Number | Class Name | Pre | Post | Gap Junc | Chart ID | Type |
|---|---|---|---|---|---|---|
| 359 | CEP | URB | URB | | 1F | S |
| 360 | CEP | URA | | | 1F | S |
| 361 | CEP | RMH | RIS | | 2T | S |
| 362 | CEP | RMH | | | 1F | S |
| 363 | CEP | RMG | | | 1F | S |
| 364 | CEP | RMD | | | 1F | S |
| 365 | CEP | RIC | URB | | 2F | S |
| 366 | CEP | RIC | RIS | | 1F | S |
| 367 | CEP | OLQ | RIH | | 2F | S |
| 368 | CEP | OLQ | RIH | | 1T | S |
| 369 | CEP | OLL | ADE | OLQ | 2F | S |
| 370 | CEP | OLL | ALM | RIH | 1T | S |
| 371 | CEP | IL1 | ALM | | 2T | S |
| 372 | CEP | IL1 | | | 1F | S |
| 373 | CEP | AVE | OLL | OLQ | 1T | S |
| 374 | CEP | AVE | OLL | RIH | 2F | S |
| 375 | CEP | | ADE | | 1T | S |
| 376 | DVA | VB/DB | | | 5F | I |
| 377 | DVA | VBn | | | 6T | I |
| 378 | DVA | SMB | | | 6T | I |
| 379 | DVA | SMB | PDE | PHC | 5T | I |
| 380 | DVA | RIR | AIZ | PVR | 5T | I |
| 381 | DVA | RIR | AIZ | PVC | 6T | I |
| 382 | DVA | PVR | PLM | | 5T | I |
| 383 | DVA | PVR | PHC | | 6F | I |
| 384 | DVA | PVC | | | 5F | I |
| 385 | DVA | PVC | PVD | | 6F | I |
| 386 | DVA | DBn | | | 6T | I |
| 387 | DVA | AVE | PLM | AVB | 6T | I |
| 388 | DVA | AVD/E | | | 5F | I |
| 389 | DVA | AVA | PDE | PVR | 6F | I |
| 390 | DVA | AVA | | AVB | 5F | I |
| 391 | DVA | AUA | | | 6T | I |
| 392 | DVA | AUA | AIZ | | 5T | I |
| 393 | DVA | AQR | SDQ | PVC | 5T | I |
| 394 | DVA | AQR | SDQ | | 6T | I |
| 395 | DVA | AIZ | | | 6T | I |
| 396 | DVA | | PHA | | 6F | I |
| 397 | DVA | | PHA | | 5T | I |
| 398 | DVA | | PVD | | 5T | I |

| Entry Number | Class Name | Pre | Post | Gap Junc | Chart ID | Type |
|---|---|---|---|---|---|---|
| 399 | DVB | (Anal) | | | 6T | I |
| 400 | DVB | RMF | | AIB | 6T | I |
| 401 | DVB | RIG | | | 6T | I |
| 402 | DVB | PDA | | AVL | 6F | I |
| 403 | DVB | NMJ | | | 6T | I |
| 404 | DVB | DVC | | AVK | 6F | I |
| 405 | DVB | DD6 | | | 6T | I |
| 406 | DVB | DA8 | | | 6T | I |
| 407 | DVB | AVL | | | 6F | I |
| 408 | DVB | AVK | | | 6F | I |
| 409 | DVB | AIB | | RIB | 6T | I |
| 410 | DVC | RMF | | | 6T | I |
| 411 | DVC | RIG | | AIB | 6T | I |
| 412 | DVC | AVA | DVB | AVL | 6F | I |
| 413 | DVC | AIB | | VD1 | 6T | I |
| 414 | DVC | | | PVP | 6F | I |
| 415 | FLP | PVC | ADE | | 2F | S |
| 416 | FLP | AVE | | RIH | 2T | S |
| 417 | FLP | AVD | | | 2F | S |
| 418 | FLP | AVB | | | 2F | S |
| 419 | FLP | AVA | | | 2F | S |
| 420 | FLP | AIB | | FLP | 2T | S |
| 421 | FLP | ADE | | | 2F | S |
| 422 | FLP | | | AVD | 2T | S |
| 423 | HSN | VUL | | | 3F | M |
| 424 | HSN | VCn | | | 3F | M |
| 425 | HSN | RIF | | | 3F | M |
| 426 | HSN | BDU | | | 3T | M |
| 427 | HSN | AWB | PLM | HSN | 3T | M |
| 428 | HSN | AVJ | AVF | | 3F | M |
| 429 | HSN | AVF | BDU | | 3F | M |
| 430 | HSN | AIZ | | | 3T | M |
| 431 | IL1 | RMD | CEP | RME | 1F | S |
| 432 | IL1 | RIP | | | 1F | S |
| 433 | IL1 | MUS1 | IL2 | | 1F | S |
| 434 | IL1 | | | IL1 | 1T | S |
| 435 | IL1 | | | OLL | 1T | S |
| 436 | IL2 | URA | | URX | 1F | S |
| 437 | IL2 | RME | | | 1F | S |
| 438 | IL2 | RIP | | | 1F | S |

| Entry Number | Class Name | Pre | Post | Gap Junc | Chart ID | Type |
|---|---|---|---|---|---|---|
| 439 | IL2 | RIH | | RMG | 1T | S |
| 440 | IL2 | OLQ | | | 1T | S |
| 441 | IL2 | IL1 | | | 1F | S |
| 442 | IL2 | ADE | | | 1T | S |
| 443 | LUA | PVC | AVD | PLM | 6F | I |
| 444 | LUA | AVJ | | AVA | 6T | I |
| 445 | LUA | AVD | | | 6F | I |
| 446 | LUA | AVA | AVA | | 6F | I |
| 447 | MUS1 | | SMD | | 1F | B |
| 448 | MUS1 | | RIV | | 1F | B |
| 449 | MUS1 | | URA | | 1F | B |
| 450 | MUS1 | | RMG | | 1F | B |
| 451 | MUS1 | | RMD | | 1F | B |
| 452 | MUS1 | | RMH | | 1F | B |
| 453 | MUS1 | | IL1 | | 1F | B |
| 454 | MUS1 | | RIM | | 1F | B |
| 455 | MUS1 | | RME | RMD | 1F | B |
| 456 | MUS1 | | SMB | | 1F | B |
| 457 | MUS1 | | RMF | | 1F | B |
| 458 | MUS5 | | RID | | 5F | B |
| 459 | MUS5 | | VD/DD | | 5F | B |
| 460 | MUS5 | | AS | | 5F | B |
| 461 | MUS5 | | VB/DB | | 5F | B |
| 462 | MUS5 | | SAB | | 5F | B |
| 463 | MUS5 | | VA/DA | | 5F | B |
| 464 | OLL | SMD | | OLL | 2T | S |
| 465 | OLL | RMD | | | 2T | S |
| 466 | OLL | RIB | | IL1 | 2T | S |
| 467 | OLL | CEP | CEP | | 2F | S |
| 468 | OLL | AVE | ADE | RIG | 2F | S |
| 469 | OLQ | SIB | | RMD | 2T | S |
| 470 | OLQ | RMD | IL2 | RIB | 2T | S |
| 471 | OLQ | RIH | CEP | RIG | 2F | S |
| 472 | OLQ | RIC | RIH | CEP | 2F | S |
| 473 | OLQ | | | URB | 2F | S |
| 474 | OLQ | | | RIH | 2F | S |
| 475 | PDA | PVN | DVB | | 6F | M |
| 476 | PDA | NMJ(DC) | | | 6T | M |
| 477 | PDA | HDC | | | 6T | M |
| 478 | PDA | DD6 | | | 6T | M |

| Entry Number | Class Name | Pre | Post | Gap Junc | Chart ID | Type |
|---|---|---|---|---|---|---|
| 479 | PDA | DA9 | | | 6T | M |
| 480 | PDE | HDC | | PVC | 6T | I |
| 481 | PDE | DVA | AVK | | 6F | I |
| 482 | PDE | AVK | PVM | PVM | 6F | I |
| 483 | PDE | | PLM | | 6F | I |
| 484 | PDE | | | PDE | 6T | I |
| 485 | PHA | PVQ | | | 6T | S |
| 486 | PHA | PHB | | PVP | 6F | S |
| 487 | PHA | PHA | PHA | PHA | 6T | S |
| 488 | PHA | DVA | | | 6F | S |
| 489 | PHA | AVH | | | 6T | S |
| 490 | PHA | AVG | | | 6T | S |
| 491 | PHA | AVF | | | 6T | S |
| 492 | PHB | VA12 | | PHB | 6T | S |
| 493 | PHB | PVC | | | 6F | S |
| 494 | PHB | AVD | | | 6F | S |
| 495 | PHB | AVA | PHA | | 6F | S |
| 496 | PHB | | | AVH | 6T | S |
| 497 | PHC | PVC | | | 6F | S |
| 498 | PHC | DVA | | | 6F | S |
| 499 | PHC | DA9 | | PHC | 6T | S |
| 500 | PHC | | | VA12 | 6T | S |
| 501 | PHC | | | DA9 | 6T | S |
| 502 | PLM | PDE | | PVR | 6F | S |
| 503 | PLM | PDE | | | 2T | S |
| 504 | PLM | HSN | | | 6T | S |
| 505 | PLM | HSN | | LUA | 2T | S |
| 506 | PLM | DVA | | | 6T | S |
| 507 | PLM | DVA | | PVR | 2T | S |
| 508 | PLM | AVD | | PVC | 2F | S |
| 509 | PLM | AVD | | LUA | 6F | S |
| 510 | PLM | AVA | | | 2F | S |
| 511 | PLM | AVA | | PVC | 6F | S |
| 512 | PLN | SMB | | | 1F | I |
| 513 | PLN | SAA | | | 1F | I |
| 514 | PQR | AVD | | | 6F | S |
| 515 | PQR | AVA | PVN | PVP | 6F | S |
| 516 | PVC | VD/DB | | | 5F | I |
| 517 | PVC | VBn | PHC | PVW | 2T | I |
| 518 | PVC | VBn | AVJ | DVA | 6T | I |

| Entry Number | Class Name | Pre | Post | Gap Junc | Chart ID | Type |
|---|---|---|---|---|---|---|
| 519 | PVC | RID | ALM | PVC | 6T | I |
| 520 | PVC | RID | | | 5F | I |
| 521 | PVC | RID | AVJ | PVC | 2T | I |
| 522 | PVC | DBn | AQR | | 6T | I |
| 523 | PVC | DBn | PHB | AVJ | 2T | I |
| 524 | PVC | AVE | FLP | | 2F | I |
| 525 | PVC | AVE | AVM | AVJ | 6T | I |
| 526 | PVC | AVD/E | DVA | AVA | 5F | I |
| 527 | PVC | AVD | PVM | AVA | 6F | I |
| 528 | PVC | AVD | ALM | PLM | 2F | I |
| 529 | PVC | AVB | AVA | | 2F | I |
| 530 | PVC | AVB | PVP | PDE | 6T | I |
| 531 | PVC | AVB | | | 5F | I |
| 532 | PVC | AVA | DVA | PVW | 6F | I |
| 533 | PVC | AVA | AVM | AVA | 2F | I |
| 534 | PVC | AVA | AVA | | 5F | I |
| 535 | PVC | | PVD | | 5T | I |
| 536 | PVC | | PVM | | 2T | I |
| 537 | PVC | | AQR | PDE | 5T | I |
| 538 | PVC | | PVN | | 5T | I |
| 539 | PVC | | PHB | DVA | 5T | I |
| 540 | PVC | | DVA | | 2T | I |
| 541 | PVC | | ALM | AVJ | 5T | I |
| 542 | PVC | | PHB | | 6F | I |
| 543 | PVC | | PVM | | 5T | I |
| 544 | PVC | | PVN | | 2T | I |
| 545 | PVC | | PHC | | 6F | I |
| 546 | PVC | | AVM | PVW | 5T | I |
| 547 | PVC | | AVJ | PLM | 5T | I |
| 548 | PVC | | PVD | PLM | 6F | I |
| 549 | PVC | | PVN | | 6F | I |
| 550 | PVC | | AVA | | 6F | I |
| 551 | PVC | | LUA | | 5T | I |
| 552 | PVC | | PVD | | 2T | I |
| 553 | PVC | | AQR | | 2F | I |
| 554 | PVC | | PHC | PHC | 5T | I |
| 555 | PVC | | VA12 | PDE | 2T | I |
| 556 | PVC | | PVP | PVC | 5T | I |
| 557 | PVC | | LUA | | 2T | I |
| 558 | PVC | | VA12 | | 5T | I |

| Entry Number | Class Name | Pre | Post | Gap Junc | Chart ID | Type |
|---|---|---|---|---|---|---|
| 559 | PVC | | VA12 | | 6T | I |
| 560 | PVC | | PVP | | 2F | I |
| 561 | PVC | | LUA | | 6F | I |
| 562 | PVD | PVC | | | 6F | I |
| 563 | PVD | DVA | | | 6F | I |
| 564 | PVD | AVA | | | 6F | I |
| 565 | PVM | PVR | | | 6F | S |
| 566 | PVM | PVC | | PDE | 6F | S |
| 567 | PVM | PDE | | | 6F | S |
| 568 | PVM | AVK | | AVL | 6F | S |
| 569 | PVN | VDn | | | 6T | I |
| 570 | PVN | VDn | | | 3T | I |
| 571 | PVN | PVW | | | 3T | I |
| 572 | PVN | PVT | | | 6T | I |
| 573 | PVN | PVT | | | 3T | I |
| 574 | PVN | PVC | | | 3T | I |
| 575 | PVN | PVC | PDA | | 6F | I |
| 576 | PVN | PQR | PDA | HSN | 3T | I |
| 577 | PVN | PQR | PVW | | 6F | I |
| 578 | PVN | NMJ(VC) | BDU | AVB | 6T | I |
| 579 | PVN | NMJ(VC) | | PVQ | 3T | I |
| 580 | PVN | DDn | | | 3T | I |
| 581 | PVN | DDn | | | 6T | I |
| 582 | PVN | BDU | | | 6T | I |
| 583 | PVN | BDU | | | 3F | I |
| 584 | PVN | AVL | | | 3T | I |
| 585 | PVN | AVL | | | 6T | I |
| 586 | PVN | AVJ | | | 6T | I |
| 587 | PVN | AVJ | BDU | | 3F | I |
| 588 | PVN | AVJ | | PVQ | 6T | I |
| 589 | PVN | AVD | | | 6F | I |
| 590 | PVN | AVD | AVG | AVB | 3T | I |
| 591 | PVN | AVA | | | 3T | I |
| 592 | PVN | AVA | AVG | HSN | 6T | I |
| 593 | PVP | RIG | | VDn | 6T | I |
| 594 | PVP | RIG | | | 2F | I |
| 595 | PVP | PVC | | AQR | 2F | I |
| 596 | PVP | PVC | RIF | PVP | 6T | I |
| 597 | PVP | HDC | | | 6T | I |
| 598 | PVP | HDC | AVH | DVC | 2T | I |

| Entry Number | Class Name | Pre | Post | Gap Junc | Chart ID | Type |
|---|---|---|---|---|---|---|
| 599 | PVP | AVL | | PQR | 2T | I |
| 600 | PVP | AVL | | PQR | 6F | I |
| 601 | PVP | AVH | RIF | PVP | 2T | I |
| 602 | PVP | AVH | | | 6T | I |
| 603 | PVP | AVB | AVH | AQR | 6T | I |
| 604 | PVP | AVB | | | 2F | I |
| 605 | PVP | AVA | | DVC | 6F | I |
| 606 | PVP | AVA | | | 2F | I |
| 607 | PVP | | | AVK | 6F | I |
| 608 | PVP | | | VDn | 2T | I |
| 609 | PVP | | | AVK | 2T | I |
| 610 | PVP | | | PHA | 6F | I |
| 611 | PVP | | | PHA | 2T | I |
| 612 | PVQ | HSN | | ADA | 4T | I |
| 613 | PVQ | AVF | PHA | PVQ | 4T | I |
| 614 | PVQ | ASK | | | 4F | I |
| 615 | PVQ | AIA | ASJ | ASK | 4F | I |
| 616 | PVR | RIP | | | 6T | S |
| 617 | PVR | AVJ | | | 6T | S |
| 618 | PVR | AVB | AVM | ALM | 6T | S |
| 619 | PVR | | PVM | DVA | 6F | S |
| 620 | PVR | | DVA | PLM | 6F | S |
| 621 | PVW | PVN | | PVC | 6F | I |
| 622 | PVW | AVD | | | 6F | I |
| 623 | RIA | SMD | RIR | | 1F | I |
| 624 | RIA | SIA | RMD | | 1F | I |
| 625 | RIA | RMD | RIB | | 1F | I |
| 626 | RIA | RIV | SMD | | 1F | I |
| 627 | RIA | | ADF | | 1T | I |
| 628 | RIA | | AIZ | | 1T | I |
| 629 | RIA | | AWB | | 1T | I |
| 630 | RIA | | BAG | | 1T | I |
| 631 | RIA | | URX | | 1F | I |
| 632 | RIA | | AQR | | 1T | I |
| 633 | RIA | | RIV | | 1F | I |
| 634 | RIA | | AUA | | 1T | I |
| 635 | RIA | | AIY | | 1T | I |
| 636 | RIA | | ASH | | 1T | I |
| 637 | RIA | | RIH | | 1T | I |
| 638 | RIB | RIA | URY | RIG | 1F | I |

| Entry Number | Class Name | Pre | Post | Gap Junc | Chart ID | Type |
|---|---|---|---|---|---|---|
| 639 | RIB | AVE | OLL | RIB | 1T | I |
| 640 | RIB | AVA | BAG | AVB | 1T | I |
| 641 | RIB | AIZ | AIY | DVB | 1T | I |
| 642 | RIB | AIN | OLQ | | 1T | I |
| 643 | RIB | AIB | AUA | SMB | 1T | I |
| 644 | RIB | | RIS | SIB | 1F | I |
| 645 | RIB | | | SMD | 1F | I |
| 646 | RIB | | AIB | | 1T | I |
| 647 | RIB | | RIH | | 1T | I |
| 648 | RIB | | RIG | SIA | 1F | I |
| 649 | RIC | SMD | URX | ASH | 2T | I |
| 650 | RIC | SMD | | AVK | 2T | I |
| 651 | RIC | SMD | URX | | 1F | I |
| 652 | RIC | SMB | CEP | | 1F | I |
| 653 | RIC | AVA | OLQ | ASH | 1T | I |
| 654 | RIC | AVA | URB | | 2F | I |
| 655 | RIC | | | AVK | 1T | I |
| 656 | RIC | | URB | | 1F | I |
| 657 | RIC | | CEP | | 2F | I |
| 658 | RIC | | OLQ | | 2F | I |
| 659 | RID | VD/DD | PVC | AVB | 5F | M |
| 660 | RID | MUS5 | | | 5F | M |
| 661 | RIF | RIM | | | 3T | I |
| 662 | RIF | PVP | | | 3T | I |
| 663 | RIF | AVJ | HSN | AVG | 3F | I |
| 664 | RIF | AVB | AIA | | 3T | I |
| 665 | RIF | ALM | | | 3T | I |
| 666 | RIG | RMH | ADE | RIG | 1T | I |
| 667 | RIG | RMH | DVC | RIB | 2T | I |
| 668 | RIG | RIR | DVB | URY | 2T | I |
| 669 | RIG | RIR | | RIB | 1F | I |
| 670 | RIG | RIB | URX | URY | 1F | I |
| 671 | RIG | RIB | | | 2T | I |
| 672 | RIG | BAG | DVB | AQR | 1T | I |
| 673 | RIG | BAG | ADE | AQR | 2F | I |
| 674 | RIG | AVK | | RIG | 2T | I |
| 675 | RIG | AVK | BAG | BAG | 1T | I |
| 676 | RIG | AVE | DVC | OLQ | 1T | I |
| 677 | RIG | AVE | PVP | BAG | 2F | I |
| 678 | RIG | AIZ | URX | AVK | 2T | I |

| Entry Number | Class Name | Pre | Post | Gap Junc | Chart ID | Type |
|---|---|---|---|---|---|---|
| 679 | RIG | AIZ | PVP | OLL | 1T | I |
| 680 | RIG | | | OLL | 2F | I |
| 681 | RIG | | BAG | OLQ | 2F | I |
| 682 | RIG | | | AVK | 1T | I |
| 683 | RIH | RIP | | | 2T | S |
| 684 | RIH | RIB | | | 2T | S |
| 685 | RIH | RIA | IL2 | ADF | 2T | S |
| 686 | RIH | OLQ | | OLQ | 2F | S |
| 687 | RIH | CEP | OLQ | CEP | 2F | S |
| 688 | RIH | AIZ | | FLP | 2T | S |
| 689 | RIM | SMD | AIB | AVE | 1T | M |
| 690 | RIM | SAA | RIS | RIS | 1F | M |
| 691 | RIM | RMD | | | 1F | M |
| 692 | RIM | MUS1 | SAA | | 1F | M |
| 693 | RIM | AVB | RIF | AVA | 1T | M |
| 694 | RIM | | AIZ | AIY | 1T | M |
| 695 | RIP | OLQ? | RIH | PHARYNX | 1T | I |
| 696 | RIP | | IL2 | RME | 1F | I |
| 697 | RIP | | IL1 | | 1F | I |
| 698 | RIP | | URA | | 1F | I |
| 699 | RIP | | PVR | | 1T | I |
| 700 | RIR | URX | DVA | | 3T | I |
| 701 | RIR | URX | RIG | | 1F | I |
| 702 | RIR | RIA | RIG | BAG | 3T | I |
| 703 | RIR | RIA | | | 1F | I |
| 704 | RIR | AIZ | ADF | | 3T | I |
| 705 | RIR | | AVH | | 3F | I |
| 706 | RIS | RMD | | | 1F | I |
| 707 | RIS | RIM | | | 1F | I |
| 708 | RIS | RIB | | RIM | 1F | I |
| 709 | RIS | CEP | | | 1F | I |
| 710 | RIS | AVK | PVC | AVJ | 1T | I |
| 711 | RIS | AVE | SDQ | SMD | 1T | I |
| 712 | RIS | | | AIB | 1T | I |
| 713 | RIV | SMD | | RIV | 1T | M |
| 714 | RIV | SIA | | SMD | 1T | M |
| 715 | RIV | SAA | | AIB | 1T | M |
| 716 | RIV | RMD | | SDQ | 1T | M |
| 717 | RIV | RIA | RIA | | 1F | M |
| 718 | RIV | MUS1 | | | 1F | M |

| Entry Number | Class Name | Pre | Post | Gap Junc | Chart ID | Type |
|---|---|---|---|---|---|---|
| 719 | RMD | URY | OLQ | RMD | 1T | M |
| 720 | RMD | SMD | RIM | SMD | 1F | M |
| 721 | RMD | RIA | RIS | SAA | 1F | M |
| 722 | RMD | MUS1 | RIA | MUS1 | 1F | M |
| 723 | RMD | | IL1 | | 1F | M |
| 724 | RMD | | CEP | | 1F | M |
| 725 | RMD | | ADE | AVE | 1T | M |
| 726 | RMD | | RMD | | 1T | M |
| 727 | RMD | | RIV | | 1T | M |
| 728 | RMD | | RMG | | 1F | M |
| 729 | RMD | | RMF | | 1F | M |
| 730 | RMD | | URA | | 1F | M |
| 731 | RMD | | OLQ | | 1T | M |
| 732 | RMD | | RMH | | 1F | M |
| 733 | RMD | | OLL | | 1T | M |
| 734 | RME | SMD | OLL | RME | 1T | M |
| 735 | RME | MUS1 | SMB | RIP | 1F | M |
| 736 | RME | | IL2 | IL1 | 1F | M |
| 737 | RME | | URA | AVE | 1T | M |
| 738 | RME | | | GLR | 1T | M |
| 739 | RMF | RMG | | RMG | 1F | M |
| 740 | RMF | RMD | | | 1F | M |
| 741 | RMF | MUS1 | | | 1F | M |
| 742 | RMF | AVK | DVB | | 1T | M |
| 743 | RMF | | DVC | | 1T | M |
| 744 | RMG | RMD | CEP | RMH | 1F | M |
| 745 | RMG | MUS1 | RMF | RMF | 1F | M |
| 746 | RMG | AVE | ADE | URX | 1T | M |
| 747 | RMG | | | AWB | 1T | M |
| 748 | RMG | | | ASH | 1T | M |
| 749 | RMG | | | IL2 | 1T | M |
| 750 | RMG | | | ADL | 1T | M |
| 751 | RMG | | | ASK | 1T | M |
| 752 | RMH | RMD | CEP | RMG | 1F | M |
| 753 | RMH | MUS1 | | | 1F | M |
| 754 | RMH | | RIG | | 1T | M |
| 755 | SAA | RIM | PLN | RMD | 1F | I |
| 756 | SAA | AVA | RIV | SMB | 1T | I |
| 757 | SAA | | RIM | | 1F | I |
| 758 | SAA | | SMB | | 1F | I |

| Entry Number | Class Name | Pre | Post | Gap Junc | Chart ID | Type |
|---|---|---|---|---|---|---|
| 759 | SAA | | ALN | | 1T | I |
| 760 | SAA | | VB1 | | 1T | I |
| 761 | SAB | MUS5 | AVA | AVA | 5F | M |
| 762 | SAB | | AVD/E | | 5F | M |
| 763 | SDQ | RIS | | | 2T | I |
| 764 | SDQ | DVA | | RIV | 2T | I |
| 765 | SDQ | AVB | | | 2F | I |
| 766 | SDQ | AVA | | AVB | 2F | I |
| 767 | SDQ | AIB | | SDQ | 2T | I |
| 768 | SIA | | RIA | RIB | 1F | I |
| 769 | SIB | | | RIB | 1F | I |
| 770 | SMB | SAA | RIC | | 1F | M |
| 771 | SMB | RME | PLN | | 1F | M |
| 772 | SMB | MUS1 | | | 1F | M |
| 773 | SMB | | ADF | | 1T | M |
| 774 | SMB | | AVH | | 1T | M |
| 775 | SMB | | AIZ | RIB | 1T | M |
| 776 | SMB | | ALN | SAA | 1T | M |
| 777 | SMB | | DVA | AVK | 1T | M |
| 778 | SMD | SMD | RIM | RIS | 1T | M |
| 779 | SMD | RIA | URY | RIB | 1F | M |
| 780 | SMD | MUS1 | RIC | RMD | 1F | M |
| 781 | SMD | | RIA | | 1F | M |
| 782 | SMD | | SMD | | 1T | M |
| 783 | SMD | | OLL | | 1T | M |
| 784 | SMD | | SAA | | 1T | M |
| 785 | SMD | | RME | SMD | 1T | M |
| 786 | SMD | | RIV | RIV | 1T | M |
| 787 | SMD | | RMD | | 1F | M |
| 788 | URA | RME | | | 1T | S |
| 789 | URA | RMD | CEP | | 1F | S |
| 790 | URA | RIP | | | 1F | S |
| 791 | URA | MUS1 | IL2 | | 1F | S |
| 792 | URB | URX | CEP | | 1F | S |
| 793 | URB | RIC | | | 1F | S |
| 794 | URB | RIC | | | 2F | S |
| 795 | URB | IL1 | | | 1F | S |
| 796 | URB | CEP | | | 1F | S |
| 797 | URB | CEP | CEP | OLQ | 2F | S |
| 798 | URX | RIG | RIR | IL2 | 1F | S |

| Entry Number | Class Name | Pre | Post | Gap Junc | Chart ID | Type |
|---|---|---|---|---|---|---|
| 799 | URX | RIC | | | 1F | S |
| 800 | URX | RIA | URB | | 1F | S |
| 801 | URX | AVE | | AUA | 1T | S |
| 802 | URX | AVB | | | 1T | S |
| 803 | URX | AUA | | RMG | 1T | S |
| 804 | URY | SMD | | RIG | 1F | S |
| 805 | URY | RIB | | | 1F | S |
| 806 | URY | AVE | | AVA | 1T | S |
| 807 | VA/DA | VD/DD | AVD/E | AVA | 5F | M |
| 808 | VA/DA | MUS5 | AVA | | 5F | M |
| 809 | VA/DA | | | VA/DA | 5T | M |
| 810 | VB/DB | VD/DD | DVA | AVB | 5F | M |
| 811 | VB/DB | MUS5 | PVC | | 5F | M |
| 812 | VB/DB | | | VB/DB | 5T | M |
| 813 | VCn | VUL | HSN | | 3F | M |
| 814 | VCn | VDn | | | 3T | M |
| 815 | VCn | VCn | VCn | VCn | 3T | M |
| 816 | VCn | NMJ(VC) | | | 3T | M |
| 817 | VCn | DDn | | | 3T | M |
| 818 | VD/DD | MUS5 | RID | | 5F | M |
| 819 | VD/DD | | | PVP | 5T | M |
| 820 | VD/DD | | AS | | 5F | M |
| 821 | VD/DD | | | VD/DD | 5T | M |
| 822 | VD/DD | | VA/DA | | 5F | M |
| 823 | VD/DD | | VB/DB | | 5F | M |
| 824 | VUL | | VCn | | 3F | B |
| 825 | VUL | | HSN | | 3F | B |

--- End of File ---

## C. GAPJUNC FILE SET

### C.1.  *GAPJUNC.TXT*

#### C.1.a.  Description

The GAPJUNC data file has four (4) columns.

1) The **Entry Number** column provides a sequential numbering system for each of the neuron pairs that form gap junctions.

2) The **Neuron A** column contains neurons that pair with neurons in the Neuron B column to form gap junctions.

3) The **Neuron B** column contains neurons that pair with neurons in the Neuron A column to form gap junctions.

4) The **Count** column contains the number of gap junctions made by any Neuron A - Neuron B pair.

#### C.1.b.  Derivation

The main reference for the GAPJUNC data file is:

White, J.G.; Southgate, E.; Thomson, J.N.; and Brenner, S. The Structure of the Nervous System of the Nematode *Caenorhabditis elegans*. Phil. Trans. R. Soc. Lond. 314(B 1165):1-340, 1986.

The framework for the entire GAPJUNC data file was derived from a manual count of all the gap junctions made by any one neuron in the diagrams printed in Appendix 1 - Connectivity Data section of the reference above (for an example, see Figures 1.6 and 1.7 on p. 6). The list was enhanced by the table of VENTRAL CORD SYNAPSES (the second column of which contains a count of gap junctions) provided at the end of the description of pertinent neurons (Figure 1.8, p. 7, top right). Text and electron micrograph pictures were used to clarify and verify connections where available (see Figure 1.8, pp. 7-8).

#### C.1.c.  Comments

Several formats of the data referenced in literature are shown in Figures 1.6 and 1.7 (p. 6) and Figure 1.8 (pp. 7-8).

The GAPJUNC data file integrates data from diagrams, textual descriptions, and electron micrograph pictures provided in the reference above.

Ambiguities in the number of gap junctions per neuron and in the laterality of the neuron which makes a gap junction [whether it belongs to XXXL (a neuron on the left side of the body) or XXXR (a neuron on the right side of the body)] were decided as far as the text and electron micrograph pictures can provide, and sometimes arbitrarily. When two different values for the Count column were encountered, the higher value was accepted (i.e., if the figure/table of Neuron A shows 4 gap junctions with Neuron B, and the figure/table of Neuron B shows only 3 gap junctions with Neuron A, then the Count column will reflect the higher value of 4, since gap junctions are symmetrical connections. Some laterality problems were settled by checking if a neuron crossed the contralateral side through a commissure, and whether or not its neuron pair (with which the gap junction is made) remained ipsilateral through its entire length (see Figure 8, p. 7, top left).

The most ambiguous cases were of those neuron groups where a "typical" neuron was given. In these cases, the "typical" neuron gap junctions were replicated accordingly (with cross reference to text and pictures). In cases when there was no other recourse but to make an educated guess, the trend of gap junction distribution was noted within the neuron group. Gap junctions were then "interpolated and distributed" within the neuron group prudently, with Count values based on the trend of the gap junction distribution.

IMPORTANT: Remember that gap junctions are symmetrical or bidirectional connections. If Neuron A is connected to Neuron B then Neuron B must be connected to Neuron A. For simplicity of presentation, however, the GAPJUNC data file presents only the gap junctions of Neuron A with Neuron B. Simple spreadsheet manipulation by proper column duplication, horizontal transposition and resorting of data should produce the symmetrical list (i.e., copy Neuron B column into the end of Neuron A column, copy Neuron A column into the end of Neuron B column, copy the Count column into the end of itself, then resort).

--- End of File ---

**C.2.** *GAPJUNC.PRN*

| Entry Number | Neuron A | Neuron B | Count |
|---|---|---|---|
| 1 | ADAL | ADAR | 1 |
| 2 | ADAL | ADFL | 1 |
| 3 | ADAL | ASHL | 1 |
| 4 | ADAL | AVDR | 2 |
| 5 | ADAL | PVQL | 1 |
| 6 | ADAR | ADFR | 1 |
| 7 | ADAR | ASHR | 1 |
| 8 | ADAR | AVDL | 2 |
| 9 | ADAR | PVQR | 1 |
| 10 | ADEL | AINL | 1 |
| 11 | ADEL | AVKR | 1 |
| 12 | ADER | AVKL | 1 |
| 13 | ADFL | ADFR | 1 |
| 14 | ADFR | AIAR | 1 |
| 15 | ADFR | RIH | 1 |
| 16 | ADLL | ADLR | 1 |
| 17 | ADLL | OLQVL | 1 |
| 18 | ADLL | RMGL | 1 |
| 19 | ADLR | OLQVR | 1 |
| 20 | AFDL | AFDR | 1 |
| 21 | AFDL | AIBL | 1 |
| 22 | AFDR | AIBR | 1 |
| 23 | AIAL | AIAR | 1 |
| 24 | AIAL | ASIL | 2 |
| 25 | AIAL | AWAL | 1 |
| 26 | AIAR | ASIR | 2 |
| 27 | AIAR | AWAR | 1 |
| 28 | AIBL | DVB | 1 |
| 29 | AIBL | DVC | 1 |
| 30 | AIBL | RIGR | 3 |
| 31 | AIBL | RIMR | 1 |
| 32 | AIBL | RIVL | 1 |
| 33 | AIBR | DVB | 1 |
| 34 | AIBR | DVC | 2 |
| 35 | AIBR | RIGL | 3 |
| 36 | AIBR | RIML | 1 |
| 37 | AIBR | RIS | 1 |
| 38 | AIBR | RIVR | 1 |

| Entry Number | Neuron A | Neuron B | Count |
|---|---|---|---|
| 39 | AIML | AIMR | 1 |
| 40 | AIML | SIBDR | 1 |
| 41 | AINL | AINR | 2 |
| 42 | AINL | ASGR | 1 |
| 43 | AINL | AUAR | 1 |
| 44 | AINR | ASGL | 1 |
| 45 | AINR | AUAL | 1 |
| 46 | AIYL | AIYR | 1 |
| 47 | AIYL | RIML | 1 |
| 48 | AIYR | RIMR | 1 |
| 49 | AIZL | AIZR | 2 |
| 50 | AIZL | ASGL | 1 |
| 51 | AIZL | ASHL | 1 |
| 52 | AIZR | ASGR | 1 |
| 53 | AIZR | ASHR | 1 |
| 54 | AIZR | AWAR | 1 |
| 55 | ALA | RID | 1 |
| 56 | ALML | AVM | 1 |
| 57 | ALML | PVR | 1 |
| 58 | ALMR | AVDR | 1 |
| 59 | ALMR | AVM | 1 |
| 60 | ALNL | SMBDR | 1 |
| 61 | AQR | AVBL | 1 |
| 62 | AQR | AVKL | 2 |
| 63 | AQR | AVKR | 1 |
| 64 | AQR | PVPL | 7 |
| 65 | AQR | PVPR | 9 |
| 66 | AQR | RIGL | 2 |
| 67 | AQR | RIGR | 1 |
| 68 | AS1 | AVAL | 2 |
| 69 | AS1 | AVAR | 2 |
| 70 | AS1 | VA3 | 1 |
| 71 | AS2 | AVAL | 2 |
| 72 | AS2 | AVAR | 1 |
| 73 | AS2 | VA3 | 1 |
| 74 | AS2 | VA5 | 2 |
| 75 | AS3 | AVAL | 2 |
| 76 | AS3 | AVAR | 1 |
| 77 | AS3 | VA4 | 1 |
| 78 | AS3 | VA6 | 2 |

| Entry Number | Neuron A | Neuron B | Count |
|---|---|---|---|
| 79 | AS4 | AVAL | 2 |
| 80 | AS4 | AVAR | 1 |
| 81 | AS4 | VA5 | 1 |
| 82 | AS4 | VA7 | 2 |
| 83 | AS5 | AVAL | 1 |
| 84 | AS5 | AVAR | 1 |
| 85 | AS5 | VA6 | 1 |
| 86 | AS5 | VA8 | 2 |
| 87 | AS6 | AVAL | 1 |
| 88 | AS6 | AVAR | 1 |
| 89 | AS6 | VA7 | 1 |
| 90 | AS6 | VA9 | 2 |
| 91 | AS7 | AVAL | 2 |
| 92 | AS7 | AVAR | 1 |
| 93 | AS7 | VA8 | 1 |
| 94 | AS7 | VA10 | 2 |
| 95 | AS8 | AVAL | 2 |
| 96 | AS8 | AVAR | 1 |
| 97 | AS8 | VA9 | 1 |
| 98 | AS8 | VA11 | 2 |
| 99 | AS9 | AVAL | 2 |
| 100 | AS9 | AVAR | 1 |
| 101 | AS9 | VA10 | 1 |
| 102 | AS9 | VA12 | 2 |
| 103 | AS10 | AVAR | 1 |
| 104 | AS10 | VA11 | 1 |
| 105 | ASHL | ASHR | 1 |
| 106 | ASHL | RICL | 2 |
| 107 | ASHL | RMGL | 1 |
| 108 | ASHR | ASKR | 1 |
| 109 | ASHR | RICR | 2 |
| 110 | ASHR | RMGR | 1 |
| 111 | ASIL | ASIR | 1 |
| 112 | ASJL | ASJR | 1 |
| 113 | ASKL | ASKR | 3 |
| 114 | ASKL | PVQL | 5 |
| 115 | ASKL | RMGL | 1 |
| 116 | ASKR | PVQR | 4 |
| 117 | ASKR | RMGR | 1 |
| 118 | AUAL | AUAR | 1 |

| Entry Number | Neuron A | Neuron B | Count |
|---|---|---|---|
| 119 | AUAL | AWBL | 1 |
| 120 | AUAL | URXL | 1 |
| 121 | AUAR | AWBR | 1 |
| 122 | AUAR | URXR | 1 |
| 123 | AVAL | AVAR | 4 |
| 124 | AVAL | AVJL | 1 |
| 125 | AVAL | LUAL | 1 |
| 126 | AVAL | PVCR | 5 |
| 127 | AVAL | RIMR | 3 |
| 128 | AVAL | SABD | 4 |
| 129 | AVAL | SABVR | 1 |
| 130 | AVAL | URYDL | 1 |
| 131 | AVAL | URYVR | 1 |
| 132 | AVAR | AVJR | 1 |
| 133 | AVAR | PVCL | 5 |
| 134 | AVAR | RIML | 3 |
| 135 | AVAR | RIMR | 1 |
| 136 | AVAR | SABVL | 2 |
| 137 | AVAR | SABVR | 1 |
| 138 | AVAR | URYDR | 1 |
| 139 | AVAR | URYVL | 1 |
| 140 | AVBL | AVBR | 2 |
| 141 | AVBL | DVA | 1 |
| 142 | AVBL | PVNR | 1 |
| 143 | AVBL | RIBL | 1 |
| 144 | AVBL | RIBR | 1 |
| 145 | AVBL | RID | 1 |
| 146 | AVBL | SDQR | 1 |
| 147 | AVBL | SIBVL | 1 |
| 148 | AVBR | DVA | 1 |
| 149 | AVBR | PVNL | 1 |
| 150 | AVBR | RIBL | 1 |
| 151 | AVBR | RIBR | 1 |
| 152 | AVBR | RID | 2 |
| 153 | AVBR | SIBVL | 1 |
| 154 | AVDL | AVJL | 1 |
| 155 | AVDL | AVM | 1 |
| 156 | AVDL | FLPL | 1 |
| 157 | AVDL | FLPR | 1 |
| 158 | AVDR | AVJR | 1 |

| Entry Number | Neuron A | Neuron B | Count |
|---|---|---|---|
| 159 | AVDR | FLPR | 1 |
| 160 | AVEL | AVER | 1 |
| 161 | AVEL | RIML | 2 |
| 162 | AVEL | RIMR | 3 |
| 163 | AVEL | RMDVR | 1 |
| 164 | AVEL | RMEV | 1 |
| 165 | AVER | RIML | 3 |
| 166 | AVER | RIMR | 2 |
| 167 | AVER | RMDVL | 1 |
| 168 | AVER | RMDVR | 1 |
| 169 | AVER | RMEV | 1 |
| 170 | AVFL | AVFR | 21 |
| 171 | AVFL | AVHR | 4 |
| 172 | AVFL | AVJR | 1 |
| 173 | AVFR | AVHL | 4 |
| 174 | AVFR | PVQR | 1 |
| 175 | AVG | RIFL | 1 |
| 176 | AVG | RIFR | 1 |
| 177 | AVHL | PHBR | 1 |
| 178 | AVJL | AVJR | 4 |
| 179 | AVJL | PVCL | 1 |
| 180 | AVJL | PVCR | 1 |
| 181 | AVJL | RIS | 2 |
| 182 | AVJR | PVCL | 1 |
| 183 | AVKL | AVKR | 2 |
| 184 | AVKL | DVB | 2 |
| 185 | AVKL | PVPL | 1 |
| 186 | AVKL | PVPR | 2 |
| 187 | AVKL | RICL | 1 |
| 188 | AVKL | RICR | 1 |
| 189 | AVKL | RIGL | 1 |
| 190 | AVKL | SMBDL | 1 |
| 191 | AVKL | SMBDR | 1 |
| 192 | AVKL | SMBVR | 1 |
| 193 | AVKR | PVPL | 2 |
| 194 | AVKR | RICL | 1 |
| 195 | AVKR | RICR | 1 |
| 196 | AVKR | RIGL | 1 |
| 197 | AVKR | RIGR | 1 |
| 198 | AVKR | SMBDL | 1 |

| Entry Number | Neuron A | Neuron B | Count |
|---|---|---|---|
| 199 | AVKR | SMBDR | 2 |
| 200 | AVKR | SMBVR | 1 |
| 201 | AVL | DVB | 1 |
| 202 | AVL | DVC | 9 |
| 203 | AVL | PVM | 1 |
| 204 | AWAL | AWAR | 1 |
| 205 | AWBL | AWBR | 1 |
| 206 | AWBL | RMGL | 1 |
| 207 | AWBR | RICL | 1 |
| 208 | AWBR | RMGR | 1 |
| 209 | BAGL | BAGR | 1 |
| 210 | BAGL | RIGR | 1 |
| 211 | BAGL | RIR | 1 |
| 212 | BAGR | RIGL | 1 |
| 213 | BAGR | RIR | 1 |
| 214 | CANL | EXCAN | 2 |
| 215 | CANR | EXCAN | 2 |
| 216 | CEPDL | OLQDL | 1 |
| 217 | CEPDL | RIH | 1 |
| 218 | CEPDR | OLQDR | 2 |
| 219 | CEPDR | RIH | 1 |
| 220 | CEPVL | OLQVL | 1 |
| 221 | CEPVL | RIH | 1 |
| 222 | CEPVR | OLQVR | 1 |
| 223 | CEPVR | RIH | 1 |
| 224 | DA1 | AVAL | 4 |
| 225 | DA1 | AVAR | 4 |
| 226 | DA1 | SABVL | 2 |
| 227 | DA1 | SABVR | 3 |
| 228 | DA2 | AVAL | 2 |
| 229 | DA2 | AVAR | 2 |
| 230 | DA2 | SABVL | 1 |
| 231 | DA2 | VA1 | 2 |
| 232 | DA3 | AVAR | 2 |
| 233 | DA4 | AVAL | 3 |
| 234 | DA4 | AVAR | 2 |
| 235 | DA4 | VA3 | 2 |
| 236 | DA5 | AVAL | 3 |
| 237 | DA5 | AVAR | 2 |
| 238 | DA5 | VA4 | 2 |

| Entry Number | Neuron A | Neuron B | Count |
|---|---|---|---|
| 239 | DA6 | AVAR | 2 |
| 240 | DA6 | VA5 | 2 |
| 241 | DA7 | AVAL | 1 |
| 242 | DA7 | AVAR | 1 |
| 243 | DA7 | VA6 | 2 |
| 244 | DA8 | AVAR | 1 |
| 245 | DA9 | PHCR | 1 |
| 246 | DB1 | AVBL | 1 |
| 247 | DB1 | AVBR | 1 |
| 248 | DB1 | DB2 | 6 |
| 249 | DB2 | AVBR | 1 |
| 250 | DB2 | DB3 | 6 |
| 251 | DB2 | VB1 | 2 |
| 252 | DB3 | AVBL | 3 |
| 253 | DB3 | AVBR | 2 |
| 254 | DB4 | AVBR | 1 |
| 255 | DB4 | DB5 | 6 |
| 256 | DB4 | VB4 | 1 |
| 257 | DB5 | AVAL | 1 |
| 258 | DB5 | AVAR | 1 |
| 259 | DB5 | AVBL | 1 |
| 260 | DB5 | AVBR | 1 |
| 261 | DB5 | DB6 | 6 |
| 262 | DB6 | AVBL | 1 |
| 263 | DB6 | AVBR | 1 |
| 264 | DB6 | DB7 | 6 |
| 265 | DB6 | VB7 | 1 |
| 266 | DB7 | AVBL | 2 |
| 267 | DB7 | AVBR | 1 |
| 268 | DD1 | DD2 | 2 |
| 269 | DD1 | VD1 | 4 |
| 270 | DD2 | DD3 | 2 |
| 271 | DD2 | VD4 | 2 |
| 272 | DD3 | DD4 | 2 |
| 273 | DD3 | VD6 | 2 |
| 274 | DD4 | DD5 | 2 |
| 275 | DD4 | VD8 | 2 |
| 276 | DD5 | DD6 | 2 |
| 277 | DD5 | VD10 | 2 |
| 278 | DD6 | VD13 | 2 |

| Entry Number | Neuron A | Neuron B | Count |
|---|---|---|---|
| 279 | DVA | PVCR | 1 |
| 280 | DVA | PVR | 2 |
| 281 | DVB | RIBL | 1 |
| 282 | DVB | RIBR | 1 |
| 283 | DVC | PVPL | 6 |
| 284 | DVC | PVPR | 6 |
| 285 | FLPL | FLPR | 2 |
| 286 | FLPR | RIH | 1 |
| 287 | GLRDL | RMED | 1 |
| 288 | GLRDR | RMED | 1 |
| 289 | GLRL | MUSCLE | 1 |
| 290 | GLRL | RMEL | 1 |
| 291 | GLRR | MUSCLE | 1 |
| 292 | GLRR | RMER | 1 |
| 293 | GLRVL | RMEV | 1 |
| 294 | GLRVR | RMEV | 1 |
| 295 | HSNL | HSNR | 1 |
| 296 | HSNR | PVNR | 1 |
| 297 | IL1DL | IL1DR | 1 |
| 298 | IL1DL | OLLL | 1 |
| 299 | IL1DL | RMEV | 1 |
| 300 | IL1DR | OLLR | 1 |
| 301 | IL1DR | RMED | 1 |
| 302 | IL1DR | RMEV | 1 |
| 303 | IL1L | IL1DL | 1 |
| 304 | IL1L | IL1VL | 1 |
| 305 | IL1R | IL1DR | 1 |
| 306 | IL1R | IL1VR | 1 |
| 307 | IL1VL | IL1VR | 1 |
| 308 | IL1VL | RMED | 1 |
| 309 | IL2L | RMGL | 1 |
| 310 | IL2R | RMGR | 1 |
| 311 | IL2R | URXR | 1 |
| 312 | LUAL | PVR | 1 |
| 313 | LUAR | PVR | 1 |
| 314 | OLLL | OLLR | 2 |
| 315 | OLLL | RIGL | 1 |
| 316 | OLLR | RIGR | 1 |
| 317 | OLQDL | RIBL | 2 |
| 318 | OLQDL | RIGL | 1 |

| Entry Number | Neuron A | Neuron B | Count |
|---|---|---|---|
| 319 | OLQDL | RMDVL | 1 |
| 320 | OLQDL | URBL | 1 |
| 321 | OLQDR | RIBR | 2 |
| 322 | OLQDR | RIGR | 1 |
| 323 | OLQDR | RIH | 1 |
| 324 | OLQDR | RMDVR | 2 |
| 325 | OLQDR | URBR | 1 |
| 326 | OLQVL | RIBL | 1 |
| 327 | OLQVL | RIGL | 1 |
| 328 | OLQVL | RMDDL | 1 |
| 329 | OLQVL | URBL | 1 |
| 330 | OLQVR | RIBR | 1 |
| 331 | OLQVR | RIGR | 1 |
| 332 | OLQVR | RMDDR | 1 |
| 333 | OLQVR | URBR | 1 |
| 334 | PDEL | PDER | 3 |
| 335 | PDER | PVCR | 1 |
| 336 | PHAL | PHAR | 2 |
| 337 | PHAL | PVPR | 1 |
| 338 | PHAR | PVPL | 2 |
| 339 | PHBL | PHBR | 3 |
| 340 | PHCL | PHCR | 1 |
| 341 | PHCR | PVCR | 1 |
| 342 | PLML | LUAL | 1 |
| 343 | PLML | PVCL | 1 |
| 344 | PLML | PVR | 1 |
| 345 | PLMR | LUAR | 1 |
| 346 | PLMR | PHCR | 1 |
| 347 | PLMR | PVCR | 1 |
| 348 | PLMR | PVR | 1 |
| 349 | PLNR | SMDVR | 1 |
| 350 | PQR | PVPL | 2 |
| 351 | PQR | PVPR | 2 |
| 352 | PVCL | PVCR | 4 |
| 353 | PVCL | PVWL | 1 |
| 354 | PVCR | PVWR | 2 |
| 355 | PVM | PDEL | 1 |
| 356 | PVM | PDER | 1 |
| 357 | PVNR | PVQR | 1 |
| 358 | PVPL | PVPR | 1 |

| Entry Number | Neuron A | Neuron B | Count |
|---|---|---|---|
| 359 | PVPL | PVT | 1 |
| 360 | PVQL | PVQR | 2 |
| 361 | RIBL | RIBL | 1 |
| 362 | RIBL | RIBR | 3 |
| 363 | RIBL | RIGL | 1 |
| 364 | RIBL | SIADL | 1 |
| 365 | RIBL | SIAVL | 1 |
| 366 | RIBL | SIBDL | 1 |
| 367 | RIBL | SIBVL | 1 |
| 368 | RIBL | SIBVR | 1 |
| 369 | RIBL | SMBDL | 1 |
| 370 | RIBL | SMDDL | 1 |
| 371 | RIBL | SMDVR | 2 |
| 372 | RIBR | RIBR | 1 |
| 373 | RIBR | RIGR | 1 |
| 374 | RIBR | SIADR | 1 |
| 375 | RIBR | SIAVR | 1 |
| 376 | RIBR | SIBDR | 1 |
| 377 | RIBR | SIBVR | 2 |
| 378 | RIBR | SMBDR | 1 |
| 379 | RIBR | SMDDL | 1 |
| 380 | RIBR | SMDDR | 1 |
| 381 | RIBR | SMDVL | 2 |
| 382 | RID | RID | 1 |
| 383 | RIGL | RIGR | 6 |
| 384 | RIGL | RMEL | 1 |
| 385 | RIGL | URYDL | 1 |
| 386 | RIGL | URYVL | 1 |
| 387 | RIGR | URYDR | 1 |
| 388 | RIGR | URYVR | 1 |
| 389 | RIML | RIS | 1 |
| 390 | RIMR | RIS | 2 |
| 391 | RIPL | IL1DL | 1 |
| 392 | RIPL | RMED | 1 |
| 393 | RIPR | IL1DR | 1 |
| 394 | RIPR | RMED | 1 |
| 395 | RIS | SMDDL | 1 |
| 396 | RIS | SMDDR | 1 |
| 397 | RIS | SMDVL | 1 |
| 398 | RIVL | RIVR | 2 |

| Entry Number | Neuron A | Neuron B | Count |
|---|---|---|---|
| 399 | RIVL | SDQR | 2 |
| 400 | RIVL | SMDVL | 1 |
| 401 | RIVR | SDQR | 1 |
| 402 | RIVR | SMDVR | 2 |
| 403 | RMDDL | RMDVL | 1 |
| 404 | RMDDL | SAADL | 1 |
| 405 | RMDDL | SMDDL | 1 |
| 406 | RMDDR | RMDVR | 1 |
| 407 | RMDDR | SAADL | 1 |
| 408 | RMDDR | SAADR | 1 |
| 409 | RMDDR | SMDDR | 1 |
| 410 | RMDL | RMDDL | 1 |
| 411 | RMDL | RMDVL | 1 |
| 412 | RMDR | RMDVR | 1 |
| 413 | RMDVL | MUSCLE | 1 |
| 414 | RMDVL | RMDVR | 1 |
| 415 | RMDVL | SAAVL | 1 |
| 416 | RMDVL | SMDVL | 1 |
| 417 | RMDVL | SMDVR | 1 |
| 418 | RMDVR | SAAVR | 1 |
| 419 | RMDVR | SMDVR | 1 |
| 420 | RMED | RMEV | 1 |
| 421 | RMEL | RMEV | 1 |
| 422 | RMER | RMEV | 1 |
| 423 | RMFL | RMGL | 1 |
| 424 | RMFL | RMGR | 1 |
| 425 | RMFL | URBR | 1 |
| 426 | RMGL | RMHL | 3 |
| 427 | RMGL | URXL | 1 |
| 428 | RMGR | RMFL | 1 |
| 429 | RMGR | RMHL | 1 |
| 430 | RMGR | RMHR | 1 |
| 431 | RMGR | URXR | 1 |
| 432 | RMHL | SIBVR | 1 |
| 433 | SAADL | SMBDL | 1 |
| 434 | SAADR | SMBDR | 1 |
| 435 | SAAVL | SMBVR | 3 |
| 436 | SAAVR | SMBVL | 2 |
| 437 | SDQL | SDQR | 1 |
| 438 | SDQR | SIBVL | 1 |

| Entry Number | Neuron A | Neuron B | Count |
|---|---|---|---|
| 439 | SIBDL | SIBVL | 1 |
| 440 | SIBDL | SIBVR | 1 |
| 441 | SIBDR | SIBVL | 1 |
| 442 | SIBDR | SIBVR | 1 |
| 443 | SMDVL | SMDVR | 1 |
| 444 | VA1 | AVAL | 2 |
| 445 | VA2 | AVAL | 1 |
| 446 | VA2 | AVAR | 2 |
| 447 | VA2 | SABD | 3 |
| 448 | VA2 | VD1 | 1 |
| 449 | VA3 | AVAL | 1 |
| 450 | VA3 | AVAR | 2 |
| 451 | VA3 | SABD | 1 |
| 452 | VA4 | AVAL | 1 |
| 453 | VA4 | AVAR | 2 |
| 454 | VA4 | SABD | 1 |
| 455 | VA4 | VA5 | 1 |
| 456 | VA5 | AVAL | 3 |
| 457 | VA5 | AVAR | 4 |
| 458 | VA5 | VA6 | 1 |
| 459 | VA6 | AVAL | 2 |
| 460 | VA6 | AVAR | 3 |
| 461 | VA6 | VA7 | 1 |
| 462 | VA7 | AVAL | 2 |
| 463 | VA7 | AVAR | 3 |
| 464 | VA7 | VA8 | 1 |
| 465 | VA8 | AVAR | 1 |
| 466 | VA8 | VA9 | 1 |
| 467 | VA9 | AVAL | 1 |
| 468 | VA9 | AVAR | 2 |
| 469 | VA9 | VA10 | 1 |
| 470 | VA10 | AVAL | 1 |
| 471 | VA10 | AVAR | 1 |
| 472 | VA10 | VA11 | 1 |
| 473 | VA11 | AVAL | 4 |
| 474 | VA11 | AVAR | 4 |
| 475 | VA12 | AVAR | 1 |
| 476 | VA12 | PHCL | 1 |
| 477 | VA12 | VB11 | 1 |
| 478 | VB1 | AVBL | 1 |

| Entry Number | Neuron A | Neuron B | Count |
|---|---|---|---|
| 479 | VB1 | VB2 | 4 |
| 480 | VB1 | VD1 | 1 |
| 481 | VB2 | AVBL | 2 |
| 482 | VB2 | AVBR | 1 |
| 483 | VB2 | RIGL | 1 |
| 484 | VB2 | VB3 | 1 |
| 485 | VB3 | AVBL | 1 |
| 486 | VB4 | AVBL | 1 |
| 487 | VB4 | AVBR | 1 |
| 488 | VB4 | VB5 | 1 |
| 489 | VB5 | AVBL | 1 |
| 490 | VB5 | VB6 | 1 |
| 491 | VB6 | AVBL | 1 |
| 492 | VB6 | AVBR | 1 |
| 493 | VB6 | VB7 | 1 |
| 494 | VB7 | AVBL | 2 |
| 495 | VB7 | AVBR | 2 |
| 496 | VB7 | VB8 | 1 |
| 497 | VB8 | AVBL | 1 |
| 498 | VB8 | AVBR | 1 |
| 499 | VB8 | VB9 | 1 |
| 500 | VB9 | AVBL | 1 |
| 501 | VB9 | AVBR | 1 |
| 502 | VB9 | VB10 | 1 |
| 503 | VB10 | AVBL | 1 |
| 504 | VB10 | AVBR | 1 |
| 505 | VB10 | VB11 | 1 |
| 506 | VB11 | AVBL | 2 |
| 507 | VB11 | AVBR | 1 |
| 508 | VC1 | VC2 | 1 |
| 509 | VC1 | VC3 | 2 |
| 510 | VC2 | VC3 | 2 |
| 511 | VC2 | VC4 | 2 |
| 512 | VC3 | VC4 | 1 |
| 513 | VC3 | VC5 | 2 |
| 514 | VC3 | vm2 | 1 |
| 515 | VC4 | AVFL | 1 |
| 516 | VC4 | AVFR | 1 |
| 517 | VC4 | VC5 | 1 |
| 518 | VC4 | VC6 | 2 |

| Entry Number | Neuron A | Neuron B | Count |
|---|---|---|---|
| 519 | VC5 | HSNL | 1 |
| 520 | VD1 | DVC | 3 |
| 521 | VD1 | RIFL | 1 |
| 522 | VD1 | RIFR | 1 |
| 523 | VD1 | RIGL | 2 |
| 524 | VD1 | RIGR | 1 |
| 525 | VD1 | SMDDR | 1 |
| 526 | VD1 | VD2 | 7 |
| 527 | VD2 | VD3 | 1 |
| 528 | VD3 | PVPL | 1 |
| 529 | VD3 | VD4 | 1 |
| 530 | VD4 | PVPR | 1 |
| 531 | VD4 | VD5 | 1 |
| 532 | VD5 | PVPR | 1 |
| 533 | VD5 | VD6 | 1 |
| 534 | VD6 | VD7 | 1 |
| 535 | VD7 | VD8 | 1 |
| 536 | VD8 | VD9 | 1 |
| 537 | VD9 | PDER | 1 |
| 538 | VD9 | VD10 | 1 |
| 539 | VD10 | VD11 | 1 |
| 540 | VD11 | VD12 | 1 |
| 541 | VD12 | VD13 | 1 |
| 542 | VD13 | PVPL | 1 |
| 543 | VD13 | PVPR | 1 |

--- End of File ---

# D. LEGENDGO FILE SET

## D.1. *LEGENDGO.TXT*

### D.1.a. Description

The LEGENDGO data file is a directory for and the prerequisite to the understanding of the PREPOSGO files. The LEGENDGO data file has 5 columns.

1) Two **Seq** columns provide a sequential numbering system for each of the possible types of cell designations in the PREPOSGO files.

2) Two **Type** columns contain all the possible types of cell designations in the PREPOS-GO files. The meaning of the two to four letter designations are explained in the Legend of Neuron Types column.

3) The **Legend of Neuron Types** column contains the meaning of each letter in the position that it occupies (i.e., as the first, second, third, or fourth letter in a sequence) in the two to four-letter designations of cell types used in the PREPOSGO files.

### D.1.b. Derivation

To be able to sort individual neurons into their respective ganglia, to sort them as to location, and to quickly sort them as to their basic function/s, an orderly method whereby neurons can be tagged with such labels was devised. These labels were used to produce the PREPOSGO data file, where the presynaptic neurons are sorted in ganglionic order. With the VENTRAL CORD NEURON GROUP (G) labeled as one "ganglion", the rest of the labels for ganglia roughly correspond to the actual anteroposterior sequence in the nematode. Exception to this sequence is the PHARYNGEAL NEURON GROUP (P), the connectivity data of the which are yet incomplete in this version.

### D.1.c. Comments

Connections of neurons to non-neural cells are seen in labels with W as the first letter, followed by an E, I, N, or M. E = external or exteroceptive, meaning receiving input from the external environment (i.e., external to the organism). I = internal or inter-oceptive, meaning receiving input from the internal environment (i.e., within the organism). N = neuroendocrine and visceral, meaning having output to the internal environment (i.e., within the organism). M = motor, meaning having output to the external environment (i.e., external to the organism). In many instances, these are conjectured, e.g., the CANL neuron.

--- End of File ---

**D.2.** *LEGENDGO.PRN*

| Seq | Type | Seq | Type | Legend of Neuron Types |
|-----|------|-----|------|------------------------|
| 1 | ALI | 37 | EUM | 1ST LETTER = GANGLION OR CELL GROUP |
| 2 | ALM | 38 | EUS | |
| 3 | ALMS | 39 | FLI | A = ANTERIOR GANGLION |
| 4 | ALS | 40 | FLM | B = DORSAL GANGLION |
| 5 | ARI | 41 | FLS | C = LATERAL GANGLION |
| 6 | ARM | 42 | FRI | D = VENTRAL GANGLION |
| 7 | ARMS | 43 | FRM | E = RETROVESICULAR GANGLION |
| 8 | ARS | 44 | FRS | F = POSTEROLATERAL GANGLION |
| 9 | AUM | 45 | FUS | G = VENTRAL CORD NEURON GROUP |
| 10 | BLS | 46 | GUM | H = PRE-ANAL GANGLION |
| 11 | BLSI | 47 | HLI | J = DORSORECTAL GANGLION |
| 12 | BRS | 48 | HRI | K = LUMBAR GANGLION |
| 13 | BRSI | 49 | HUI | P = PHARYNGEAL NEURON GROUP ** |
| 14 | BUI | 50 | HUM | W = NON-NEURAL CELL GROUP |
| 15 | BUM | 51 | JUI | |
| 16 | CLI | 52 | JUM | 2ND LETTER = LOCATION |
| 17 | CLM | 53 | KLI | |
| 18 | CLMI | 54 | KLS | L = LEFT SIDE        E = EXTERNAL |
| 19 | CLS | 55 | KRI | R = RIGHT SIDE  I = INTEROCEPTIVE |
| 20 | CRI | 56 | KRS | U = UNPAIRED    N = NEUROENDOCRINE |
| 21 | CRM | 57 | KUS |                              M = MOTOR |
| 22 | CRMI | 58 | PLI | |
| 23 | CRS | 59 | PLIS | 3RD OR 4TH LETTER = FUNCTION TYPE |
| 24 | DLI | 60 | PLM | |
| 25 | DLM | 61 | PLSM | S = SENSORY |
| 26 | DRI | 62 | PRI | M = MOTOR |
| 27 | DRM | 63 | PRIS | I = INTERNEURON |
| 28 | DUI | 64 | PRM | |
| 29 | DUIM | 65 | PRSM | |
| 30 | ELI | 66 | PUIM | |
| 31 | ELM | 67 | PUIS | ** CONNECTIONS TO AND OF THE PHA- |
| 32 | ELS | 68 | PUM | RYNGEAL NEURONS ARE YET INCOM- |
| 33 | ERI | 69 | WE | PLETE. THIS DESIGNATION IS TO |
| 34 | ERM | 70 | WI | FACILITATE RAPID ISOLATION OR |
| 35 | ERS | 71 | WM | REMOVAL FROM THE LIST OF COMPLE- |
| 36 | EUI | 72 | WN | TED CONNECTIONS. |

--- End of File ---

### E. PREPOSGO FILE SET

#### E.1. *PREPOSGO.TXT*

#### E.1.a. Description

The LEGENDGO file set is a prerequisite to the understanding of the PREPOSGO file set. The PREPOSGO data file has seven (7) columns.

1) The **Entry Number** column provides a sequential numbering system for each of the pre- to postsynaptic neuron pairs.

2) In the **Type** column are sorting tags for the neurons in the Pre column. These tags or designations are fully explained in the LEGENDGO files.

3) The **Pre** column contains the neurons that are presynaptic to the neurons in the Post column. Neurons in this column also have corresponding tags in the Type column.

4) The **Post** column contains the neurons and non-neurons that are postsynaptic to the neurons in the Pre column. Non-neurons are labeled in the WE, WI, WM, or WM format as explained in the LEGENDGO files.

5) The **Ref** column contains data for non-neurons as they appear in literature, prior to their conversion into the non-neuron format found in the Post column.

6) The **Den** column contains the number of synapses between any two pre- to post synaptic neuron pair (Pre column neuron to Post column neuron pair).

7) The **Uniq** column provides a list where all neurons of the hermaphroditic *C. elegans* (in the Pre column) are listed once. The neurons in this column are sequenced according to their first appearance in the Pre column.

IMPORTANT: There are two (2) blocks of data in the PREPOSGO data file. The first is from Entry Numbers 1 to 2367 (cell Types A to P), and the second is from Entry Numbers 2368 to 2463 (cell Types A to P again). The reason for the short, second block of data is given in the Comments section of this text file.

In the PREPOSGO.PRN file in Disk 1, all column fields containing integers are written in a four-character field length (I4), and all column fields containing letters are written in a seven-character field length (A7). In the left to right appearance of the seven columns, the PREPOSGO file is thus written in an I4,A7,A7,A7,A7,I4,A7 format by FORTRAN computer programming convention.

**E.1.b. Derivation**

The references for the PREPOSGO data file are:

1) White, J.G.; Southgate, E.; Thomson, J.N.; and Brenner, S. The Structure of the Nervous System of the Nematode *Caenorhabditis elegans*. Phil. Trans. R. Soc. Lond. 314(B 1165):1-340, 1986.

2) White, J.G.; Southgate, E.; Thomson, J.N.; and Brenner, S. The Structure of the Ventral Nerve Cord of *Caenorhabditis elegans*. Phil. Trans. R. Soc. Lond. 275(B):28-348, 1976.

The framework for the entire PREPOSGO data file was derived from a manual count of all the presynaptic and postsynaptic connections made by any one neuron in the diagrams printed in appendices of reference (1) above (pp. 63-340). Unlike the GAP-JUNC data file, only limited cross counting for consistency was performed. Counts in the Den column were based primarily on efferent symbols (see Figure 1.7, p. 6). The list was enhanced by the table of VENTRAL CORD SYNAPSES provided at the end of the description of pertinent neurons (Figure 1.8, p. 7, top right). Text and electron micrograph pictures were used to clarify and verify connections where available (see Figure 1.8, pp. 7-8).

**E.1.c. Comments**

Several formats of the data referenced in literature are shown in Figures 1.6 and 1.7 (p. 6), and Figure 1.8 (pp. 7-8).

The PREPOSGO data file integrates data from diagrams, textual descriptions, and electron micrograph pictures provided in the references above.

When a neuron in the Pre column made a synaptic connection with a non-neural cell in the Post column, the corresponding non-neural cell in the Post column was designated according to the representation of a non-neural cell in the LEGENDGO files (WE, WI, WM, or WN). Research motivation for conversion to this type of format was discussed in Chapter II, pp. 27-28. The Ref column, however, preserves the data prior to conversion, and uses the following conventions:

1) NMJ - NeuroMuscular Junction

2) CEPshVR, CEPshVL, CEPshDR, and CEPshDL - CEPhalic sheath cells, sheet-like processes that envelop the neuropil of the ring and part of the ventral ganglion [12, p. 419]

   Note: In Entry Number 1539, Pre = BDUL, Ref is CEPshVL. No symmetric entry is found where Pre = BDUR, Ref = CEPshVR. For purposes of computation, however, the following entry may be inserted between Entry Numbers 1611 and 1612: Pre = BDUR, Post = WE, Ref = CEPshVR, Den = 1, as it was done for pertinent computations in Chapter II.

3) HDC - HypoDermal Cell

4) NONE1 - No presynaptic contacts have been seen on these cells except for occasional regions that show typical presynaptic specializations but no vesicles. In one place, the process of an SIBD cell was seen in the motor endplate region of the nerve ring, but again, no vesicles were seen [11, p. 308].

5) NONE2 - SIA is unusual in that it appears to have no synaptic outputs except, perhaps, via gap junctions to RIB [11, p. 305].

6) NONE3 - No synaptic output has been seen from SAB in adults; however, in the L1 larval stage, SABs have several NMJs in the anterior ventral cord [11, p. 299].

7) MDE - Muscle-Dorsal-Excitatory, inferred from text, instead of using NMJ

8) MDI - Muscle-Dorsal-Inhibitory, inferred from text, instead of using NMJ

9) MVE - Muscle-Ventral-Excitatory, inferred from text, instead of using NMJ

10) MVI - Muscle-Ventral-Inhibitory, inferred from text, instead of using NMJ

11) NONE4 - A single synapse onto the lateral hypodermis has been seen on one of the processes. Apart from a few rather unconvincing gap junctions to the excretory canal, no other synapses can be ambiguously assigned to CAN. Laser ablation experiments have, however, shown CAN to be essential to the survival of the animal [11, p. 169].

12) MUE - Muscle-Uterine-Excitatory, inferred from text, instead of using VM2 or vm2 (vulval muscles)

13) NONE5 - No synaptic outputs have been seen from this neuron [11, p. 244].

14) PHARYN1 - Pharyngeal motoneurons [12, p. 421]

15) PHARYN2 - Pharyngeal sensory/motoneurons [12, p. 421]

16) PHARYN3 - Pharyngeal neurons that synapse onto the marginal cells [12, p. 421]

17) PHARYN4 - Pharyngeal motoneuron/interneuron [12, p. 421]

18) PHARYN5 - Pharyngeal neurosecretory motoneurons that contain serotonin [12, p. 421].

Even if data in literature indicated no synaptic connections (e.g., NONEn in the Ref column), a minimum of Den = 1 was assumed for logical consistency.

In the Type P (pharyngeal) neuron group, the Post and Den column entries of Entry Numbers 2348 to 2355 were left blank due to lack of data available to the editors. The pharyngeal connectivity data are yet incomplete in this version, and one may opt to proceed with computation without them, and without the second data block of PREPOSGO (see below).

By strict ganglionic classification, neurons of the Posterior Lateral (or Posterolateral, cell Type F) ganglion are PDEL, PDER, PVDL, PVDR, PVM, and SDQL only, excluding the sheath and socket cells illustrated in Figure 4, page 14, of reference (1) above. Since the rest of the neurons in the referenced figure do not belong to any of the named ganglia, they were included with the Posterolateral ganglion. The neurons included are ALML, ALMR, AVM, BDUL, BDUR, CANL, CANR, HSNL, HSNR, and SDQR.

Ambiguities in the number of synapses per neuron and in the laterality of the neuron which makes the synapse [whether it belongs to XXXL (a neuron on the left side of the body) or XXXR (a neuron on the right side of the body)] were decided as far as the text and electron micrograph pictures can provide, and sometimes arbitrarily. When two different values for the Den column were encountered, the higher value was accepted (i.e., if the figure/table of the neuron in the Pre column shows 4 presynaptic connections with a neuron in the Post column, and the figure/table of the corresponding neuron pair in the Post column shows only 3 postsynaptic connections with the neuron in the Pre column, then the Dens column will reflect the higher value of 4). Some laterality problems were decided by checking if a neuron crossed the contralateral side through a commissure, and whether or not its neuron pair (with which the synapse is made) remained ipsilateral through its entire length (see Figure 8, p. 7, top left).

The most ambiguous cases were of those neuron groups where a "typical" neuron was given. In these cases, the "typical" neuron synapses were replicated accordingly (with cross reference to text and pictures). In these cases, when there was no other recourse but to make an educated guess, the trend of synaptic distribution was noted within the neuron group. Pre- and postsynaptic connections were then "interpolated and distributed" within the neuron group prudently, with Dens values based on the trend of the synaptic distribution.

Neurons that were previously presynaptic in the first data block are postsynaptic in the second data block of PREPOSGO. The presynaptic elements in the second data block are non-neural cells (WE, designation explained in the LEGENDGO files). For example, neuron BAGL is presynaptic to neuron RIBR in the first data block (Entry Number 32), but is postsynaptic to WE in the second data block (Entry Number 2368). From the text of references, it is known that these neurons receive input from non-neural cells. The basis for the external environment designation ("E" after the "W") of the non-neural cell is conjectural, and will be improved or specified as further data are reviewed or available.

--- End of File ---

**E.2.** *PREPOSGO.PRN*

| Entry Number | Type | Pre | Post | Ref | Den | Uniq |
|---|---|---|---|---|---|---|
| 1 | ALI | RIPL | OLQDL | | 2 | RIPL |
| 2 | ALM | RMEL | WM | NMJ | 11 | RMEL |
| 3 | ALMS | IL1DL | RMDVL | | 6 | IL1DL |
| 4 | ALMS | IL1DL | PVR | | 1 | IL1L |
| 5 | ALMS | IL1DL | RIPL | | 3 | IL1VL |
| 6 | ALMS | IL1DL | WM | NMJ | 4 | URADL |
| 7 | ALMS | IL1DL | IL1L | | 1 | URAVL |
| 8 | ALMS | IL1L | RMDR | | 3 | BAGL |
| 9 | ALMS | IL1L | RMDL | | 1 | CEPVL |
| 10 | ALMS | IL1L | RMER | | 1 | IL2DL |
| 11 | ALMS | IL1L | IL1DL | | 1 | IL2L |
| 12 | ALMS | IL1L | RMDVR | | 2 | IL2VL |
| 13 | ALMS | IL1L | WM | NMJ | 10 | OLLL |
| 14 | ALMS | IL1L | RMDVL | | 4 | OLQDL |
| 15 | ALMS | IL1L | RMDDL | | 3 | OLQVL |
| 16 | ALMS | IL1L | AVER | | 2 | URBL |
| 17 | ALMS | IL1VL | URYVL | | 1 | URYDL |
| 18 | ALMS | IL1VL | WM | NMJ | 10 | URYVL |
| 19 | ALMS | IL1VL | RMDDL | | 11 | RMED |
| 20 | ALMS | IL1VL | RIPL | | 7 | RMEV |
| 21 | ALMS | IL1VL | IL1L | | 1 | RIPR |
| 22 | ALMS | URADL | RIPL | | 3 | RMER |
| 23 | ALMS | URADL | IL1DL | | 2 | IL1DR |
| 24 | ALMS | URADL | WM | NMJ | 6 | IL1R |
| 25 | ALMS | URADL | RMEL | | 1 | IL1VR |
| 26 | ALMS | URADL | RMDDR | | 1 | URADR |
| 27 | ALMS | URAVL | RMER | | 1 | URAVR |
| 28 | ALMS | URAVL | RMEV | | 2 | BAGR |
| 29 | ALMS | URAVL | RIPL | | 3 | CEPVR |
| 30 | ALMS | URAVL | RMEL | | 1 | IL2DR |
| 31 | ALMS | URAVL | WM | NMJ | 9 | IL2R |
| 32 | ALS | BAGL | RIBR | | 6 | IL2VR |
| 33 | ALS | BAGL | RIAR | | 6 | OLLR |
| 34 | ALS | BAGL | RIGR | | 4 | OLQDR |
| 35 | ALS | BAGL | AVER | | 2 | OLQVR |
| 36 | ALS | BAGL | AIBL | | 1 | URBR |
| 37 | ALS | BAGL | AVEL | | 1 | URYDR |
| 38 | ALS | BAGL | RIGL | | 1 | URYVR |

| Entry Number | Type | Pre | Post | Ref | Den | Uniq |
|---|---|---|---|---|---|---|
| 39 | ALS | BAGL | BAGR | | 1 | ALA |
| 40 | ALS | BAGL | AVAR | | 2 | CEPDL |
| 41 | ALS | CEPVL | OLQVL | | 5 | URXL |
| 42 | ALS | CEPVL | AVER | | 3 | RID |
| 43 | ALS | CEPVL | RICL | | 6 | CEPDR |
| 44 | ALS | CEPVL | ADLL | | 1 | URXR |
| 45 | ALS | CEPVL | RMHL | | 1 | AIBL |
| 46 | ALS | CEPVL | URAVL | | 2 | AINL |
| 47 | ALS | CEPVL | RIPL | | 1 | AIZL |
| 48 | ALS | CEPVL | OLLL | | 4 | AVAL |
| 49 | ALS | CEPVL | RICR | | 4 | AVBL |
| 50 | ALS | CEPVL | IL1VL | | 2 | AVDL |
| 51 | ALS | CEPVL | RMDDL | | 3 | AVEL |
| 52 | ALS | CEPVL | SIAVL | | 2 | AVHL |
| 53 | ALS | IL2DL | AUAL | | 1 | AVJL |
| 54 | ALS | IL2DL | OLQDL | | 2 | RIAL |
| 55 | ALS | IL2DL | RIPL | | 9 | RIBL |
| 56 | ALS | IL2DL | RIBL | | 1 | RICL |
| 57 | ALS | IL2DL | URADL | | 3 | SAAVL |
| 58 | ALS | IL2DL | IL1DL | | 7 | SIBDL |
| 59 | ALS | IL2DL | RMEL | | 4 | RIML |
| 60 | ALS | IL2DL | RMER | | 4 | RMDL |
| 61 | ALS | IL2L | ADEL | | 2 | RMDVL |
| 62 | ALS | IL2L | AVEL | | 1 | SMDVL |
| 63 | ALS | IL2L | RICL | | 1 | RIVL |
| 64 | ALS | IL2L | RMDR | | 1 | ADFL |
| 65 | ALS | IL2L | OLQVL | | 8 | ADLL |
| 66 | ALS | IL2L | RIH | | 9 | AFDL |
| 67 | ALS | IL2L | IL1L | | 1 | ASEL |
| 68 | ALS | IL2L | RMEV | | 2 | ASGL |
| 69 | ALS | IL2L | RMDL | | 3 | ASHL |
| 70 | ALS | IL2L | URXL | | 2 | ASIL |
| 71 | ALS | IL2L | RMER | | 2 | ASJL |
| 72 | ALS | IL2L | OLQDL | | 6 | ASKL |
| 73 | ALS | IL2VL | RIAL | | 1 | AUAL |
| 74 | ALS | IL2VL | IL1VL | | 6 | AWAL |
| 75 | ALS | IL2VL | RMEL | | 2 | AWBL |
| 76 | ALS | IL2VL | RMEV | | 1 | AWCL |
| 77 | ALS | IL2VL | RIPL | | 11 | AIBR |
| 78 | ALS | IL2VL | BAGR | | 1 | AINR |

| Entry Number | Type | Pre | Post | Ref | Den | Uniq |
|---|---|---|---|---|---|---|
| 79 | ALS | IL2VL | RIH | | 2 | AIZR |
| 80 | ALS | IL2VL | URAVL | | 4 | AUAR |
| 81 | ALS | IL2VL | OLQVL | | 1 | AVAR |
| 82 | ALS | IL2VL | RMER | | 5 | AVBR |
| 83 | ALS | OLLL | URYDL | | 1 | AVDR |
| 84 | ALS | OLLL | SMDVR | | 4 | AVER |
| 85 | ALS | OLLL | RMDDL | | 7 | AVHR |
| 86 | ALS | OLLL | SMDDR | | 4 | AVJR |
| 87 | ALS | OLLL | RIBL | | 7 | RIAR |
| 88 | ALS | OLLL | SMDDL | | 3 | RIBR |
| 89 | ALS | OLLL | AVER | | 19 | RICR |
| 90 | ALS | OLLL | RMDVL | | 1 | SAAVR |
| 91 | ALS | OLLL | CEPDL | | 2 | SIBDR |
| 92 | ALS | OLLL | SAADR | | 1 | RIMR |
| 93 | ALS | OLLL | RMDL | | 2 | RMDR |
| 94 | ALS | OLLL | RMEL | | 2 | RMDVR |
| 95 | ALS | OLLL | CEPVL | | 2 | SMDVR |
| 96 | ALS | OLLL | IL1VL | | 1 | RIVR |
| 97 | ALS | OLLL | RMDDR | | 1 | ADFR |
| 98 | ALS | OLQDL | SIBVL | | 3 | ADLR |
| 99 | ALS | OLQDL | RICR | | 1 | AFDR |
| 100 | ALS | OLQDL | RMDDR | | 4 | ASER |
| 101 | ALS | OLQVL | IL2VL | | 1 | ASGR |
| 102 | ALS | OLQVL | RMDVR | | 4 | ASHR |
| 103 | ALS | OLQVL | RIH | | 2 | ASIR |
| 104 | ALS | OLQVL | SIBDL | | 3 | ASJR |
| 105 | ALS | OLQVL | RICL | | 1 | ASKR |
| 106 | ALS | OLQVL | RIPL | | 2 | AWAR |
| 107 | ALS | OLQVL | IL1VL | | 2 | AWBR |
| 108 | ALS | URBL | RICR | | 1 | AWCR |
| 109 | ALS | URBL | URXL | | 4 | RIH |
| 110 | ALS | URBL | CEPDL | | 1 | RIR |
| 111 | ALS | URBL | IL1L | | 1 | RIS |
| 112 | ALS | URBL | SMBDR | | 1 | AVL |
| 113 | ALS | URBL | RMDDR | | 1 | AIAL |
| 114 | ALS | URBL | SIAVL | | 1 | AIML |
| 115 | ALS | URBL | AVBL | | 1 | AIYL |
| 116 | ALS | URYDL | RIBL | | 1 | AVKL |
| 117 | ALS | URYDL | RMDDR | | 4 | SAADL |
| 118 | ALS | URYDL | SMDDL | | 1 | SIADL |

106

| Entry Number | Type | Pre | Post | Ref | Den | Uniq |
|---|---|---|---|---|---|---|
| 119 | ALS | URYDL | RMDVL | | 5 | SIAVL |
| 120 | ALS | URYDL | AVER | | 3 | SIBVL |
| 121 | ALS | URYDL | SMDDR | | 2 | RMDDL |
| 122 | ALS | URYVL | RIAL | | 1 | RMFL |
| 123 | ALS | URYVL | RIH | | 1 | RMHL |
| 124 | ALS | URYVL | RMDDL | | 4 | SMBDL |
| 125 | ALS | URYVL | AVER | | 5 | SMBVL |
| 126 | ALS | URYVL | RIS | | 1 | SMDDL |
| 127 | ALS | URYVL | AVBR | | 1 | AIAR |
| 128 | ALS | URYVL | SMDVR | | 4 | AIMR |
| 129 | ALS | URYVL | IL1VL | | 1 | AIYR |
| 130 | ALS | URYVL | RMDVR | | 2 | AVKR |
| 131 | ALS | URYVL | SIBVL | | 1 | SAADR |
| 132 | ALS | URYVL | RIBL | | 2 | SIADR |
| 133 | AUM | RMED | RIBR | | 1 | SIAVR |
| 134 | AUM | RMED | WM | NMJ | 10 | SIBVR |
| 135 | AUM | RMED | RIBL | | 1 | RMDDR |
| 136 | AUM | RMEV | SMDDR | | 1 | RMFR |
| 137 | AUM | RMEV | WM | NMJ | 5 | RMHR |
| 138 | ARI | RIPR | OLQDR | | 2 | SMBDR |
| 139 | ARM | RMER | WM | NMJ | 14 | SMBVR |
| 140 | ARMS | IL1DR | WM | NMJ | 8 | SMDDR |
| 141 | ARMS | IL1DR | RMDVR | | 8 | SABD |
| 142 | ARMS | IL1DR | RIPR | | 7 | ADAL |
| 143 | ARMS | IL1R | RMDR | | 1 | AVFL |
| 144 | ARMS | IL1R | AVEL | | 1 | RIFL |
| 145 | ARMS | IL1R | RMDL | | 4 | RIGL |
| 146 | ARMS | IL1R | URXR | | 1 | SABVL |
| 147 | ARMS | IL1R | IL1DR | | 1 | RMGL |
| 148 | ARMS | IL1R | RMDDR | | 4 | ADEL |
| 149 | ARMS | IL1R | RMDVL | | 1 | FLPL |
| 150 | ARMS | IL1R | RMEL | | 2 | AS1 |
| 151 | ARMS | IL1R | RMDDL | | 3 | DA1 |
| 152 | ARMS | IL1R | WM | NMJ | 12 | DB1 |
| 153 | ARMS | IL1R | RMDVR | | 4 | DB2 |
| 154 | ARMS | IL1VR | RIPR | | 8 | DD1 |
| 155 | ARMS | IL1VR | IL1R | | 1 | VA1 |
| 156 | ARMS | IL1VR | IL2VR | | 1 | VB1 |
| 157 | ARMS | IL1VR | WM | NMJ | 10 | VB2 |
| 158 | ARMS | IL1VR | RMDDR | | 11 | VD1 |

| Entry Number | Type | Pre | Post | Ref | Den | Uniq |
|---|---|---|---|---|---|---|
| 159 | ARMS | URADR | WM | NMJ | 6 | VD2 |
| 160 | ARMS | URADR | URYDR | | 1 | ADAR |
| 161 | ARMS | URADR | RMDVR | | 1 | AVFR |
| 162 | ARMS | URADR | RMED | | 1 | RIFR |
| 163 | ARMS | URADR | RMER | | 1 | RIGR |
| 164 | ARMS | URADR | RIPR | | 3 | SABVR |
| 165 | ARMS | URAVR | IL1R | | 1 | RMGR |
| 166 | ARMS | URAVR | RMER | | 2 | ADER |
| 167 | ARMS | URAVR | WM | NMJ | 9 | FLPR |
| 168 | ARMS | URAVR | RMDVL | | 1 | AQR |
| 169 | ARMS | URAVR | RMEV | | 2 | AVG |
| 170 | ARMS | URAVR | RIPR | | 4 | BDUL |
| 171 | ARS | BAGR | RIBL | | 4 | CANL |
| 172 | ARS | BAGR | RIAL | | 4 | PVDL |
| 173 | ARS | BAGR | AIYL | | 1 | SDQL |
| 174 | ARS | BAGR | RIGL | | 7 | HSNL |
| 175 | ARS | BAGR | AVEL | | 4 | ALML |
| 176 | ARS | BAGR | AVAL | | 3 | PDEL |
| 177 | ARS | BAGR | AUAL | | 1 | BDUR |
| 178 | ARS | CEPVR | SIAVR | | 2 | CANR |
| 179 | ARS | CEPVR | RMHR | | 2 | PVDR |
| 180 | ARS | CEPVR | RIPR | | 1 | SDQR |
| 181 | ARS | CEPVR | RICL | | 2 | HSNR |
| 182 | ARS | CEPVR | OLQVR | | 3 | ALMR |
| 183 | ARS | CEPVR | OLLR | | 5 | PDER |
| 184 | ARS | CEPVR | RMDDR | | 2 | AVM |
| 185 | ARS | CEPVR | IL1VR | | 2 | PVM |
| 186 | ARS | CEPVR | RIVL | | 1 | AS10 |
| 187 | ARS | CEPVR | RICR | | 4 | AS2 |
| 188 | ARS | CEPVR | WE | CEPshVR | 1 | AS3 |
| 189 | ARS | CEPVR | AVEL | | 4 | AS4 |
| 190 | ARS | CEPVR | URAVR | | 1 | AS5 |
| 191 | ARS | CEPVR | ASGR | | 1 | AS6 |
| 192 | ARS | CEPVR | IL2VR | | 1 | AS7 |
| 193 | ARS | IL2DR | RMEL | | 3 | AS8 |
| 194 | ARS | IL2DR | RIPR | | 12 | AS9 |
| 195 | ARS | IL2DR | RMED | | 1 | DA2 |
| 196 | ARS | IL2DR | RICR | | 1 | DA3 |
| 197 | ARS | IL2DR | URADR | | 3 | DA4 |
| 198 | ARS | IL2DR | IL1DR | | 7 | DA5 |

| Entry Number | Type | Pre | Post | Ref | Den | Uniq |
|---|---|---|---|---|---|---|
| 199 | ARS | IL2DR | RMER | | 2 | DA6 |
| 200 | ARS | IL2DR | RIH | | 1 | DA7 |
| 201 | ARS | IL2DR | CEPDR | | 1 | DB3 |
| 202 | ARS | IL2R | RMEV | | 1 | DB4 |
| 203 | ARS | IL2R | RIH | | 5 | DB5 |
| 204 | ARS | IL2R | OLLR | | 2 | DB6 |
| 205 | ARS | IL2R | URAVR | | 1 | DB7 |
| 206 | ARS | IL2R | IL1R | | 1 | DD2 |
| 207 | ARS | IL2R | OLQVR | | 7 | DD3 |
| 208 | ARS | IL2R | RMEL | | 2 | DD4 |
| 209 | ARS | IL2R | URBR | | 1 | DD5 |
| 210 | ARS | IL2R | ADER | | 2 | VA10 |
| 211 | ARS | IL2R | OLQDR | | 1 | VA11 |
| 212 | ARS | IL2R | RMDL | | 1 | VA2 |
| 213 | ARS | IL2VR | RIPR | | 14 | VA3 |
| 214 | ARS | IL2VR | RIAR | | 2 | VA4 |
| 215 | ARS | IL2VR | RIH | | 4 | VA5 |
| 216 | ARS | IL2VR | OLQVR | | 2 | VA6 |
| 217 | ARS | IL2VR | URAVR | | 3 | VA7 |
| 218 | ARS | IL2VR | URXR | | 1 | VA8 |
| 219 | ARS | IL2VR | RMEL | | 3 | VA9 |
| 220 | ARS | IL2VR | IL1VR | | 5 | VB10 |
| 221 | ARS | IL2VR | RMEV | | 3 | VB11 |
| 222 | ARS | IL2VR | RMER | | 2 | VB3 |
| 223 | ARS | OLLR | CEPDR | | 2 | VB4 |
| 224 | ARS | OLLR | RIBR | | 10 | VB5 |
| 225 | ARS | OLLR | IL2R | | 1 | VB6 |
| 226 | ARS | OLLR | RMDVR | | 3 | VB7 |
| 227 | ARS | OLLR | IL1VR | | 1 | VB8 |
| 228 | ARS | OLLR | SMDDL | | 1 | VB9 |
| 229 | ARS | OLLR | IL1DR | | 3 | VC1 |
| 230 | ARS | OLLR | SMDVL | | 3 | VC2 |
| 231 | ARS | OLLR | CEPVR | | 6 | VC3 |
| 232 | ARS | OLLR | RMDDR | | 10 | VC4 |
| 233 | ARS | OLLR | AVER | | 1 | VC5 |
| 234 | ARS | OLLR | RMER | | 2 | VC6 |
| 235 | ARS | OLLR | SMDDR | | 1 | VD10 |
| 236 | ARS | OLLR | RMDL | | 2 | VD11 |
| 237 | ARS | OLLR | AVEL | | 14 | VD3 |
| 238 | ARS | OLLR | RMDR | | 1 | VD4 |

| Entry Number | Type | Pre | Post | Ref | Den | Uniq |
|---|---|---|---|---|---|---|
| 239 | ARS | OLLR | SMDVR | | 3 | VD5 |
| 240 | ARS | OLQDR | RICR | | 1 | VD6 |
| 241 | ARS | OLQDR | SIBVR | | 2 | VD7 |
| 242 | ARS | OLQDR | RMHR | | 1 | VD8 |
| 243 | ARS | OLQDR | RICL | | 1 | VD9 |
| 244 | ARS | OLQDR | RMDDL | | 3 | PVT |
| 245 | ARS | OLQVR | RIPR | | 1 | PVPL |
| 246 | ARS | OLQVR | IL1VR | | 1 | AS11 |
| 247 | ARS | OLQVR | RMDVL | | 3 | DA8 |
| 248 | ARS | OLQVR | RICR | | 1 | DA9 |
| 249 | ARS | OLQVR | RMER | | 1 | DD6 |
| 250 | ARS | OLQVR | IL2VR | | 1 | PDA |
| 251 | ARS | OLQVR | RIH | | 1 | PDB |
| 252 | ARS | OLQVR | SIBDR | | 3 | VA12 |
| 253 | ARS | URBR | RMGR | | 1 | VD12 |
| 254 | ARS | URBR | SMBDL | | 1 | VD13 |
| 255 | ARS | URBR | RMDR | | 1 | PVPR |
| 256 | ARS | URBR | CEPDR | | 1 | DVA |
| 257 | ARS | URBR | SIAVR | | 1 | DVC |
| 258 | ARS | URBR | RMDL | | 1 | DVB |
| 259 | ARS | URBR | SIADL | | 1 | LUAL |
| 260 | ARS | URBR | IL2R | | 1 | PVCL |
| 261 | ARS | URBR | AVBR | | 1 | PVNL |
| 262 | ARS | URBR | RICR | | 1 | PVQL |
| 263 | ARS | URBR | IL1R | | 3 | PVWL |
| 264 | ARS | URBR | URXR | | 7 | ALNL |
| 265 | ARS | URYDR | RMDDL | | 2 | PHAL |
| 266 | ARS | URYDR | RIBR | | 1 | PHBL |
| 267 | ARS | URYDR | AVER | | 2 | PHCL |
| 268 | ARS | URYDR | SMDDR | | 1 | PLML |
| 269 | ARS | URYDR | AVEL | | 2 | PLNL |
| 270 | ARS | URYDR | SMDDL | | 4 | LUAR |
| 271 | ARS | URYDR | RMDVR | | 4 | PVCR |
| 272 | ARS | URYVR | IL1VR | | 1 | PVNR |
| 273 | ARS | URYVR | AVEL | | 6 | PVQR |
| 274 | ARS | URYVR | RIAR | | 1 | PVWR |
| 275 | ARS | URYVR | SIBVL | | 1 | ALNR |
| 276 | ARS | URYVR | AVAL | | 1 | PHAR |
| 277 | ARS | URYVR | RIBR | | 1 | PHBR |
| 278 | ARS | URYVR | SMDVL | | 3 | PHCR |

| Entry Number | Type | Pre | Post | Ref | Den | Uniq |
|---|---|---|---|---|---|---|
| 279 | ARS | URYVR | RMDDL | | 1 | PLMR |
| 280 | ARS | URYVR | RMDDR | | 4 | PLNR |
| 281 | ARS | URYVR | RMDVL | | 3 | PQR |
| 282 | ARS | URYVR | SIBVR | | 1 | PVR |
| 283 | BUI | ALA | WE | CEPshDL | 2 | I1L |
| 284 | BUI | ALA | RMDR | | 1 | I1R |
| 285 | BUI | ALA | WE | CEPshVR | 1 | I2L |
| 286 | BUI | ALA | AVER | | 2 | I2R |
| 287 | BUI | ALA | AVEL | | 2 | I3 |
| 288 | BLS | CEPDL | RICR | | 3 | I4 |
| 289 | BLS | CEPDL | URBL | | 2 | I5 |
| 290 | BLS | CEPDL | RMGL | | 3 | I6 |
| 291 | BLS | CEPDL | RIPL | | 2 | M1 |
| 292 | BLS | CEPDL | URYDL | | 2 | M2L |
| 293 | BLS | CEPDL | RMDVL | | 3 | M2R |
| 294 | BLS | CEPDL | IL1DL | | 4 | M3L |
| 295 | BLS | CEPDL | SMBDL | | 1 | M3R |
| 296 | BLS | CEPDL | SIADR | | 1 | M4 |
| 297 | BLS | CEPDL | RMHR | | 4 | M5 |
| 298 | BLS | CEPDL | OLLL | | 2 | MCL |
| 299 | BLS | CEPDL | RIS | | 1 | MCR |
| 300 | BLS | CEPDL | OLQDL | | 6 | MI |
| 301 | BLS | CEPDL | AVER | | 5 | NSML |
| 302 | BLS | CEPDL | RIBL | | 2 | NSMR |
| 303 | BLS | CEPDL | RICL | | 1 | |
| 304 | BLS | CEPDL | URADL | | 2 | |
| 305 | BLS | CEPDL | IL1L | | 1 | |
| 306 | BLSI | URXL | AVJR | | 2 | |
| 307 | BLSI | URXL | AVEL | | 3 | |
| 308 | BLSI | URXL | RIGL | | 2 | |
| 309 | BLSI | URXL | AUAL | | 7 | |
| 310 | BLSI | URXL | RIAL | | 8 | |
| 311 | BLSI | URXL | RICL | | 1 | |
| 312 | BLSI | URXL | AVBL | | 2 | |
| 313 | BUM | RID | RMED | | 1 | |
| 314 | BUM | RID | WM | NMJ | 1 | |
| 315 | BUM | RID | DD6 | | 1 | |
| 316 | BRS | CEPDR | AVEL | | 5 | |
| 317 | BRS | CEPDR | URADR | | 1 | |
| 318 | BRS | CEPDR | RMGR | | 1 | |

| Entry Number | Type | Pre | Post | Ref | Den | Uniq |
|---|---|---|---|---|---|---|
| 319 | BRS | CEPDR | RIBR | | 1 | |
| 320 | BRS | CEPDR | IL1DR | | 5 | |
| 321 | BRS | CEPDR | URBR | | 2 | |
| 322 | BRS | CEPDR | IL1R | | 1 | |
| 323 | BRS | CEPDR | RICL | | 4 | |
| 324 | BRS | CEPDR | OLLR | | 6 | |
| 325 | BRS | CEPDR | RMHL | | 4 | |
| 326 | BRS | CEPDR | OLQDR | | 5 | |
| 327 | BRS | CEPDR | SMBDR | | 1 | |
| 328 | BRS | CEPDR | URYDR | | 1 | |
| 329 | BRS | CEPDR | RMDDL | | 1 | |
| 330 | BRS | CEPDR | SIADL | | 1 | |
| 331 | BRS | CEPDR | RICR | | 2 | |
| 332 | BRS | CEPDR | BDUR | | 1 | |
| 333 | BRS | CEPDR | RMDVR | | 2 | |
| 334 | BRSI | URXR | RIGR | | 2 | |
| 335 | BRSI | URXR | RIAR | | 4 | |
| 336 | BRSI | URXR | RIPR | | 3 | |
| 337 | BRSI | URXR | AVBR | | 1 | |
| 338 | BRSI | URXR | RMDR | | 1 | |
| 339 | BRSI | URXR | RICR | | 1 | |
| 340 | BRSI | URXR | OLQVR | | 1 | |
| 341 | BRSI | URXR | AVBL | | 1 | |
| 342 | BRSI | URXR | SIAVR | | 1 | |
| 343 | BRSI | URXR | AVER | | 1 | |
| 344 | BRSI | URXR | AUAR | | 4 | |
| 345 | CLI | AIBL | SAADR | | 2 | |
| 346 | CLI | AIBL | ASER | | 1 | |
| 347 | CLI | AIBL | AVAL | | 2 | |
| 348 | CLI | AIBL | AIYL | | 1 | |
| 349 | CLI | AIBL | AVBL | | 3 | |
| 350 | CLI | AIBL | SMDDR | | 1 | |
| 351 | CLI | AIBL | FLPL | | 1 | |
| 352 | CLI | AIBL | SAADL | | 1 | |
| 353 | CLI | AIBL | RIMR | | 12 | |
| 354 | CLI | AIBL | RIBR | | 2 | |
| 355 | CLI | AINL | ASGR | | 1 | |
| 356 | CLI | AINL | RIBR | | 2 | |
| 357 | CLI | AINL | ASEL | | 2 | |
| 358 | CLI | AINL | RIBL | | 1 | |

| Entry Number | Type | Pre | Post | Ref | Den | Uniq |
|---|---|---|---|---|---|---|
| 359 | CLI | AINL | AFDR | | 5 | |
| 360 | CLI | AINL | BAGL | | 3 | |
| 361 | CLI | AINL | AUAR | | 1 | |
| 362 | CLI | AINL | WE | CEPshVR | 1 | |
| 363 | CLI | AIZL | SMBDL | | 4 | |
| 364 | CLI | AIZL | DVA | | 1 | |
| 365 | CLI | AIZL | AIBL | | 4 | |
| 366 | CLI | AIZL | ASIL | | 1 | |
| 367 | CLI | AIZL | ASEL | | 1 | |
| 368 | CLI | AIZL | ADFL | | 2 | |
| 369 | CLI | AIZL | SMBVL | | 7 | |
| 370 | CLI | AIZL | AIAL | | 2 | |
| 371 | CLI | AIZL | RIAL | | 7 | |
| 372 | CLI | AIZL | RIML | | 3 | |
| 373 | CLI | AIZL | AVER | | 4 | |
| 374 | CLI | AIZL | RIH | | 1 | |
| 375 | CLI | AIZL | AIBR | | 7 | |
| 376 | CLI | AVAL | DB3 | | 1 | |
| 377 | CLI | AVAL | DB5 | | 2 | |
| 378 | CLI | AVAL | AS2 | | 3 | |
| 379 | CLI | AVAL | DA7 | | 7 | |
| 380 | CLI | AVAL | VD13 | | 2 | |
| 381 | CLI | AVAL | DA6 | | 2 | |
| 382 | CLI | AVAL | AS11 | | 9 | |
| 383 | CLI | AVAL | DA5 | | 9 | |
| 384 | CLI | AVAL | VA10 | | 10 | |
| 385 | CLI | AVAL | DA4 | | 12 | |
| 386 | CLI | AVAL | AS10 | | 4 | |
| 387 | CLI | AVAL | DA3 | | 9 | |
| 388 | CLI | AVAL | DA9 | | 5 | |
| 389 | CLI | AVAL | DA2 | | 4 | |
| 390 | CLI | AVAL | PDEL | | 2 | |
| 391 | CLI | AVAL | DA1 | | 4 | |
| 392 | CLI | AVAL | VA3 | | 4 | |
| 393 | CLI | AVAL | AVDL | | 2 | |
| 394 | CLI | AVAL | AS1 | | 1 | |
| 395 | CLI | AVAL | AVBL | | 2 | |
| 396 | CLI | AVAL | ADLL | | 1 | |
| 397 | CLI | AVAL | AVAR | | 3 | |
| 398 | CLI | AVAL | VA6 | | 4 | |

| Entry Number | Type | Pre | Post | Ref | Den | Uniq |
|---|---|---|---|---|---|---|
| 399 | CLI | AVAL | AS6 | | 2 | |
| 400 | CLI | AVAL | VA4 | | 4 | |
| 401 | CLI | AVAL | AS5 | | 4 | |
| 402 | CLI | AVAL | VA2 | | 3 | |
| 403 | CLI | AVAL | AS4 | | 2 | |
| 404 | CLI | AVAL | VA11 | | 14 | |
| 405 | CLI | AVAL | AS3 | | 2 | |
| 406 | CLI | AVAL | SABVR | | 2 | |
| 407 | CLI | AVAL | DA8 | | 14 | |
| 408 | CLI | AVAL | PVCR | | 28 | |
| 409 | CLI | AVAL | VA12 | | 2 | |
| 410 | CLI | AVAL | LUAL | | 5 | |
| 411 | CLI | AVAL | DB3 | | 2 | |
| 412 | CLI | AVAL | SABD | | 1 | |
| 413 | CLI | AVAL | VA5 | | 6 | |
| 414 | CLI | AVAL | AVEL | | 4 | |
| 415 | CLI | AVBL | AS4 | | 2 | |
| 416 | CLI | AVBL | AVL | | 1 | |
| 417 | CLI | AVBL | AVER | | 3 | |
| 418 | CLI | AVBL | AS6 | | 2 | |
| 419 | CLI | AVBL | AVDR | | 3 | |
| 420 | CLI | AVBL | VA10 | | 1 | |
| 421 | CLI | AVBL | AS10 | | 2 | |
| 422 | CLI | AVBL | AS5 | | 1 | |
| 423 | CLI | AVBL | AVBR | | 2 | |
| 424 | CLI | AVBL | DA5 | | 1 | |
| 425 | CLI | AVBL | AVAR | | 25 | |
| 426 | CLI | AVBL | HSNR | | 1 | |
| 427 | CLI | AVBL | VA4 | | 1 | |
| 428 | CLI | AVBL | VA7 | | 1 | |
| 429 | CLI | AVBL | WN | HDC | 6 | |
| 430 | CLI | AVBL | AS3 | | 2 | |
| 431 | CLI | AVBL | AS1 | | 1 | |
| 432 | CLI | AVBL | VD3 | | 1 | |
| 433 | CLI | AVDL | LUAL | | 3 | |
| 434 | CLI | AVDL | AVAL | | 70 | |
| 435 | CLI | AVDL | DVC | | 1 | |
| 436 | CLI | AVDL | DA4 | | 4 | |
| 437 | CLI | AVDL | AS5 | | 1 | |
| 438 | CLI | AVDL | DA3 | | 6 | |

| Entry Number | Type | Pre | Post | Ref | Den | Uniq |
|---|---|---|---|---|---|---|
| 439 | CLI | AVDL | DA9 | | 1 | |
| 440 | CLI | AVDL | DA2 | | 2 | |
| 441 | CLI | AVDL | AS1 | | 1 | |
| 442 | CLI | AVDL | DA1 | | 3 | |
| 443 | CLI | AVDL | VA2 | | 1 | |
| 444 | CLI | AVDL | AVDR | | 2 | |
| 445 | CLI | AVDL | AS4 | | 1 | |
| 446 | CLI | AVDL | AVBL | | 1 | |
| 447 | CLI | AVDL | DA8 | | 1 | |
| 448 | CLI | AVDL | DA5 | | 3 | |
| 449 | CLI | AVDL | AS11 | | 2 | |
| 450 | CLI | AVDL | DB4 | | 1 | |
| 451 | CLI | AVDL | SABVL | | 6 | |
| 452 | CLI | AVDL | VA11 | | 1 | |
| 453 | CLI | AVDL | PQR | | 1 | |
| 454 | CLI | AVDL | VA6 | | 2 | |
| 455 | CLI | AVDL | VA5 | | 1 | |
| 456 | CLI | AVDL | SABD | | 3 | |
| 457 | CLI | AVDL | PVCL | | 1 | |
| 458 | CLI | AVDL | VA3 | | 3 | |
| 459 | CLI | AVDL | AS10 | | 2 | |
| 460 | CLI | AVEL | DA2 | | 6 | |
| 461 | CLI | AVEL | VA2 | | 2 | |
| 462 | CLI | AVEL | AVDL | | 1 | |
| 463 | CLI | AVEL | DA4 | | 1 | |
| 464 | CLI | AVEL | DA1 | | 12 | |
| 465 | CLI | AVEL | VA3 | | 5 | |
| 466 | CLI | AVEL | AVAL | | 44 | |
| 467 | CLI | AVEL | VA1 | | 6 | |
| 468 | CLI | AVEL | AS1 | | 5 | |
| 469 | CLI | AVEL | SABD | | 8 | |
| 470 | CLI | AVEL | VD2 | | 1 | |
| 471 | CLI | AVEL | PVCL | | 1 | |
| 472 | CLI | AVEL | VA4 | | 1 | |
| 473 | CLI | AVEL | AS3 | | 1 | |
| 474 | CLI | AVEL | AS2 | | 2 | |
| 475 | CLI | AVEL | DA3 | | 4 | |
| 476 | CLI | AVEL | VA6 | | 1 | |
| 477 | CLI | AVEL | DB3 | | 2 | |
| 478 | CLI | AVEL | DVB | | 1 | |

| Entry Number | Type | Pre | Post | Ref | Den | Uniq |
|---|---|---|---|---|---|---|
| 479 | CLI | AVEL | SABVL | | 17 | |
| 480 | CLI | AVEL | VD3 | | 1 | |
| 481 | CLI | AVHL | AVBR | | 2 | |
| 482 | CLI | AVHL | AVJL | | 4 | |
| 483 | CLI | AVHL | AVDL | | 1 | |
| 484 | CLI | AVHL | PVQR | | 2 | |
| 485 | CLI | AVHL | AVHR | | 2 | |
| 486 | CLI | AVHL | PVPR | | 1 | |
| 487 | CLI | AVHL | SMBVR | | 2 | |
| 488 | CLI | AVHL | AWBR | | 1 | |
| 489 | CLI | AVHL | AVJR | | 1 | |
| 490 | CLI | AVHL | RIMR | | 1 | |
| 491 | CLI | AVHL | RIR | | 3 | |
| 492 | CLI | AVHL | VD1 | | 1 | |
| 493 | CLI | AVHL | ADFR | | 3 | |
| 494 | CLI | AVHL | SMBDR | | 2 | |
| 495 | CLI | AVHL | AVFR | | 5 | |
| 496 | CLI | AVJL | AVHR | | 1 | |
| 497 | CLI | AVJL | PVQR | | 1 | |
| 498 | CLI | AVJL | SABVR | | 1 | |
| 499 | CLI | AVJL | PVCL | | 1 | |
| 500 | CLI | AVJL | RIFR | | 1 | |
| 501 | CLI | AVJL | PVCR | | 3 | |
| 502 | CLI | AVJL | AVER | | 4 | |
| 503 | CLI | AVJL | PVNR | | 2 | |
| 504 | CLI | AVJL | AVBR | | 5 | |
| 505 | CLI | AVJL | HSNR | | 1 | |
| 506 | CLI | AVJL | RIS | | 2 | |
| 507 | CLI | AVJL | AVDL | | 3 | |
| 508 | CLI | AVJL | AVBL | | 1 | |
| 509 | CLI | AVJL | AVFR | | 1 | |
| 510 | CLI | AVJL | AVAL | | 2 | |
| 511 | CLI | RIAL | RIAR | | 1 | |
| 512 | CLI | RIAL | RMDR | | 5 | |
| 513 | CLI | RIAL | SMDVL | | 3 | |
| 514 | CLI | RIAL | SMDVR | | 12 | |
| 515 | CLI | RIAL | RIVL | | 2 | |
| 516 | CLI | RIAL | RMDDR | | 6 | |
| 517 | CLI | RIAL | RMDL | | 6 | |
| 518 | CLI | RIAL | RIVR | | 3 | |

116

| Entry Number | Type | Pre | Post | Ref | Den | Uniq |
|---|---|---|---|---|---|---|
| 519 | CLI | RIAL | SMDDL | | 9 | |
| 520 | CLI | RIAL | RMDVR | | 10 | |
| 521 | CLI | RIAL | RMDDL | | 10 | |
| 522 | CLI | RIAL | RMDVL | | 7 | |
| 523 | CLI | RIAL | SMDDR | | 8 | |
| 524 | CLI | RIAL | SIADL | | 1 | |
| 525 | CLI | RIBL | RIAL | | 3 | |
| 526 | CLI | RIBL | AIBR | | 1 | |
| 527 | CLI | RIBL | AUAL | | 1 | |
| 528 | CLI | RIBL | AVEL | | 2 | |
| 529 | CLI | RIBL | AVBR | | 1 | |
| 530 | CLI | RIBL | AVDR | | 1 | |
| 531 | CLI | RIBL | AVAL | | 1 | |
| 532 | CLI | RIBL | SMDVR | | 2 | |
| 533 | CLI | RIBL | AVER | | 5 | |
| 534 | CLI | RICL | SMDDL | | 3 | |
| 535 | CLI | RICL | AVAL | | 5 | |
| 536 | CLI | RICL | SMDVR | | 1 | |
| 537 | CLI | RICL | SMBDL | | 2 | |
| 538 | CLI | RICL | RMFR | | 1 | |
| 539 | CLI | RICL | SMDDR | | 4 | |
| 540 | CLI | RICL | RIVR | | 1 | |
| 541 | CLI | RICL | AVAR | | 6 | |
| 542 | CLI | RICL | RIML | | 1 | |
| 543 | CLI | RICL | RIMR | | 2 | |
| 544 | CLI | SAAVL | OLLR | | 1 | |
| 545 | CLI | SAAVL | RIML | | 2 | |
| 546 | CLI | SAAVL | SMDDR | | 6 | |
| 547 | CLI | SAAVL | RIMR | | 9 | |
| 548 | CLI | SAAVL | RMFR | | 2 | |
| 549 | CLI | SAAVL | AVAL | | 15 | |
| 550 | CLI | SIBDL | WN | NONE1 | 1 | |
| 551 | CLM | RIML | RMDR | | 3 | |
| 552 | CLM | RIML | RIBL | | 1 | |
| 553 | CLM | RIML | RMDL | | 1 | |
| 554 | CLM | RIML | AVBL | | 2 | |
| 555 | CLM | RIML | RIS | | 1 | |
| 556 | CLM | RIML | SAAVL | | 3 | |
| 557 | CLM | RIML | WM | NMJ | 4 | |
| 558 | CLM | RIML | SMDDR | | 5 | |

| Entry Number | Type | Pre | Post | Ref | Den | Uniq |
|---|---|---|---|---|---|---|
| 559 | CLM | RIML | AVBR | | 3 | |
| 560 | CLM | RIML | RMFR | | 1 | |
| 561 | CLM | RIML | SAAVR | | 2 | |
| 562 | CLM | RIML | SMDVL | | 1 | |
| 563 | CLM | RIML | SAADR | | 1 | |
| 564 | CLM | RIML | AVAL | | 1 | |
| 565 | CLM | RMDL | RMFL | | 1 | |
| 566 | CLM | RMDL | OLLR | | 1 | |
| 567 | CLM | RMDL | RMDR | | 2 | |
| 568 | CLM | RMDL | RIAR | | 3 | |
| 569 | CLM | RMDL | RMDDR | | 1 | |
| 570 | CLM | RMDL | WM | NMJ | 11 | |
| 571 | CLM | RMDL | RIAL | | 5 | |
| 572 | CLM | RMDVL | RMDDR | | 6 | |
| 573 | CLM | RMDVL | RIAR | | 1 | |
| 574 | CLM | RMDVL | WM | NMJ | 7 | |
| 575 | CLM | RMDVL | RMDDL | | 1 | |
| 576 | CLM | SMDVL | PVR | | 1 | |
| 577 | CLM | SMDVL | RIAR | | 7 | |
| 578 | CLM | SMDVL | RMDDR | | 1 | |
| 579 | CLM | SMDVL | RIAL | | 3 | |
| 580 | CLM | SMDVL | RIVL | | 1 | |
| 581 | CLM | SMDVL | SMDDR | | 4 | |
| 582 | CLM | SMDVL | WM | NMJ | 2 | |
| 583 | CLMI | RIVL | SAADR | | 3 | |
| 584 | CLMI | RIVL | SIAVR | | 3 | |
| 585 | CLMI | RIVL | SMBVR | | 1 | |
| 586 | CLMI | RIVL | WM | NMJ | 4 | |
| 587 | CLMI | RIVL | RMDVL | | 1 | |
| 588 | CLMI | RIVL | RMDL | | 2 | |
| 589 | CLMI | RIVL | SMDDR | | 1 | |
| 590 | CLMI | RIVL | RIAL | | 1 | |
| 591 | CLMI | RIVL | RIAR | | 1 | |
| 592 | CLS | ADFL | RIR | | 2 | |
| 593 | CLS | ADFL | OLQVL | | 1 | |
| 594 | CLS | ADFL | SMBVL | | 2 | |
| 595 | CLS | ADFL | RIAL | | 12 | |
| 596 | CLS | ADFL | AIZL | | 11 | |
| 597 | CLS | ADFL | AUAL | | 3 | |
| 598 | CLS | ADFL | ADAL | | 1 | |

| Entry Number | Type | Pre | Post | Ref | Den | Uniq |
|---|---|---|---|---|---|---|
| 599 | CLS | ADLL | ASER | | 3 | |
| 600 | CLS | ADLL | AVAL | | 2 | |
| 601 | CLS | ADLL | AIBR | | 1 | |
| 602 | CLS | ADLL | AVDR | | 4 | |
| 603 | CLS | ADLL | ASHL | | 2 | |
| 604 | CLS | ADLL | AVDL | | 1 | |
| 605 | CLS | ADLL | AWBL | | 2 | |
| 606 | CLS | ADLL | AVBL | | 1 | |
| 607 | CLS | ADLL | RIPL | | 1 | |
| 608 | CLS | ADLL | AVAR | | 4 | |
| 609 | CLS | ADLL | ALA | | 1 | |
| 610 | CLS | ADLL | AVJL | | 1 | |
| 611 | CLS | ADLL | AIAL | | 5 | |
| 612 | CLS | ADLL | SMDDL | | 2 | |
| 613 | CLS | ADLL | AVJR | | 2 | |
| 614 | CLS | ADLL | AIBL | | 6 | |
| 615 | CLS | ADLL | OLQVL | | 1 | |
| 616 | CLS | AFDL | AIYL | | 7 | |
| 617 | CLS | AFDL | AINR | | 1 | |
| 618 | CLS | ASEL | AWCR | | 2 | |
| 619 | CLS | ASEL | AIYR | | 6 | |
| 620 | CLS | ASEL | AIYL | | 12 | |
| 621 | CLS | ASEL | AIBL | | 7 | |
| 622 | CLS | ASEL | RIAR | | 1 | |
| 623 | CLS | ASEL | AIAL | | 4 | |
| 624 | CLS | ASEL | AIBR | | 3 | |
| 625 | CLS | ASEL | AWCL | | 4 | |
| 626 | CLS | ASEL | ADFR | | 1 | |
| 627 | CLS | ASGL | ASKL | | 1 | |
| 628 | CLS | ASGL | AINR | | 1 | |
| 629 | CLS | ASGL | AIAL | | 9 | |
| 630 | CLS | ASGL | AIBL | | 3 | |
| 631 | CLS | ASHL | RIML | | 1 | |
| 632 | CLS | ASHL | RIAL | | 4 | |
| 633 | CLS | ASHL | ASKL | | 1 | |
| 634 | CLS | ASHL | AIAL | | 7 | |
| 635 | CLS | ASHL | RIPL | | 1 | |
| 636 | CLS | ASHL | ADFL | | 3 | |
| 637 | CLS | ASHL | AVDR | | 4 | |
| 638 | CLS | ASHL | AVBL | | 6 | |

| Entry Number | Type | Pre | Post | Ref | Den | Uniq |
|---|---|---|---|---|---|---|
| 639 | CLS | ASHL | AVAL | | 2 | |
| 640 | CLS | ASHL | RIR | | 1 | |
| 641 | CLS | ASHL | AVDL | | 2 | |
| 642 | CLS | ASHL | AIBL | | 5 | |
| 643 | CLS | ASIL | AIBL | | 1 | |
| 644 | CLS | ASIL | AWCL | | 2 | |
| 645 | CLS | ASIL | ASKL | | 2 | |
| 646 | CLS | ASIL | ASER | | 1 | |
| 647 | CLS | ASIL | AWCR | | 2 | |
| 648 | CLS | ASIL | AIZL | | 1 | |
| 649 | CLS | ASIL | AIYL | | 2 | |
| 650 | CLS | ASIL | ASEL | | 1 | |
| 651 | CLS | ASJL | ASKL | | 4 | |
| 652 | CLS | ASJL | PVQL | | 13 | |
| 653 | CLS | ASKL | AIAL | | 11 | |
| 654 | CLS | ASKL | AIML | | 2 | |
| 655 | CLS | ASKL | ASJL | | 1 | |
| 656 | CLS | ASKL | AIBL | | 3 | |
| 657 | CLS | AUAL | AVAL | | 2 | |
| 658 | CLS | AUAL | AVEL | | 4 | |
| 659 | CLS | AUAL | AVDR | | 1 | |
| 660 | CLS | AUAL | RIAL | | 6 | |
| 661 | CLS | AUAL | RIBL | | 9 | |
| 662 | CLS | AWAL | ASGL | | 1 | |
| 663 | CLS | AWAL | ASEL | | 4 | |
| 664 | CLS | AWAL | ADAL | | 1 | |
| 665 | CLS | AWAL | AIZL | | 10 | |
| 666 | CLS | AWAL | AWBL | | 1 | |
| 667 | CLS | AWAL | AIYL | | 1 | |
| 668 | CLS | AWAL | AFDL | | 5 | |
| 669 | CLS | AWBL | SMBDR | | 1 | |
| 670 | CLS | AWBL | RIAL | | 3 | |
| 671 | CLS | AWBL | AIZL | | 7 | |
| 672 | CLS | AWBL | AVBL | | 1 | |
| 673 | CLS | AWBL | AIBR | | 1 | |
| 674 | CLS | AWBL | ADFL | | 8 | |
| 675 | CLS | AWCL | AIBR | | 1 | |
| 676 | CLS | AWCL | AIAR | | 4 | |
| 677 | CLS | AWCL | AVAL | | 1 | |
| 678 | CLS | AWCL | RIAL | | 2 | |

| Entry Number | Type | Pre | Post | Ref | Den | Uniq |
|---|---|---|---|---|---|---|
| 679 | CLS | AWCL | AIBL | | 1 | |
| 680 | CLS | AWCL | AIYL | | 13 | |
| 681 | CLS | AWCL | AIAL | | 2 | |
| 682 | CRI | AIBR | AVEL | | 1 | |
| 683 | CRI | AIBR | SMDDL | | 3 | |
| 684 | CRI | AIBR | VB1 | | 1 | |
| 685 | CRI | AIBR | SAADL | | 1 | |
| 686 | CRI | AIBR | RIMR | | 1 | |
| 687 | CRI | AIBR | RIAL | | 1 | |
| 688 | CRI | AIBR | RIML | | 14 | |
| 689 | CRI | AIBR | AVAR | | 1 | |
| 690 | CRI | AIBR | AVBR | | 3 | |
| 691 | CRI | AIBR | RIBL | | 4 | |
| 692 | CRI | AINR | AIAL | | 2 | |
| 693 | CRI | AINR | AFDL | | 5 | |
| 694 | CRI | AINR | ASER | | 3 | |
| 695 | CRI | AINR | AIBL | | 2 | |
| 696 | CRI | AINR | BAGR | | 3 | |
| 697 | CRI | AINR | RID | | 1 | |
| 698 | CRI | AINR | WE | CEPshVL | 1 | |
| 699 | CRI | AINR | RIBL | | 1 | |
| 700 | CRI | AIZR | DVA | | 2 | |
| 701 | CRI | AIZR | SMBVR | | 3 | |
| 702 | CRI | AIZR | AVEL | | 4 | |
| 703 | CRI | AIZR | RIMR | | 4 | |
| 704 | CRI | AIZR | AVER | | 1 | |
| 705 | CRI | AIZR | SMBDR | | 5 | |
| 706 | CRI | AIZR | AIBL | | 9 | |
| 707 | CRI | AIZR | AIAR | | 1 | |
| 708 | CRI | AIZR | AIBR | | 1 | |
| 709 | CRI | AIZR | RIAR | | 7 | |
| 710 | CRI | AUAR | AVER | | 3 | |
| 711 | CRI | AUAR | RIAR | | 10 | |
| 712 | CRI | AUAR | URXR | | 1 | |
| 713 | CRI | AUAR | RIBR | | 12 | |
| 714 | CRI | AUAR | AVAR | | 1 | |
| 715 | CRI | AUAR | AIYR | | 1 | |
| 716 | CRI | AVAR | SABD | | 1 | |
| 717 | CRI | AVAR | VA2 | | 3 | |
| 718 | CRI | AVAR | AS11 | | 9 | |

| Entry Number | Type | Pre | Post | Ref | Den | Uniq |
|---|---|---|---|---|---|---|
| 719 | CRI | AVAR | VA6 | | 4 | |
| 720 | CRI | AVAR | DA6 | | 2 | |
| 721 | CRI | AVAR | VD13 | | 2 | |
| 722 | CRI | AVAR | AS10 | | 4 | |
| 723 | CRI | AVAR | AVER | | 4 | |
| 724 | CRI | AVAR | RIGL | | 1 | |
| 725 | CRI | AVAR | DA1 | | 4 | |
| 726 | CRI | AVAR | AS1 | | 1 | |
| 727 | CRI | AVAR | DA2 | | 4 | |
| 728 | CRI | AVAR | VA4 | | 4 | |
| 729 | CRI | AVAR | DA3 | | 9 | |
| 730 | CRI | AVAR | VA11 | | 14 | |
| 731 | CRI | AVAR | AVDR | | 2 | |
| 732 | CRI | AVAR | DB3 | | 2 | |
| 733 | CRI | AVAR | AVBR | | 2 | |
| 734 | CRI | AVAR | ADER | | 1 | |
| 735 | CRI | AVAR | AVAL | | 3 | |
| 736 | CRI | AVAR | DA4 | | 12 | |
| 737 | CRI | AVAR | AS6 | | 2 | |
| 738 | CRI | AVAR | DA7 | | 7 | |
| 739 | CRI | AVAR | AS5 | | 4 | |
| 740 | CRI | AVAR | DA9 | | 5 | |
| 741 | CRI | AVAR | AS4 | | 2 | |
| 742 | CRI | AVAR | DB5 | | 2 | |
| 743 | CRI | AVAR | AS3 | | 2 | |
| 744 | CRI | AVAR | PVCL | | 28 | |
| 745 | CRI | AVAR | AS2 | | 3 | |
| 746 | CRI | AVAR | LUAR | | 5 | |
| 747 | CRI | AVAR | DA5 | | 9 | |
| 748 | CRI | AVAR | VA10 | | 10 | |
| 749 | CRI | AVAR | PDER | | 2 | |
| 750 | CRI | AVAR | VA12 | | 2 | |
| 751 | CRI | AVAR | VA5 | | 6 | |
| 752 | CRI | AVAR | SABVL | | 2 | |
| 753 | CRI | AVAR | DA8 | | 14 | |
| 754 | CRI | AVAR | VA3 | | 4 | |
| 755 | CRI | AVBR | AS4 | | 2 | |
| 756 | CRI | AVBR | HSNL | | 1 | |
| 757 | CRI | AVBR | VA10 | | 1 | |
| 758 | CRI | AVBR | VA7 | | 1 | |

| Entry Number | Type | Pre | Post | Ref | Den | Uniq |
|---|---|---|---|---|---|---|
| 759 | CRI | AVBR | VA4 | | 1 | |
| 760 | CRI | AVBR | AVEL | | 3 | |
| 761 | CRI | AVBR | AS5 | | 1 | |
| 762 | CRI | AVBR | VD3 | | 1 | |
| 763 | CRI | AVBR | AS6 | | 2 | |
| 764 | CRI | AVBR | WN | HDC | 6 | |
| 765 | CRI | AVBR | AVAL | | 25 | |
| 766 | CRI | AVBR | AVBL | | 2 | |
| 767 | CRI | AVBR | AS3 | | 2 | |
| 768 | CRI | AVBR | AVL | | 1 | |
| 769 | CRI | AVBR | DA5 | | 1 | |
| 770 | CRI | AVBR | AVDL | | 3 | |
| 771 | CRI | AVBR | AS10 | | 2 | |
| 772 | CRI | AVBR | AS1 | | 1 | |
| 773 | CRI | AVDR | DA9 | | 1 | |
| 774 | CRI | AVDR | VA11 | | 1 | |
| 775 | CRI | AVDR | DA4 | | 4 | |
| 776 | CRI | AVDR | AS11 | | 2 | |
| 777 | CRI | AVDR | AS1 | | 1 | |
| 778 | CRI | AVDR | DA3 | | 6 | |
| 779 | CRI | AVDR | AS4 | | 1 | |
| 780 | CRI | AVDR | VA5 | | 1 | |
| 781 | CRI | AVDR | AVBR | | 1 | |
| 782 | CRI | AVDR | VA6 | | 2 | |
| 783 | CRI | AVDR | PQR | | 1 | |
| 784 | CRI | AVDR | DA2 | | 2 | |
| 785 | CRI | AVDR | AVAR | | 70 | |
| 786 | CRI | AVDR | DA1 | | 3 | |
| 787 | CRI | AVDR | DA5 | | 3 | |
| 788 | CRI | AVDR | AVDL | | 2 | |
| 789 | CRI | AVDR | AS5 | | 1 | |
| 790 | CRI | AVDR | LUAR | | 3 | |
| 791 | CRI | AVDR | DVC | | 1 | |
| 792 | CRI | AVDR | PVCR | | 1 | |
| 793 | CRI | AVDR | AS10 | | 2 | |
| 794 | CRI | AVDR | DA8 | | 1 | |
| 795 | CRI | AVDR | VA2 | | 1 | |
| 796 | CRI | AVDR | DB4 | | 1 | |
| 797 | CRI | AVDR | SABD | | 3 | |
| 798 | CRI | AVDR | SABVR | | 6 | |

123

| Entry Number | Type | Pre | Post | Ref | Den | Uniq |
|---|---|---|---|---|---|---|
| 799 | CRI | AVDR | VA3 | | 3 | |
| 800 | CRI | AVER | DA4 | | 1 | |
| 801 | CRI | AVER | AS1 | | 5 | |
| 802 | CRI | AVER | DB3 | | 2 | |
| 803 | CRI | AVER | AS3 | | 1 | |
| 804 | CRI | AVER | SABVR | | 17 | |
| 805 | CRI | AVER | AVAR | | 44 | |
| 806 | CRI | AVER | DA3 | | 4 | |
| 807 | CRI | AVER | AVDR | | 1 | |
| 808 | CRI | AVER | VA4 | | 1 | |
| 809 | CRI | AVER | DVB | | 1 | |
| 810 | CRI | AVER | DA2 | | 6 | |
| 811 | CRI | AVER | VD3 | | 1 | |
| 812 | CRI | AVER | VA2 | | 2 | |
| 813 | CRI | AVER | VA1 | | 6 | |
| 814 | CRI | AVER | VD2 | | 1 | |
| 815 | CRI | AVER | VA3 | | 5 | |
| 816 | CRI | AVER | PVCR | | 1 | |
| 817 | CRI | AVER | AS2 | | 2 | |
| 818 | CRI | AVER | VA6 | | 1 | |
| 819 | CRI | AVER | SABD | | 8 | |
| 820 | CRI | AVER | DA1 | | 12 | |
| 821 | CRI | AVHR | AVDR | | 1 | |
| 822 | CRI | AVHR | SMBDL | | 1 | |
| 823 | CRI | AVHR | AVBL | | 1 | |
| 824 | CRI | AVHR | VD1 | | 1 | |
| 825 | CRI | AVHR | SMBVL | | 1 | |
| 826 | CRI | AVHR | RIGL | | 1 | |
| 827 | CRI | AVHR | ADLR | | 2 | |
| 828 | CRI | AVHR | RIR | | 3 | |
| 829 | CRI | AVHR | AVJL | | 1 | |
| 830 | CRI | AVHR | PVQL | | 2 | |
| 831 | CRI | AVHR | PVPL | | 4 | |
| 832 | CRI | AVHR | AVHL | | 2 | |
| 833 | CRI | AVHR | AQR | | 1 | |
| 834 | CRI | AVHR | AVJR | | 2 | |
| 835 | CRI | AVHR | AVFL | | 5 | |
| 836 | CRI | AVJR | AVM | | 1 | |
| 837 | CRI | AVJR | SABVL | | 1 | |
| 838 | CRI | AVJR | AVDR | | 4 | |

| Entry Number | Type | Pre | Post | Ref | Den | Uniq |
|---|---|---|---|---|---|---|
| 839 | CRI | AVJR | AVAR | | 2 | |
| 840 | CRI | AVJR | AVHL | | 1 | |
| 841 | CRI | AVJR | AVBL | | 4 | |
| 842 | CRI | AVJR | PVCL | | 2 | |
| 843 | CRI | AVJR | AVBR | | 1 | |
| 844 | CRI | AVJR | AVEL | | 4 | |
| 845 | CRI | AVJR | PVQL | | 1 | |
| 846 | CRI | AVJR | PVNL | | 1 | |
| 847 | CRI | AVJR | HSNL | | 1 | |
| 848 | CRI | AVJR | AVJL | | 1 | |
| 849 | CRI | AVJR | PVCR | | 4 | |
| 850 | CRI | AVJR | AVEL | | 1 | |
| 851 | CRI | RIAR | CEPVR | | 1 | |
| 852 | CRI | RIAR | RIVL | | 2 | |
| 853 | CRI | RIAR | RMDVL | | 4 | |
| 854 | CRI | RIAR | RMDDL | | 12 | |
| 855 | CRI | RIAR | SMDDL | | 6 | |
| 856 | CRI | RIAR | RMDDR | | 8 | |
| 857 | CRI | RIAR | SIADL | | 1 | |
| 858 | CRI | RIAR | SMDDR | | 10 | |
| 859 | CRI | RIAR | IL1R | | 1 | |
| 860 | CRI | RIAR | SMDVL | | 13 | |
| 861 | CRI | RIAR | SIADR | | 1 | |
| 862 | CRI | RIAR | SMDVR | | 7 | |
| 863 | CRI | RIAR | SIAVL | | 1 | |
| 864 | CRI | RIAR | SAADR | | 1 | |
| 865 | CRI | RIAR | RMDR | | 7 | |
| 866 | CRI | RIAR | RMDL | | 5 | |
| 867 | CRI | RIAR | RMDVR | | 9 | |
| 868 | CRI | RIBR | AVEL | | 4 | |
| 869 | CRI | RIBR | AVAR | | 2 | |
| 870 | CRI | RIBR | RIH | | 1 | |
| 871 | CRI | RIBR | RIAR | | 2 | |
| 872 | CRI | RIBR | AIBL | | 1 | |
| 873 | CRI | RIBR | SMDDL | | 1 | |
| 874 | CRI | RICR | ADAR | | 1 | |
| 875 | CRI | RICR | SMDVL | | 2 | |
| 876 | CRI | RICR | SMBDR | | 1 | |
| 877 | CRI | RICR | SMDDR | | 3 | |
| 878 | CRI | RICR | AVAL | | 5 | |

| Entry Number | Type | Pre | Post | Ref | Den | Uniq |
|---|---|---|---|---|---|---|
| 879 | CRI | RICR | AVAR | | 3 | |
| 880 | CRI | RICR | SMDVR | | 1 | |
| 881 | CRI | RICR | SMDDL | | 2 | |
| 882 | CRI | SAAVR | AVAR | | 11 | |
| 883 | CRI | SAAVR | SMDDL | | 6 | |
| 884 | CRI | SAAVR | RIML | | 4 | |
| 885 | CRI | SAAVR | RIMR | | 1 | |
| 886 | CRI | SIBDR | WN | NONE1 | 1 | |
| 887 | CRM | RIMR | AVBR | | 5 | |
| 888 | CRM | RIMR | WM | NMJ | 4 | |
| 889 | CRM | RIMR | RMDL | | 2 | |
| 890 | CRM | RIMR | SMDDR | | 3 | |
| 891 | CRM | RIMR | AVKL | | 1 | |
| 892 | CRM | RIMR | SAAVL | | 3 | |
| 893 | CRM | RIMR | RIS | | 1 | |
| 894 | CRM | RIMR | RMFR | | 2 | |
| 895 | CRM | RIMR | ADAR | | 1 | |
| 896 | CRM | RIMR | RMDR | | 1 | |
| 897 | CRM | RIMR | AIBL | | 4 | |
| 898 | CRM | RIMR | AVBL | | 2 | |
| 899 | CRM | RIMR | AVJL | | 1 | |
| 900 | CRM | RIMR | SAAVR | | 3 | |
| 901 | CRM | RIMR | AVAL | | 2 | |
| 902 | CRM | RIMR | SMDDL | | 2 | |
| 903 | CRM | RIMR | RIBR | | 1 | |
| 904 | CRM | RMDR | WM | NMJ | 4 | |
| 905 | CRM | RMDR | RIAL | | 5 | |
| 906 | CRM | RMDR | RMDL | | 1 | |
| 907 | CRM | RMDR | RIAR | | 7 | |
| 908 | CRM | RMDR | RMDDL | | 1 | |
| 909 | CRM | RMDR | AVKL | | 1 | |
| 910 | CRM | RMDVR | SIBVR | | 1 | |
| 911 | CRM | RMDVR | RMDDL | | 10 | |
| 912 | CRM | RMDVR | WM | NMJ | 9 | |
| 913 | CRM | SMDVR | WM | NMJ | 3 | |
| 914 | CRM | SMDVR | SMDDL | | 2 | |
| 915 | CRM | SMDVR | RIVR | | 1 | |
| 916 | CRM | SMDVR | RMDDL | | 2 | |
| 917 | CRM | SMDVR | RIAL | | 8 | |
| 918 | CRM | SMDVR | RIAR | | 4 | |

| Entry Number | Type | Pre | Post | Ref | Den | Uniq |
|---|---|---|---|---|---|---|
| 919 | CRMI | RIVR | SMDDL | | 2 | |
| 920 | CRMI | RIVR | RMEV | | 1 | |
| 921 | CRMI | RIVR | RIAR | | 2 | |
| 922 | CRMI | RIVR | RMDR | | 1 | |
| 923 | CRMI | RIVR | SMDVR | | 2 | |
| 924 | CRMI | RIVR | RMDVR | | 1 | |
| 925 | CRMI | RIVR | SIAVL | | 2 | |
| 926 | CRMI | RIVR | RMDDL | | 1 | |
| 927 | CRMI | RIVR | WM | NMJ | 5 | |
| 928 | CRMI | RIVR | RIAL | | 2 | |
| 929 | CRMI | RIVR | SAADL | | 2 | |
| 930 | CRS | ADFR | PVPR | | 1 | |
| 931 | CRS | ADFR | RIGR | | 2 | |
| 932 | CRS | ADFR | ASHR | | 1 | |
| 933 | CRS | ADFR | AIYR | | 1 | |
| 934 | CRS | ADFR | SMBVR | | 2 | |
| 935 | CRS | ADFR | ADAR | | 1 | |
| 936 | CRS | ADFR | AWBR | | 2 | |
| 937 | CRS | ADFR | AIZR | | 8 | |
| 938 | CRS | ADFR | RIR | | 3 | |
| 939 | CRS | ADFR | URXR | | 1 | |
| 940 | CRS | ADFR | RIAR | | 12 | |
| 941 | CRS | ADFR | SMBDL | | 1 | |
| 942 | CRS | ADFR | AUAR | | 3 | |
| 943 | CRS | ADLR | PVCL | | 2 | |
| 944 | CRS | ADLR | AVDR | | 2 | |
| 945 | CRS | ADLR | AIBR | | 11 | |
| 946 | CRS | ADLR | AVJR | | 2 | |
| 947 | CRS | ADLR | ASHR | | 3 | |
| 948 | CRS | ADLR | AVDL | | 5 | |
| 949 | CRS | ADLR | ASER | | 1 | |
| 950 | CRS | ADLR | OLLR | | 1 | |
| 951 | CRS | ADLR | AVHR | | 1 | |
| 952 | CRS | ADLR | AVBL | | 1 | |
| 953 | CRS | ADLR | AWCR | | 3 | |
| 954 | CRS | ADLR | AIAR | | 10 | |
| 955 | CRS | ADLR | AVBR | | 2 | |
| 956 | CRS | ADLR | AVAR | | 2 | |
| 957 | CRS | ADLR | RICR | | 1 | |
| 958 | CRS | ADLR | RICL | | 1 | |

| Entry Number | Type | Pre | Post | Ref | Den | Uniq |
|---|---|---|---|---|---|---|
| 959 | CRS | AFDR | AIYR | | 11 | |
| 960 | CRS | AFDR | ASER | | 1 | |
| 961 | CRS | ASER | AIYL | | 4 | |
| 962 | CRS | ASER | AWCL | | 1 | |
| 963 | CRS | ASER | AWCR | | 2 | |
| 964 | CRS | ASER | AIYR | | 12 | |
| 965 | CRS | ASER | AIBL | | 1 | |
| 966 | CRS | ASER | AFDR | | 1 | |
| 967 | CRS | ASER | AIBR | | 7 | |
| 968 | CRS | ASER | AFDL | | 3 | |
| 969 | CRS | ASER | AIAR | | 3 | |
| 970 | CRS | ASER | AIAL | | 4 | |
| 971 | CRS | ASER | AWAR | | 1 | |
| 972 | CRS | ASGR | AIAR | | 10 | |
| 973 | CRS | ASGR | AIBR | | 2 | |
| 974 | CRS | ASHR | HSNR | | 1 | |
| 975 | CRS | ASHR | AIBR | | 3 | |
| 976 | CRS | ASHR | PVPR | | 1 | |
| 977 | CRS | ASHR | AVBR | | 3 | |
| 978 | CRS | ASHR | AVER | | 2 | |
| 979 | CRS | ASHR | AVDR | | 1 | |
| 980 | CRS | ASHR | AVDL | | 5 | |
| 981 | CRS | ASHR | ADAR | | 1 | |
| 982 | CRS | ASHR | AIAR | | 9 | |
| 983 | CRS | ASHR | RMGR | | 2 | |
| 984 | CRS | ASHR | AVAR | | 5 | |
| 985 | CRS | ASHR | ADFR | | 2 | |
| 986 | CRS | ASHR | RIAR | | 2 | |
| 987 | CRS | ASIR | AWCR | | 2 | |
| 988 | CRS | ASIR | AIAR | | 3 | |
| 989 | CRS | ASIR | ASEL | | 2 | |
| 990 | CRS | ASIR | ASHR | | 1 | |
| 991 | CRS | ASIR | AWCL | | 1 | |
| 992 | CRS | ASJR | ASKR | | 4 | |
| 993 | CRS | ASJR | PVQR | | 13 | |
| 994 | CRS | ASJR | HSNR | | 1 | |
| 995 | CRS | ASKR | AWAR | | 1 | |
| 996 | CRS | ASKR | AIAR | | 11 | |
| 997 | CRS | ASKR | AIBR | | 1 | |
| 998 | CRS | ASKR | AIMR | | 1 | |

| Entry Number | Type | Pre | Post | Ref | Den | Uniq |
|---|---|---|---|---|---|---|
| 999 | CRS | AWAR | AFDR | | 9 | |
| 1000 | CRS | AWAR | RIR | | 1 | |
| 1001 | CRS | AWAR | AIYR | | 2 | |
| 1002 | CRS | AWAR | AWBR | | 2 | |
| 1003 | CRS | AWAR | ADFR | | 3 | |
| 1004 | CRS | AWAR | ASER | | 2 | |
| 1005 | CRS | AWAR | AIZR | | 7 | |
| 1006 | CRS | AWAR | RIFR | | 2 | |
| 1007 | CRS | AWAR | ASEL | | 1 | |
| 1008 | CRS | AWBR | ADFR | | 4 | |
| 1009 | CRS | AWBR | RIR | | 2 | |
| 1010 | CRS | AWBR | AIZR | | 1 | |
| 1011 | CRS | AWBR | AVBR | | 2 | |
| 1012 | CRS | AWBR | ASGR | | 1 | |
| 1013 | CRS | AWBR | SMBVR | | 1 | |
| 1014 | CRS | AWBR | RIAR | | 1 | |
| 1015 | CRS | AWCR | AIAR | | 2 | |
| 1016 | CRS | AWCR | AIYL | | 4 | |
| 1017 | CRS | AWCR | AIYR | | 9 | |
| 1018 | CRS | AWCR | AIBL | | 1 | |
| 1019 | CRS | AWCR | AWCL | | 2 | |
| 1020 | CRS | AWCR | ADLR | | 1 | |
| 1021 | CRS | AWCR | AIBR | | 3 | |
| 1022 | DUI | RIH | RIPL | | 5 | |
| 1023 | DUI | RIH | OLQVR | | 7 | |
| 1024 | DUI | RIH | ADFR | | 1 | |
| 1025 | DUI | RIH | AIZR | | 4 | |
| 1026 | DUI | RIH | CEPVL | | 1 | |
| 1027 | DUI | RIH | WE | CEPshVR | 1 | |
| 1028 | DUI | RIH | RIAR | | 9 | |
| 1029 | DUI | RIH | CEPVR | | 2 | |
| 1030 | DUI | RIH | OLQDR | | 3 | |
| 1031 | DUI | RIH | OLQDL | | 3 | |
| 1032 | DUI | RIH | AIZL | | 5 | |
| 1033 | DUI | RIH | OLQVL | | 3 | |
| 1034 | DUI | RIH | RIBL | | 5 | |
| 1035 | DUI | RIH | RIBR | | 4 | |
| 1036 | DUI | RIH | URYVR | | 1 | |
| 1037 | DUI | RIH | CEPDL | | 1 | |
| 1038 | DUI | RIH | BAGR | | 1 | |

| Entry Number | Type | Pre | Post | Ref | Den | Uniq |
|---|---|---|---|---|---|---|
| 1039 | DUI | RIH | RIPR | | 5 | |
| 1040 | DUI | RIH | AUAR | | 1 | |
| 1041 | DUI | RIH | RMER | | 1 | |
| 1042 | DUI | RIH | CEPDR | | 1 | |
| 1043 | DUI | RIH | RIAL | | 12 | |
| 1044 | DUI | RIH | IL2R | | 1 | |
| 1045 | DUI | RIH | RMEV | | 1 | |
| 1046 | DUI | RIR | DVA | | 3 | |
| 1047 | DUI | RIR | URXL | | 5 | |
| 1048 | DUI | RIR | AIZL | | 6 | |
| 1049 | DUI | RIR | RIAL | | 6 | |
| 1050 | DUI | RIR | BAGR | | 1 | |
| 1051 | DUI | RIR | RIAR | | 1 | |
| 1052 | DUI | RIR | AQR | | 1 | |
| 1053 | DUI | RIR | HSNL | | 1 | |
| 1054 | DUI | RIR | AIZR | | 5 | |
| 1055 | DUI | RIR | AUAL | | 1 | |
| 1056 | DUI | RIR | URXR | | 2 | |
| 1057 | DUI | RIS | RMDDL | | 1 | |
| 1058 | DUI | RIS | SMDDL | | 1 | |
| 1059 | DUI | RIS | RMDR | | 5 | |
| 1060 | DUI | RIS | RIBL | | 3 | |
| 1061 | DUI | RIS | RMDL | | 2 | |
| 1062 | DUI | RIS | AVL | | 2 | |
| 1063 | DUI | RIS | CEPDL | | 1 | |
| 1064 | DUI | RIS | OLLR | | 1 | |
| 1065 | DUI | RIS | RIMR | | 4 | |
| 1066 | DUI | RIS | SMDVR | | 1 | |
| 1067 | DUI | RIS | RIML | | 2 | |
| 1068 | DUI | RIS | CEPVR | | 1 | |
| 1069 | DUI | RIS | RIBR | | 6 | |
| 1070 | DUI | RIS | AVKL | | 1 | |
| 1071 | DUI | RIS | AVKR | | 3 | |
| 1072 | DUI | RIS | SMDDR | | 1 | |
| 1073 | DUI | RIS | URYVR | | 1 | |
| 1074 | DUI | RIS | CEPDR | | 2 | |
| 1075 | DUI | RIS | AVER | | 6 | |
| 1076 | DUI | RIS | SMDVL | | 1 | |
| 1077 | DUI | RIS | AVEL | | 8 | |
| 1078 | DUI | RIS | CEPVL | | 2 | |

130

| Entry Number | Type | Pre | Post | Ref | Den | Uniq |
|---|---|---|---|---|---|---|
| 1079 | DUIM | AVL | PVPR | | 1 | |
| 1080 | DUIM | AVL | PVWR | | 1 | |
| 1081 | DUIM | AVL | AS1 | | 1 | |
| 1082 | DUIM | AVL | AVER | | 1 | |
| 1083 | DUIM | AVL | SABVR | | 7 | |
| 1084 | DUIM | AVL | DD6 | | 2 | |
| 1085 | DUIM | AVL | SABD | | 5 | |
| 1086 | DUIM | AVL | DA2 | | 1 | |
| 1087 | DUIM | AVL | VD12 | | 3 | |
| 1088 | DUIM | AVL | DD1 | | 1 | |
| 1089 | DUIM | AVL | AVFR | | 1 | |
| 1090 | DUIM | AVL | WM | NMJ | 6 | |
| 1091 | DLI | AIAL | RIFL | | 1 | |
| 1092 | DLI | AIAL | HSNL | | 1 | |
| 1093 | DLI | AIAL | AWCR | | 1 | |
| 1094 | DLI | AIAL | AIZL | | 2 | |
| 1095 | DLI | AIAL | ASKL | | 3 | |
| 1096 | DLI | AIAL | AIBL | | 11 | |
| 1097 | DLI | AIAL | ASGL | | 1 | |
| 1098 | DLI | AIAL | AIML | | 2 | |
| 1099 | DLI | AIAL | ADAL | | 1 | |
| 1100 | DLI | AIAL | ASER | | 2 | |
| 1101 | DLI | AIML | PVQL | | 1 | |
| 1102 | DLI | AIML | WE | CEPshVL | 1 | |
| 1103 | DLI | AIML | RIFL | | 1 | |
| 1104 | DLI | AIML | ASKL | | 2 | |
| 1105 | DLI | AIML | SMBVL | | 1 | |
| 1106 | DLI | AIML | AVFR | | 1 | |
| 1107 | DLI | AIML | AVDL | | 1 | |
| 1108 | DLI | AIML | ASGL | | 2 | |
| 1109 | DLI | AIML | AVER | | 1 | |
| 1110 | DLI | AIML | WE | CEPshDL | 1 | |
| 1111 | DLI | AIML | AVBR | | 2 | |
| 1112 | DLI | AIML | AVHR | | 2 | |
| 1113 | DLI | AIML | AVFL | | 4 | |
| 1114 | DLI | AIML | AVDR | | 1 | |
| 1115 | DLI | AIML | AVJL | | 1 | |
| 1116 | DLI | AIML | AVHL | | 1 | |
| 1117 | DLI | AIML | AIAL | | 5 | |
| 1118 | DLI | AIML | ALML | | 1 | |

| Entry Number | Type | Pre | Post | Ref | Den | Uniq |
|---|---|---|---|---|---|---|
| 1119 | DLI | AIYL | RIAL | | 5 | |
| 1120 | DLI | AIYL | AIZL | | 13 | |
| 1121 | DLI | AIYL | AWCL | | 1 | |
| 1122 | DLI | AIYL | HSNR | | 1 | |
| 1123 | DLI | AIYL | AWCR | | 1 | |
| 1124 | DLI | AIYL | AWAL | | 3 | |
| 1125 | DLI | AIYL | RIBL | | 5 | |
| 1126 | DLI | AVKL | AVM | | 1 | |
| 1127 | DLI | AVKL | SAADR | | 1 | |
| 1128 | DLI | AVKL | WN | HDC | 5 | |
| 1129 | DLI | AVKL | AVER | | 1 | |
| 1130 | DLI | AVKL | PVM | | 1 | |
| 1131 | DLI | AVKL | RIML | | 2 | |
| 1132 | DLI | AVKL | RIMR | | 1 | |
| 1133 | DLI | AVKL | DVA | | 1 | |
| 1134 | DLI | AVKL | SIAVR | | 1 | |
| 1135 | DLI | AVKL | PVQR | | 1 | |
| 1136 | DLI | AVKL | AVEL | | 2 | |
| 1137 | DLI | AVKL | AVBL | | 1 | |
| 1138 | DLI | AVKL | PDER | | 4 | |
| 1139 | DLI | SAADL | RIML | | 3 | |
| 1140 | DLI | SAADL | RIMR | | 5 | |
| 1141 | DLI | SAADL | AIBL | | 1 | |
| 1142 | DLI | SAADL | RMGL | | 1 | |
| 1143 | DLI | SAADL | AVAL | | 5 | |
| 1144 | DLI | SIADL | WN | NONE2 | 1 | |
| 1145 | DLI | SIAVL | WN | NONE2 | 1 | |
| 1146 | DLI | SIBVL | WN | NONE1 | 1 | |
| 1147 | DLM | RMDDL | RMDVR | | 7 | |
| 1148 | DLM | RMDDL | WM | NMJ | 8 | |
| 1149 | DLM | RMFL | RMDL | | 5 | |
| 1150 | DLM | RMFL | AVKR | | 4 | |
| 1151 | DLM | RMFL | RMDR | | 1 | |
| 1152 | DLM | RMFL | RMGR | | 1 | |
| 1153 | DLM | RMFL | AIBL | | 1 | |
| 1154 | DLM | RMFL | WM | NMJ | 5 | |
| 1155 | DLM | RMFL | AVKL | | 1 | |
| 1156 | DLM | RMHL | RMDR | | 1 | |
| 1157 | DLM | RMHL | WM | NMJ | 8 | |
| 1158 | DLM | SMBDL | WM | NMJ | 9 | |

| Entry Number | Type | Pre | Post | Ref | Den | Uniq |
|---|---|---|---|---|---|---|
| 1159 | DLM | SMBDL | AVAR | | 1 | |
| 1160 | DLM | SMBDL | RMED | | 4 | |
| 1161 | DLM | SMBDL | SAAVR | | 2 | |
| 1162 | DLM | SMBVL | RMEV | | 5 | |
| 1163 | DLM | SMBVL | SAADL | | 3 | |
| 1164 | DLM | SMBVL | PLNR | | 1 | |
| 1165 | DLM | SMBVL | WM | NMJ | 4 | |
| 1166 | DLM | SMDDL | SMDVR | | 1 | |
| 1167 | DLM | SMDDL | WM | NMJ | 3 | |
| 1168 | DLM | SMDDL | RIAL | | 1 | |
| 1169 | DLM | SMDDL | RIAR | | 1 | |
| 1170 | DRI | AIAR | ADAR | | 1 | |
| 1171 | DRI | AIAR | AWCR | | 1 | |
| 1172 | DRI | AIAR | RIFR | | 2 | |
| 1173 | DRI | AIAR | ADLR | | 1 | |
| 1174 | DRI | AIAR | AIBR | | 12 | |
| 1175 | DRI | AIAR | AWAR | | 1 | |
| 1176 | DRI | AIAR | ASER | | 1 | |
| 1177 | DRI | AIAR | AIZR | | 1 | |
| 1178 | DRI | AIMR | AVFL | | 3 | |
| 1179 | DRI | AIMR | AVDR | | 1 | |
| 1180 | DRI | AIMR | AVFR | | 2 | |
| 1181 | DRI | AIMR | HSNL | | 1 | |
| 1182 | DRI | AIMR | AVJR | | 1 | |
| 1183 | DRI | AIMR | RIFR | | 1 | |
| 1184 | DRI | AIMR | ASGR | | 2 | |
| 1185 | DRI | AIMR | AIAR | | 4 | |
| 1186 | DRI | AIMR | ASKR | | 2 | |
| 1187 | DRI | AIMR | RMGR | | 2 | |
| 1188 | DRI | AIMR | HSNR | | 2 | |
| 1189 | DRI | AIMR | ASJR | | 2 | |
| 1190 | DRI | AIMR | WE | CEPshVR | 1 | |
| 1191 | DRI | AIYR | HSNL | | 1 | |
| 1192 | DRI | AIYR | RIAR | | 4 | |
| 1193 | DRI | AIYR | AIZR | | 6 | |
| 1194 | DRI | AIYR | ADFR | | 1 | |
| 1195 | DRI | AIYR | RIBR | | 4 | |
| 1196 | DRI | AIYR | AWAR | | 1 | |
| 1197 | DRI | AVKR | RIML | | 3 | |
| 1198 | DRI | AVKR | SMBDR | | 1 | |

| Entry Number | Type | Pre | Post | Ref | Den | Uniq |
|---|---|---|---|---|---|---|
| 1199 | DRI | AVKR | PVPL | | 1 | |
| 1200 | DRI | AVKR | DVA | | 1 | |
| 1201 | DRI | AVKR | SMDVR | | 1 | |
| 1202 | DRI | AVKR | PDEL | | 4 | |
| 1203 | DRI | AVKR | SAADL | | 1 | |
| 1204 | DRI | AVKR | PVQL | | 1 | |
| 1205 | DRI | AVKR | SMDDL | | 1 | |
| 1206 | DRI | AVKR | RIMR | | 3 | |
| 1207 | DRI | AVKR | WN | HDC | 5 | |
| 1208 | DRI | AVKR | SMDDR | | 2 | |
| 1209 | DRI | AVKR | PVM | | 1 | |
| 1210 | DRI | SAADR | AVAR | | 3 | |
| 1211 | DRI | SAADR | AIBR | | 1 | |
| 1212 | DRI | SAADR | RIML | | 4 | |
| 1213 | DRI | SAADR | RMFL | | 1 | |
| 1214 | DRI | SAADR | RIMR | | 5 | |
| 1215 | DRI | SAADR | RMGL | | 1 | |
| 1216 | DRI | SAADR | OLLL | | 1 | |
| 1217 | DRI | SIADR | WN | NONE2 | 1 | |
| 1218 | DRI | SIAVR | WN | NONE2 | 1 | |
| 1219 | DRI | SIBVR | WN | NONE1 | 1 | |
| 1220 | DRM | RMDDR | RMDVL | | 10 | |
| 1221 | DRM | RMDDR | WM | NMJ | 10 | |
| 1222 | DRM | RMFR | AVKL | | 1 | |
| 1223 | DRM | RMFR | AIBR | | 1 | |
| 1224 | DRM | RMFR | AVKR | | 3 | |
| 1225 | DRM | RMFR | WM | NMJ | 4 | |
| 1226 | DRM | RMFR | RMDR | | 2 | |
| 1227 | DRM | RMHR | RMGL | | 1 | |
| 1228 | DRM | RMHR | RMER | | 1 | |
| 1229 | DRM | RMHR | WM | NMJ | 7 | |
| 1230 | DRM | SMBDR | AVAL | | 1 | |
| 1231 | DRM | SMBDR | SAAVL | | 2 | |
| 1232 | DRM | SMBDR | RMED | | 3 | |
| 1233 | DRM | SMBDR | WM | NMJ | 9 | |
| 1234 | DRM | SMBVR | PLNL | | 3 | |
| 1235 | DRM | SMBVR | SAADR | | 4 | |
| 1236 | DRM | SMBVR | RMEV | | 3 | |
| 1237 | DRM | SMBVR | WM | NMJ | 4 | |
| 1238 | DRM | SMDDR | RIAL | | 2 | |

| Entry Number | Type | Pre | Post | Ref | Den | Uniq |
|---|---|---|---|---|---|---|
| 1239 | DRM | SMDDR | WM | NMJ | 3 | |
| 1240 | DRM | SMDDR | RIAR | | 1 | |
| 1241 | EUI | SABD | VA2 | | 1 | |
| 1242 | ELI | ADAL | SMDVL | | 2 | |
| 1243 | ELI | ADAL | RIML | | 3 | |
| 1244 | ELI | ADAL | AVBR | | 5 | |
| 1245 | ELI | ADAL | FLPR | | 1 | |
| 1246 | ELI | ADAL | AVJR | | 4 | |
| 1247 | ELI | ADAL | AIBL | | 1 | |
| 1248 | ELI | ADAL | AVEL | | 1 | |
| 1249 | ELI | ADAL | AVAR | | 1 | |
| 1250 | ELI | ADAL | RIPL | | 1 | |
| 1251 | ELI | ADAL | RIPL | | 1 | |
| 1252 | ELI | ADAL | AVBL | | 4 | |
| 1253 | ELI | ADAL | AIBR | | 2 | |
| 1254 | ELI | AVFL | AVG | | 2 | |
| 1255 | ELI | AVFL | AVL | | 1 | |
| 1256 | ELI | AVFL | VB1 | | 1 | |
| 1257 | ELI | AVFL | AVFR | | 5 | |
| 1258 | ELI | AVFL | AVHR | | 6 | |
| 1259 | ELI | AVFL | VD11 | | 1 | |
| 1260 | ELI | AVFL | AVJR | | 4 | |
| 1261 | ELI | AVFL | WM | NMJ | 2 | |
| 1262 | ELI | AVFL | HSNL | | 1 | |
| 1263 | ELI | AVFL | PDEL | | 1 | |
| 1264 | ELI | AVFL | AVBL | | 5 | |
| 1265 | ELI | AVFL | PVQL | | 2 | |
| 1266 | ELI | RIFL | RIMR | | 1 | |
| 1267 | ELI | RIFL | AVBL | | 10 | |
| 1268 | ELI | RIFL | AVHR | | 1 | |
| 1269 | ELI | RIFL | AVBR | | 1 | |
| 1270 | ELI | RIFL | RIML | | 2 | |
| 1271 | ELI | RIFL | PVPL | | 3 | |
| 1272 | ELI | RIFL | ALML | | 2 | |
| 1273 | ELI | RIFL | AVJR | | 2 | |
| 1274 | ELI | RIGL | RMHR | | 3 | |
| 1275 | ELI | RIGL | AIZR | | 1 | |
| 1276 | ELI | RIGL | BAGR | | 2 | |
| 1277 | ELI | RIGL | AVKL | | 1 | |
| 1278 | ELI | RIGL | AVER | | 2 | |

| Entry Number | Type | Pre | Post | Ref | Den | Uniq |
|---|---|---|---|---|---|---|
| 1279 | ELI | RIGL | DVC | | 1 | |
| 1280 | ELI | RIGL | RIBL | | 1 | |
| 1281 | ELI | RIGL | RIR | | 2 | |
| 1282 | ELI | RIGL | AIZL | | 2 | |
| 1283 | ELI | RIGL | AVEL | | 1 | |
| 1284 | ELI | RIGL | RMFL | | 1 | |
| 1285 | ELI | RIGL | ALNL | | 1 | |
| 1286 | ELI | SABVL | WN | NONE3 | 1 | |
| 1287 | ELM | RMGL | AVBR | | 2 | |
| 1288 | ELM | RMGL | ADAL | | 1 | |
| 1289 | ELM | RMGL | CEPDL | | 1 | |
| 1290 | ELM | RMGL | ALNL | | 1 | |
| 1291 | ELM | RMGL | AVEL | | 2 | |
| 1292 | ELM | RMGL | SIAVL | | 1 | |
| 1293 | ELM | RMGL | WM | NMJ | 5 | |
| 1294 | ELM | RMGL | ALML | | 1 | |
| 1295 | ELM | RMGL | AVAL | | 1 | |
| 1296 | ELM | RMGL | RMDL | | 1 | |
| 1297 | ELM | RMGL | ASKL | | 1 | |
| 1298 | ELM | RMGL | RMDVL | | 4 | |
| 1299 | ELM | RMGL | SIBVL | | 2 | |
| 1300 | ELM | RMGL | SMBVL | | 2 | |
| 1301 | ELM | RMGL | RMDR | | 3 | |
| 1302 | ELM | RMGL | RMHR | | 1 | |
| 1303 | ELM | RMGL | URXL | | 1 | |
| 1304 | ELM | RMGL | AIBR | | 1 | |
| 1305 | ELS | ADEL | RIGR | | 6 | |
| 1306 | ELS | ADEL | IL1L | | 1 | |
| 1307 | ELS | ADEL | FLPL | | 2 | |
| 1308 | ELS | ADEL | RIH | | 2 | |
| 1309 | ELS | ADEL | IL2L | | 2 | |
| 1310 | ELS | ADEL | BDUL | | 1 | |
| 1311 | ELS | ADEL | RIGL | | 5 | |
| 1312 | ELS | ADEL | SMBDR | | 1 | |
| 1313 | ELS | ADEL | RIFL | | 1 | |
| 1314 | ELS | ADEL | ADAL | | 1 | |
| 1315 | ELS | ADEL | RIAL | | 1 | |
| 1316 | ELS | ADEL | RMGL | | 3 | |
| 1317 | ELS | ADEL | OLLL | | 2 | |
| 1318 | ELS | ADEL | AVAL | | 2 | |

| Entry Number | Type | Pre | Post | Ref | Den | Uniq |
|---|---|---|---|---|---|---|
| 1319 | ELS | ADEL | CEPDL | | 1 | |
| 1320 | ELS | ADEL | AVEL | | 1 | |
| 1321 | ELS | ADEL | URBL | | 1 | |
| 1322 | ELS | ADEL | RIVR | | 1 | |
| 1323 | ELS | ADEL | RMHL | | 1 | |
| 1324 | ELS | ADEL | SIADR | | 1 | |
| 1325 | ELS | ADEL | RIVL | | 1 | |
| 1326 | ELS | ADEL | RMDL | | 2 | |
| 1327 | ELS | ADEL | SIBDR | | 1 | |
| 1328 | ELS | ADEL | AVAR | | 3 | |
| 1329 | ELS | FLPL | ADER | | 2 | |
| 1330 | ELS | FLPL | AIBL | | 1 | |
| 1331 | ELS | FLPL | FLPR | | 1 | |
| 1332 | ELS | FLPL | AVBL | | 4 | |
| 1333 | ELS | FLPL | AVAR | | 18 | |
| 1334 | ELS | FLPL | AIBR | | 2 | |
| 1335 | ELS | FLPL | AVDL | | 6 | |
| 1336 | ELS | FLPL | AVBR | | 5 | |
| 1337 | ELS | FLPL | DVA | | 1 | |
| 1338 | ELS | FLPL | ADEL | | 1 | |
| 1339 | ELS | FLPL | AVAL | | 13 | |
| 1340 | ELS | FLPL | AVDR | | 14 | |
| 1341 | EUM | AS1 | WM | MDE | 1 | |
| 1342 | EUM | AS1 | DA1 | | 1 | |
| 1343 | EUM | AS1 | VD1 | | 1 | |
| 1344 | EUM | DA1 | VD2 | | 1 | |
| 1345 | EUM | DA1 | VD1 | | 1 | |
| 1346 | EUM | DA1 | WM | MDE | 10 | |
| 1347 | EUM | DB1 | WM | MDE | 22 | |
| 1348 | EUM | DB1 | DD1 | | 4 | |
| 1349 | EUM | DB1 | VD2 | | 9 | |
| 1350 | EUM | DB1 | VD4 | | 7 | |
| 1351 | EUM | DB1 | AS2 | | 1 | |
| 1352 | EUM | DB1 | VD1 | | 1 | |
| 1353 | EUM | DB1 | VD3 | | 22 | |
| 1354 | EUM | DB1 | AS1 | | 1 | |
| 1355 | EUM | DB2 | AS3 | | 1 | |
| 1356 | EUM | DB2 | VD4 | | 22 | |
| 1357 | EUM | DB2 | VD5 | | 7 | |
| 1358 | EUM | DB2 | WM | MDE | 22 | |

| Entry Number | Type | Pre | Post | Ref | Den | Uniq |
|---|---|---|---|---|---|---|
| 1359 | EUM | DB2 | VD3 | | 9 | |
| 1360 | EUM | DB2 | VD2 | | 1 | |
| 1361 | EUM | DB2 | DD2 | | 4 | |
| 1362 | EUM | DB2 | DD1 | | 1 | |
| 1363 | EUM | DD1 | DA1 | | 2 | |
| 1364 | EUM | DD1 | VD2 | | 1 | |
| 1365 | EUM | DD1 | WM | MDI | 29 | |
| 1366 | EUM | DD1 | VD1 | | 1 | |
| 1367 | EUM | VA1 | VD1 | | 3 | |
| 1368 | EUM | VA1 | DD1 | | 8 | |
| 1369 | EUM | VA1 | WM | MVE | 10 | |
| 1370 | EUM | VA1 | VA2 | | 1 | |
| 1371 | EUM | VA1 | VB1 | | 1 | |
| 1372 | EUM | VB1 | VD2 | | 1 | |
| 1373 | EUM | VB1 | DD1 | | 1 | |
| 1374 | EUM | VB1 | WM | MVE | 5 | |
| 1375 | EUM | VB1 | RIML | | 1 | |
| 1376 | EUM | VB1 | VA1 | | 3 | |
| 1377 | EUM | VB1 | VD1 | | 2 | |
| 1378 | EUM | VB1 | SABD | | 1 | |
| 1379 | EUM | VB1 | SAADR | | 4 | |
| 1380 | EUM | VB1 | RMFL | | 2 | |
| 1381 | EUM | VB1 | SAADL | | 9 | |
| 1382 | EUM | VB1 | AIBR | | 1 | |
| 1383 | EUM | VB1 | VA2 | | 1 | |
| 1384 | EUM | VB1 | VA3 | | 1 | |
| 1385 | EUM | VB2 | VD2 | | 6 | |
| 1386 | EUM | VB2 | DD1 | | 19 | |
| 1387 | EUM | VB2 | WM | MVE | 23 | |
| 1388 | EUM | VB2 | VC2 | | 1 | |
| 1389 | EUM | VB2 | DD2 | | 1 | |
| 1390 | EUM | VB2 | VD3 | | 4 | |
| 1391 | EUM | VB2 | VA2 | | 1 | |
| 1392 | EUM | VD1 | VA1 | | 2 | |
| 1393 | EUM | VD1 | WM | MVI | 8 | |
| 1394 | EUM | VD1 | DD1 | | 1 | |
| 1395 | EUM | VD2 | WM | MVI | 8 | |
| 1396 | EUM | VD2 | AS3 | | 2 | |
| 1397 | EUM | VD2 | VB2 | | 2 | |
| 1398 | ERI | ADAR | AVEL | | 1 | |

| Entry Number | Type | Pre | Post | Ref | Den | Uniq |
|---|---|---|---|---|---|---|
| 1399 | ERI | ADAR | AVBR | | 4 | |
| 1400 | ERI | ADAR | SMDVR | | 2 | |
| 1401 | ERI | ADAR | AVBL | | 3 | |
| 1402 | ERI | ADAR | RIMR | | 4 | |
| 1403 | ERI | ADAR | RIVR | | 1 | |
| 1404 | ERI | ADAR | AVJL | | 3 | |
| 1405 | ERI | ADAR | RIPR | | 1 | |
| 1406 | ERI | AVFR | AVG | | 2 | |
| 1407 | ERI | AVFR | AVBR | | 8 | |
| 1408 | ERI | AVFR | PDER | | 1 | |
| 1409 | ERI | AVFR | AVL | | 1 | |
| 1410 | ERI | AVFR | HSNR | | 2 | |
| 1411 | ERI | AVFR | PVQL | | 1 | |
| 1412 | ERI | AVFR | AVFL | | 4 | |
| 1413 | ERI | AVFR | WM | NMJ | 2 | |
| 1414 | ERI | AVFR | AVJL | | 4 | |
| 1415 | ERI | AVFR | VD11 | | 1 | |
| 1416 | ERI | AVFR | AVHR | | 5 | |
| 1417 | ERI | AVFR | AVBL | | 1 | |
| 1418 | ERI | RIFR | AVJL | | 1 | |
| 1419 | ERI | RIFR | PVCR | | 1 | |
| 1420 | ERI | RIFR | RIMR | | 4 | |
| 1421 | ERI | RIFR | HSNR | | 1 | |
| 1422 | ERI | RIFR | ASHR | | 2 | |
| 1423 | ERI | RIFR | AVFL | | 1 | |
| 1424 | ERI | RIFR | PVPR | | 4 | |
| 1425 | ERI | RIFR | AVHL | | 1 | |
| 1426 | ERI | RIFR | PVCL | | 1 | |
| 1427 | ERI | RIFR | AVBR | | 16 | |
| 1428 | ERI | RIFR | AVJR | | 2 | |
| 1429 | ERI | RIFR | AVBL | | 1 | |
| 1430 | ERI | RIFR | RIPR | | 1 | |
| 1431 | ERI | RIGR | AVER | | 2 | |
| 1432 | ERI | RIGR | RMHL | | 4 | |
| 1433 | ERI | RIGR | RIBR | | 1 | |
| 1434 | ERI | RIGR | BAGR | | 1 | |
| 1435 | ERI | RIGR | AVKL | | 4 | |
| 1436 | ERI | RIGR | AVKR | | 1 | |
| 1437 | ERI | RIGR | RIR | | 1 | |
| 1438 | ERI | RIGR | ALNR | | 1 | |

| Entry Number | Type | Pre | Post | Ref | Den | Uniq |
|---|---|---|---|---|---|---|
| 1439 | ERI | RIGR | AIZR | | 2 | |
| 1440 | ERI | SABVR | WN | NONE3 | 1 | |
| 1441 | ERM | RMGR | URXR | | 1 | |
| 1442 | ERM | RMGR | RMDR | | 3 | |
| 1443 | ERM | RMGR | ADAR | | 1 | |
| 1444 | ERM | RMGR | AVBR | | 1 | |
| 1445 | ERM | RMGR | WM | NMJ | 3 | |
| 1446 | ERM | RMGR | AVDL | | 1 | |
| 1447 | ERM | RMGR | RMDVL | | 1 | |
| 1448 | ERM | RMGR | AVER | | 3 | |
| 1449 | ERM | RMGR | AVJL | | 1 | |
| 1450 | ERM | RMGR | AVAR | | 1 | |
| 1451 | ERM | RMGR | RMDL | | 3 | |
| 1452 | ERM | RMGR | RMDVR | | 4 | |
| 1453 | ERM | RMGR | ASHR | | 1 | |
| 1454 | ERM | RMGR | RIR | | 1 | |
| 1455 | ERM | RMGR | ALNR | | 1 | |
| 1456 | ERS | ADER | ADAR | | 1 | |
| 1457 | ERS | ADER | FLPL | | 1 | |
| 1458 | ERS | ADER | FLPR | | 2 | |
| 1459 | ERS | ADER | RMDR | | 2 | |
| 1460 | ERS | ADER | OLLR | | 2 | |
| 1461 | ERS | ADER | AVAR | | 1 | |
| 1462 | ERS | ADER | RIGL | | 7 | |
| 1463 | ERS | ADER | RMGR | | 2 | |
| 1464 | ERS | ADER | RIGR | | 4 | |
| 1465 | ERS | ADER | AVKR | | 1 | |
| 1466 | ERS | ADER | RIH | | 1 | |
| 1467 | ERS | ADER | AVER | | 1 | |
| 1468 | ERS | ADER | AVAL | | 5 | |
| 1469 | ERS | ADER | ADEL | | 2 | |
| 1470 | ERS | ADER | AVJR | | 1 | |
| 1471 | ERS | ADER | AVDR | | 1 | |
| 1472 | ERS | ADER | CEPDR | | 1 | |
| 1473 | ERS | ADER | SAAVR | | 1 | |
| 1474 | ERS | FLPR | AVBR | | 2 | |
| 1475 | ERS | FLPR | AVAR | | 4 | |
| 1476 | ERS | FLPR | ADER | | 2 | |
| 1477 | ERS | FLPR | AVDR | | 2 | |
| 1478 | ERS | FLPR | AVAL | | 10 | |

| Entry Number | Type | Pre | Post | Ref | Den | Uniq |
|---|---|---|---|---|---|---|
| 1479 | ERS | FLPR | AVDL | | 10 | |
| 1480 | ERS | FLPR | FLPL | | 1 | |
| 1481 | ERS | FLPR | AVBL | | 4 | |
| 1482 | ERS | FLPR | PVCL | | 2 | |
| 1483 | ERS | FLPR | AVEL | | 3 | |
| 1484 | ERS | FLPR | AVER | | 2 | |
| 1485 | ERS | FLPR | WN | HDC | 3 | |
| 1486 | ERS | FLPR | VB1 | | 1 | |
| 1487 | ERS | FLPR | DVA | | 1 | |
| 1488 | ERS | FLPR | AIBR | | 1 | |
| 1489 | EUS | AQR | BAGR | | 2 | |
| 1490 | EUS | AQR | AVBR | | 5 | |
| 1491 | EUS | AQR | BAGL | | 2 | |
| 1492 | EUS | AQR | RIAR | | 2 | |
| 1493 | EUS | AQR | AVJL | | 1 | |
| 1494 | EUS | AQR | AVBL | | 4 | |
| 1495 | EUS | AQR | AVDR | | 1 | |
| 1496 | EUS | AQR | PVCL | | 1 | |
| 1497 | EUS | AQR | AVDL | | 1 | |
| 1498 | EUS | AQR | PVPL | | 1 | |
| 1499 | EUS | AQR | AVAL | | 5 | |
| 1500 | EUS | AQR | URXL | | 1 | |
| 1501 | EUS | AQR | PVCR | | 2 | |
| 1502 | EUS | AQR | RIAL | | 3 | |
| 1503 | EUS | AQR | DVA | | 1 | |
| 1504 | EUS | AQR | AVAR | | 3 | |
| 1505 | EUS | AVG | AVER | | 1 | |
| 1506 | EUS | AVG | PHAL | | 2 | |
| 1507 | EUS | AVG | PHAR | | 1 | |
| 1508 | EUS | AVG | PVQL | | 1 | |
| 1509 | EUS | AVG | PVCL | | 1 | |
| 1510 | EUS | AVG | AVFL | | 1 | |
| 1511 | EUS | AVG | PVCR | | 1 | |
| 1512 | EUS | AVG | PVQR | | 1 | |
| 1513 | EUS | AVG | PVNL | | 2 | |
| 1514 | EUS | AVG | AVBL | | 4 | |
| 1515 | EUS | AVG | PVNR | | 2 | |
| 1516 | EUS | AVG | VA11 | | 1 | |
| 1517 | EUS | AVG | PVPL | | 1 | |
| 1518 | EUS | AVG | AVAL | | 3 | |

| Entry Number | Type | Pre | Post | Ref | Den | Uniq |
|---|---|---|---|---|---|---|
| 1519 | EUS | AVG | PVPR | | 1 | |
| 1520 | EUS | AVG | VA11 | | 1 | |
| 1521 | EUS | AVG | AVJR | | 1 | |
| 1522 | EUS | AVG | AVDR | | 1 | |
| 1523 | EUS | AVG | AVDL | | 1 | |
| 1524 | EUS | AVG | DA8 | | 1 | |
| 1525 | EUS | AVG | WN | HDC | 1 | |
| 1526 | EUS | AVG | DVB | | 1 | |
| 1527 | EUS | AVG | WN | HDC | 1 | |
| 1528 | EUS | AVG | DA8 | | 1 | |
| 1529 | EUS | AVG | AVAR | | 1 | |
| 1530 | FLI | BDUL | AVHL | | 1 | |
| 1531 | FLI | BDUL | AVAR | | 2 | |
| 1532 | FLI | BDUL | HSNL | | 4 | |
| 1533 | FLI | BDUL | PVNR | | 2 | |
| 1534 | FLI | BDUL | URADL | | 1 | |
| 1535 | FLI | BDUL | PVNL | | 1 | |
| 1536 | FLI | BDUL | SAADL | | 2 | |
| 1537 | FLI | BDUL | PVCL | | 2 | |
| 1538 | FLI | BDUL | AVJR | | 1 | |
| 1539 | FLI | BDUL | WE | CEPshVL | 1 | |
| 1540 | FLI | BDUL | ADEL | | 4 | |
| 1541 | FLI | CANL | WN | NONE4 | 1 | |
| 1542 | FLI | PVDL | AVAL | | 27 | |
| 1543 | FLI | PVDL | PVCR | | 28 | |
| 1544 | FLI | PVDL | PVDR | | 1 | |
| 1545 | FLI | PVDL | DVA | | 3 | |
| 1546 | FLI | PVDL | WN | HDC | 1 | |
| 1547 | FLI | SDQL | RICR | | 1 | |
| 1548 | FLI | SDQL | ALML | | 1 | |
| 1549 | FLI | SDQL | RMFL | | 1 | |
| 1550 | FLI | SDQL | AVEL | | 1 | |
| 1551 | FLI | SDQL | AIBR | | 1 | |
| 1552 | FLI | SDQL | AVAR | | 3 | |
| 1553 | FLI | SDQL | RIS | | 2 | |
| 1554 | FLI | SDQL | FLPL | | 1 | |
| 1555 | FLI | SDQL | AVAL | | 2 | |
| 1556 | FLM | HSNL | VA6 | | 1 | |
| 1557 | FLM | HSNL | AIAL | | 2 | |
| 1558 | FLM | HSNL | AVFL | | 6 | |

| Entry Number | Type | Pre | Post | Ref | Den | Uniq |
|---|---|---|---|---|---|---|
| 1559 | FLM | HSNL | WM | MUE | 22 | |
| 1560 | FLM | HSNL | AVJL | | 2 | |
| 1561 | FLM | HSNL | ASHR | | 2 | |
| 1562 | FLM | HSNL | RIML | | 2 | |
| 1563 | FLM | HSNL | AIZR | | 1 | |
| 1564 | FLM | HSNL | SABD | | 1 | |
| 1565 | FLM | HSNL | AWBR | | 2 | |
| 1566 | FLM | HSNL | SABVL | | 3 | |
| 1567 | FLM | HSNL | VC5 | | 10 | |
| 1568 | FLM | HSNL | ASKL | | 1 | |
| 1569 | FLM | HSNL | ASJR | | 1 | |
| 1570 | FLM | HSNL | ASHL | | 1 | |
| 1571 | FLM | HSNL | AS5 | | 1 | |
| 1572 | FLM | HSNL | WM | NMJ | 3 | |
| 1573 | FLM | HSNL | DA5 | | 1 | |
| 1574 | FLM | HSNL | AVDL | | 3 | |
| 1575 | FLM | HSNL | RIFL | | 3 | |
| 1576 | FLM | HSNL | VC2 | | 2 | |
| 1577 | FLM | HSNL | AWBL | | 2 | |
| 1578 | FLM | HSNL | HSNR | | 3 | |
| 1579 | FLM | HSNL | AIZL | | 2 | |
| 1580 | FLS | ALML | PVCR | | 2 | |
| 1581 | FLS | ALML | RMGL | | 1 | |
| 1582 | FLS | ALML | CEPVL | | 1 | |
| 1583 | FLS | ALML | AVDR | | 1 | |
| 1584 | FLS | ALML | BDUL | | 5 | |
| 1585 | FLS | ALML | RMDDR | | 1 | |
| 1586 | FLS | ALML | CEPDL | | 3 | |
| 1587 | FLS | ALML | AVEL | | 1 | |
| 1588 | FLS | ALML | PVCL | | 4 | |
| 1589 | FLS | ALML | SDQL | | 1 | |
| 1590 | FLS | PDEL | PVM | | 1 | |
| 1591 | FLS | PDEL | WN | HDC | 4 | |
| 1592 | FLS | PDEL | AVKL | | 22 | |
| 1593 | FLS | PDEL | PVCL | | 2 | |
| 1594 | FLS | PDEL | PVR | | 2 | |
| 1595 | FLS | PDEL | DVA | | 61 | |
| 1596 | FLS | PDEL | PDER | | 1 | |
| 1597 | FLS | PDEL | VA9 | | 1 | |
| 1598 | FRI | BDUR | AVJL | | 2 | |

| Entry Number | Type | Pre | Post | Ref | Den | Uniq |
|---|---|---|---|---|---|---|
| 1599 | FRI | BDUR | PVNR | | 1 | |
| 1600 | FRI | BDUR | AVAL | | 5 | |
| 1601 | FRI | BDUR | ALMR | | 1 | |
| 1602 | FRI | BDUR | PVNL | | 4 | |
| 1603 | FRI | BDUR | URADR | | 1 | |
| 1604 | FRI | BDUR | ADER | | 1 | |
| 1605 | FRI | BDUR | SAADR | | 1 | |
| 1606 | FRI | BDUR | AVHL | | 1 | |
| 1607 | FRI | BDUR | ADEL | | 1 | |
| 1608 | FRI | BDUR | SDQR | | 1 | |
| 1609 | FRI | BDUR | PVCL | | 2 | |
| 1610 | FRI | BDUR | PVCR | | 2 | |
| 1611 | FRI | BDUR | HSNR | | 7 | |
| 1612 | FRI | CANR | WN | NONE4 | 1 | |
| 1613 | FRI | PVDR | AVAR | | 27 | |
| 1614 | FRI | PVDR | PVDL | | 1 | |
| 1615 | FRI | PVDR | PVCL | | 28 | |
| 1616 | FRI | PVDR | DVA | | 3 | |
| 1617 | FRI | PVDR | WN | HDC | 1 | |
| 1618 | FRI | SDQR | AVBL | | 5 | |
| 1619 | FRI | SDQR | AVBR | | 4 | |
| 1620 | FRI | SDQR | DVA | | 3 | |
| 1621 | FRI | SDQR | ADLL | | 1 | |
| 1622 | FRI | SDQR | RMHL | | 2 | |
| 1623 | FRI | SDQR | AVAL | | 2 | |
| 1624 | FRI | SDQR | AIBL | | 2 | |
| 1625 | FRI | SDQR | RMHR | | 1 | |
| 1626 | FRM | HSNR | AIBR | | 1 | |
| 1627 | FRM | HSNR | BDUR | | 2 | |
| 1628 | FRM | HSNR | WM | NMJ | 3 | |
| 1629 | FRM | HSNR | AVFL | | 2 | |
| 1630 | FRM | HSNR | DA6 | | 1 | |
| 1631 | FRM | HSNR | DA5 | | 2 | |
| 1632 | FRM | HSNR | VC3 | | 2 | |
| 1633 | FRM | HSNR | AIZL | | 1 | |
| 1634 | FRM | HSNR | SABVL | | 3 | |
| 1635 | FRM | HSNR | AIZR | | 1 | |
| 1636 | FRM | HSNR | PVQR | | 1 | |
| 1637 | FRM | HSNR | AS5 | | 1 | |
| 1638 | FRM | HSNR | DB5 | | 1 | |

| Entry Number | Type | Pre | Post | Ref | Den | Uniq |
|---|---|---|---|---|---|---|
| 1639 | FRM | HSNR | ASHL | | 2 | |
| 1640 | FRM | HSNR | AVDL | | 3 | |
| 1641 | FRM | HSNR | RMGR | | 1 | |
| 1642 | FRM | HSNR | AIBL | | 1 | |
| 1643 | FRM | HSNR | WM | MUE | 21 | |
| 1644 | FRM | HSNR | SABD | | 1 | |
| 1645 | FRM | HSNR | AVJL | | 1 | |
| 1646 | FRM | HSNR | VA5 | | 1 | |
| 1647 | FRM | HSNR | AWBL | | 1 | |
| 1648 | FRM | HSNR | VC2 | | 4 | |
| 1649 | FRM | HSNR | PVNR | | 2 | |
| 1650 | FRM | HSNR | VC5 | | 5 | |
| 1651 | FRM | HSNR | VA6 | | 1 | |
| 1652 | FRM | HSNR | HSNL | | 1 | |
| 1653 | FRM | HSNR | RIMR | | 1 | |
| 1654 | FRM | HSNR | RIFR | | 3 | |
| 1655 | FRS | ALMR | BDUR | | 5 | |
| 1656 | FRS | ALMR | CEPVR | | 1 | |
| 1657 | FRS | ALMR | RMDDL | | 1 | |
| 1658 | FRS | ALMR | PVCR | | 2 | |
| 1659 | FRS | ALMR | CEPDR | | 1 | |
| 1660 | FRS | ALMR | SIADL | | 1 | |
| 1661 | FRS | PDER | PVM | | 1 | |
| 1662 | FRS | PDER | PVCR | | 2 | |
| 1663 | FRS | PDER | WN | HDC | 4 | |
| 1664 | FRS | PDER | PDEL | | 1 | |
| 1665 | FRS | PDER | DVA | | 61 | |
| 1666 | FRS | PDER | PVR | | 2 | |
| 1667 | FRS | PDER | VA9 | | 1 | |
| 1668 | FRS | PDER | AVKR | | 22 | |
| 1669 | FUS | AVM | ADER | | 3 | |
| 1670 | FUS | AVM | SIBVL | | 1 | |
| 1671 | FUS | AVM | DA1 | | 1 | |
| 1672 | FUS | AVM | PVCL | | 4 | |
| 1673 | FUS | AVM | BDUR | | 4 | |
| 1674 | FUS | AVM | RID | | 1 | |
| 1675 | FUS | AVM | DA1 | | 1 | |
| 1676 | FUS | AVM | PVCR | | 6 | |
| 1677 | FUS | AVM | AVBR | | 5 | |
| 1678 | FUS | AVM | BDUL | | 3 | |

| Entry Number | Type | Pre | Post | Ref | Den | Uniq |
|---|---|---|---|---|---|---|
| 1679 | FUS | AVM | AVBL | | 6 | |
| 1680 | FUS | AVM | AVJR | | 1 | |
| 1681 | FUS | AVM | PVR | | 3 | |
| 1682 | FUS | PVM | PVCL | | 2 | |
| 1683 | FUS | PVM | PDEL | | 10 | |
| 1684 | FUS | PVM | AVKL | | 11 | |
| 1685 | FUS | PVM | PVR | | 2 | |
| 1686 | FUS | PVM | WN | HDC | 1 | |
| 1687 | FUS | PVM | AVM | | 1 | |
| 1688 | FUS | PVM | DVA | | 3 | |
| 1689 | GUM | AS10 | WM | MDE | 15 | |
| 1690 | GUM | AS10 | VD10 | | 16 | |
| 1691 | GUM | AS10 | DA7 | | 1 | |
| 1692 | GUM | AS10 | VD9 | | 2 | |
| 1693 | GUM | AS2 | WM | MDE | 15 | |
| 1694 | GUM | AS2 | DA2 | | 1 | |
| 1695 | GUM | AS2 | VD2 | | 1 | |
| 1696 | GUM | AS3 | DA3 | | 1 | |
| 1697 | GUM | AS3 | WM | MDE | 15 | |
| 1698 | GUM | AS3 | VD2 | | 2 | |
| 1699 | GUM | AS3 | VD3 | | 16 | |
| 1700 | GUM | AS4 | WM | MDE | 15 | |
| 1701 | GUM | AS4 | VD3 | | 2 | |
| 1702 | GUM | AS4 | VD4 | | 16 | |
| 1703 | GUM | AS5 | WM | MDE | 15 | |
| 1704 | GUM | AS5 | DA4 | | 1 | |
| 1705 | GUM | AS5 | VD4 | | 2 | |
| 1706 | GUM | AS5 | VD5 | | 16 | |
| 1707 | GUM | AS6 | DA5 | | 1 | |
| 1708 | GUM | AS6 | WM | MDE | 15 | |
| 1709 | GUM | AS6 | VD5 | | 2 | |
| 1710 | GUM | AS6 | VD6 | | 16 | |
| 1711 | GUM | AS7 | WM | MDE | 15 | |
| 1712 | GUM | AS7 | VD6 | | 2 | |
| 1713 | GUM | AS7 | VD7 | | 16 | |
| 1714 | GUM | AS8 | WM | MDE | 15 | |
| 1715 | GUM | AS8 | VD8 | | 16 | |
| 1716 | GUM | AS8 | DA6 | | 1 | |
| 1717 | GUM | AS8 | VD7 | | 2 | |
| 1718 | GUM | AS9 | VD8 | | 2 | |

| Entry Number | Type | Pre | Post | Ref | Den | Uniq |
|---|---|---|---|---|---|---|
| 1719 | GUM | AS9 | VD9 | | 16 | |
| 1720 | GUM | AS9 | WM | MDE | 15 | |
| 1721 | GUM | DA2 | VD2 | | 23 | |
| 1722 | GUM | DA2 | WM | MDE | 29 | |
| 1723 | GUM | DA2 | DD1 | | 1 | |
| 1724 | GUM | DA2 | VD1 | | 1 | |
| 1725 | GUM | DA2 | VD3 | | 6 | |
| 1726 | GUM | DA3 | VD3 | | 23 | |
| 1727 | GUM | DA3 | DA4 | | 1 | |
| 1728 | GUM | DA3 | WM | MDE | 29 | |
| 1729 | GUM | DA3 | DD2 | | 1 | |
| 1730 | GUM | DA3 | VD4 | | 6 | |
| 1731 | GUM | DA4 | DA5 | | 1 | |
| 1732 | GUM | DA4 | WM | MDE | 29 | |
| 1733 | GUM | DA4 | VD5 | | 6 | |
| 1734 | GUM | DA4 | VD4 | | 23 | |
| 1735 | GUM | DA4 | DD3 | | 1 | |
| 1736 | GUM | DA4 | VD6 | | 1 | |
| 1737 | GUM | DA5 | DD4 | | 1 | |
| 1738 | GUM | DA5 | DA6 | | 1 | |
| 1739 | GUM | DA5 | WM | MDE | 29 | |
| 1740 | GUM | DA5 | VD6 | | 23 | |
| 1741 | GUM | DA5 | VD7 | | 1 | |
| 1742 | GUM | DA6 | DA7 | | 1 | |
| 1743 | GUM | DA6 | WM | MDE | 29 | |
| 1744 | GUM | DA6 | VD8 | | 23 | |
| 1745 | GUM | DA6 | DD4 | | 1 | |
| 1746 | GUM | DA6 | VD9 | | 1 | |
| 1747 | GUM | DA7 | VD11 | | 1 | |
| 1748 | GUM | DA7 | DA8 | | 1 | |
| 1749 | GUM | DA7 | WM | MDE | 29 | |
| 1750 | GUM | DA7 | VD10 | | 23 | |
| 1751 | GUM | DA7 | DD6 | | 1 | |
| 1752 | GUM | DB3 | VD6 | | 7 | |
| 1753 | GUM | DB3 | DD3 | | 9 | |
| 1754 | GUM | DB3 | VD4 | | 9 | |
| 1755 | GUM | DB3 | VD3 | | 1 | |
| 1756 | GUM | DB3 | VD5 | | 22 | |
| 1757 | GUM | DB3 | DD2 | | 4 | |
| 1758 | GUM | DB3 | WM | MDE | 22 | |

| Entry Number | Type | Pre | Post | Ref | Den | Uniq |
|---|---|---|---|---|---|---|
| 1759 | GUM | DB3 | AS4 | | 1 | |
| 1760 | GUM | DB4 | VD4 | | 1 | |
| 1761 | GUM | DB4 | WM | MDE | 22 | |
| 1762 | GUM | DB4 | DD4 | | 9 | |
| 1763 | GUM | DB4 | VD5 | | 9 | |
| 1764 | GUM | DB4 | AS5 | | 1 | |
| 1765 | GUM | DB4 | VD6 | | 22 | |
| 1766 | GUM | DB4 | DD3 | | 4 | |
| 1767 | GUM | DB4 | AS6 | | 1 | |
| 1768 | GUM | DB4 | VD7 | | 7 | |
| 1769 | GUM | DB5 | AS8 | | 1 | |
| 1770 | GUM | DB5 | WM | MDE | 22 | |
| 1771 | GUM | DB5 | DD5 | | 9 | |
| 1772 | GUM | DB5 | VD6 | | 9 | |
| 1773 | GUM | DB5 | VD7 | | 22 | |
| 1774 | GUM | DB5 | AS7 | | 1 | |
| 1775 | GUM | DB5 | DD4 | | 4 | |
| 1776 | GUM | DB5 | VD5 | | 1 | |
| 1777 | GUM | DB5 | VD8 | | 7 | |
| 1778 | GUM | DB6 | AS9 | | 1 | |
| 1779 | GUM | DB6 | VD6 | | 1 | |
| 1780 | GUM | DB6 | VD9 | | 7 | |
| 1781 | GUM | DB6 | DD5 | | 4 | |
| 1782 | GUM | DB6 | VD8 | | 22 | |
| 1783 | GUM | DB6 | VD7 | | 9 | |
| 1784 | GUM | DB6 | WM | MDE | 22 | |
| 1785 | GUM | DB6 | DD6 | | 9 | |
| 1786 | GUM | DB7 | VD11 | | 7 | |
| 1787 | GUM | DB7 | AS10 | | 1 | |
| 1788 | GUM | DB7 | AS11 | | 1 | |
| 1789 | GUM | DB7 | VD8 | | 1 | |
| 1790 | GUM | DB7 | VD9 | | 9 | |
| 1791 | GUM | DB7 | DD6 | | 4 | |
| 1792 | GUM | DB7 | VD10 | | 22 | |
| 1793 | GUM | DB7 | WM | MDE | 22 | |
| 1794 | GUM | DD2 | WM | MDI | 29 | |
| 1795 | GUM | DD2 | DA3 | | 2 | |
| 1796 | GUM | DD2 | VD3 | | 1 | |
| 1797 | GUM | DD2 | VD4 | | 1 | |
| 1798 | GUM | DD3 | WM | MDI | 29 | |

| Entry Number | Type | Pre | Post | Ref | Den | Uniq |
|---|---|---|---|---|---|---|
| 1799 | GUM | DD3 | DA4 | | 2 | |
| 1800 | GUM | DD3 | VD5 | | 1 | |
| 1801 | GUM | DD3 | VD6 | | 1 | |
| 1802 | GUM | DD4 | VD7 | | 1 | |
| 1803 | GUM | DD4 | VD8 | | 1 | |
| 1804 | GUM | DD4 | WM | MDI | 29 | |
| 1805 | GUM | DD4 | DA6 | | 2 | |
| 1806 | GUM | DD5 | DA7 | | 2 | |
| 1807 | GUM | DD5 | VD10 | | 1 | |
| 1808 | GUM | DD5 | VD9 | | 1 | |
| 1809 | GUM | DD5 | WM | MDI | 29 | |
| 1810 | GUM | VA10 | DD5 | | 8 | |
| 1811 | GUM | VA10 | WM | MVE | 10 | |
| 1812 | GUM | VA10 | VA11 | | 1 | |
| 1813 | GUM | VA11 | VA12 | | 1 | |
| 1814 | GUM | VA11 | VD11 | | 2 | |
| 1815 | GUM | VA11 | VD10 | | 3 | |
| 1816 | GUM | VA11 | WM | MVE | 10 | |
| 1817 | GUM | VA11 | DD6 | | 8 | |
| 1818 | GUM | VA2 | VD1 | | 1 | |
| 1819 | GUM | VA2 | WM | MVE | 10 | |
| 1820 | GUM | VA2 | VA3 | | 1 | |
| 1821 | GUM | VA2 | DD1 | | 8 | |
| 1822 | GUM | VA2 | VD2 | | 3 | |
| 1823 | GUM | VA3 | DD2 | | 11 | |
| 1824 | GUM | VA3 | VD2 | | 3 | |
| 1825 | GUM | VA3 | VA4 | | 1 | |
| 1826 | GUM | VA3 | WM | MVE | 31 | |
| 1827 | GUM | VA3 | VD3 | | 2 | |
| 1828 | GUM | VA3 | DD1 | | 18 | |
| 1829 | GUM | VA4 | WM | MVE | 10 | |
| 1830 | GUM | VA4 | VD4 | | 1 | |
| 1831 | GUM | VA4 | DD2 | | 8 | |
| 1832 | GUM | VA4 | VA5 | | 1 | |
| 1833 | GUM | VA5 | VA6 | | 1 | |
| 1834 | GUM | VA5 | DD3 | | 8 | |
| 1835 | GUM | VA5 | WM | MVE | 10 | |
| 1836 | GUM | VA5 | VD5 | | 2 | |
| 1837 | GUM | VA5 | VD4 | | 3 | |
| 1838 | GUM | VA6 | WM | MVE | 10 | |

| Entry Number | Type | Pre | Post | Ref | Den | Uniq |
|---|---|---|---|---|---|---|
| 1839 | GUM | VA6 | DD3 | | 8 | |
| 1840 | GUM | VA6 | VA7 | | 1 | |
| 1841 | GUM | VA7 | VD6 | | 3 | |
| 1842 | GUM | VA7 | WM | MVE | 10 | |
| 1843 | GUM | VA7 | VA8 | | 1 | |
| 1844 | GUM | VA7 | DD4 | | 8 | |
| 1845 | GUM | VA7 | VD7 | | 2 | |
| 1846 | GUM | VA8 | WM | MVE | 10 | |
| 1847 | GUM | VA8 | VA9 | | 1 | |
| 1848 | GUM | VA8 | DD4 | | 8 | |
| 1849 | GUM | VA9 | VD8 | | 3 | |
| 1850 | GUM | VA9 | DD5 | | 8 | |
| 1851 | GUM | VA9 | WM | MVE | 30 | |
| 1852 | GUM | VA9 | VA10 | | 1 | |
| 1853 | GUM | VA9 | VD9 | | 2 | |
| 1854 | GUM | VB10 | DD5 | | 6 | |
| 1855 | GUM | VB10 | DD6 | | 15 | |
| 1856 | GUM | VB10 | WM | MVE | 18 | |
| 1857 | GUM | VB11 | WM | MVE | 18 | |
| 1858 | GUM | VB11 | DD6 | | 15 | |
| 1859 | GUM | VB11 | VA12 | | 1 | |
| 1860 | GUM | VB3 | WM | MVE | 18 | |
| 1861 | GUM | VB3 | DD2 | | 15 | |
| 1862 | GUM | VB4 | DD3 | | 15 | |
| 1863 | GUM | VB4 | WM | MVE | 18 | |
| 1864 | GUM | VB4 | DD2 | | 6 | |
| 1865 | GUM | VB5 | DD3 | | 15 | |
| 1866 | GUM | VB5 | WM | MVE | 18 | |
| 1867 | GUM | VB6 | DD3 | | 6 | |
| 1868 | GUM | VB6 | DD4 | | 15 | |
| 1869 | GUM | VB6 | WM | MVE | 18 | |
| 1870 | GUM | VB7 | DD4 | | 15 | |
| 1871 | GUM | VB7 | WM | MVE | 18 | |
| 1872 | GUM | VB8 | WM | MVE | 18 | |
| 1873 | GUM | VB8 | DD4 | | 6 | |
| 1874 | GUM | VB8 | DD5 | | 15 | |
| 1875 | GUM | VB9 | WM | MVE | 18 | |
| 1876 | GUM | VB9 | DD5 | | 15 | |
| 1877 | GUM | VC1 | WM | MUE | 2 | |
| 1878 | GUM | VC1 | VC2 | | 1 | |

150

| Entry Number | Type | Pre | Post | Ref | Den | Uniq |
|---|---|---|---|---|---|---|
| 1879 | GUM | VC1 | DD1 | | 1 | |
| 1880 | GUM | VC1 | VD1 | | 1 | |
| 1881 | GUM | VC1 | WM | MVE | 2 | |
| 1882 | GUM | VC1 | HSNR | | 1 | |
| 1883 | GUM | VC2 | HSNR | | 1 | |
| 1884 | GUM | VC2 | VD1 | | 1 | |
| 1885 | GUM | VC2 | WM | MVE | 2 | |
| 1886 | GUM | VC2 | VD4 | | 2 | |
| 1887 | GUM | VC2 | WM | MUE | 8 | |
| 1888 | GUM | VC2 | DD2 | | 4 | |
| 1889 | GUM | VC3 | VD5 | | 4 | |
| 1890 | GUM | VC3 | DD3 | | 4 | |
| 1891 | GUM | VC3 | VD1 | | 1 | |
| 1892 | GUM | VC3 | VC4 | | 1 | |
| 1893 | GUM | VC3 | DD4 | | 10 | |
| 1894 | GUM | VC3 | VC1 | | 2 | |
| 1895 | GUM | VC3 | WM | MUE | 8 | |
| 1896 | GUM | VC3 | HSNR | | 1 | |
| 1897 | GUM | VC3 | PVT | | 1 | |
| 1898 | GUM | VC3 | DD1 | | 2 | |
| 1899 | GUM | VC3 | VC2 | | 1 | |
| 1900 | GUM | VC3 | VD3 | | 1 | |
| 1901 | GUM | VC3 | VD4 | | 2 | |
| 1902 | GUM | VC3 | VD7 | | 5 | |
| 1903 | GUM | VC3 | PVQR | | 3 | |
| 1904 | GUM | VC3 | VD2 | | 1 | |
| 1905 | GUM | VC3 | DD2 | | 4 | |
| 1906 | GUM | VC3 | AVL | | 1 | |
| 1907 | GUM | VC3 | VD6 | | 3 | |
| 1908 | GUM | VC4 | AVFL | | 1 | |
| 1909 | GUM | VC4 | DVB | | 2 | |
| 1910 | GUM | VC4 | WM | MUE | 8 | |
| 1911 | GUM | VC4 | AVFR | | 1 | |
| 1912 | GUM | VC4 | VC5 | | 2 | |
| 1913 | GUM | VC4 | DVC | | 2 | |
| 1914 | GUM | VC4 | DD4 | | 2 | |
| 1915 | GUM | VC5 | WM | MUE | 10 | |
| 1916 | GUM | VC5 | DVB | | 1 | |
| 1917 | GUM | VC5 | VC3 | | 1 | |
| 1918 | GUM | VC5 | AVFR | | 1 | |

| Entry Number | Type | Pre | Post | Ref | Den | Uniq |
|---|---|---|---|---|---|---|
| 1919 | GUM | VC5 | VC4 | | 1 | |
| 1920 | GUM | VC5 | DVC | | 2 | |
| 1921 | GUM | VC5 | AVFL | | 1 | |
| 1922 | GUM | VC6 | DD4 | | 1 | |
| 1923 | GUM | VC6 | VD9 | | 1 | |
| 1924 | GUM | VC6 | VD11 | | 1 | |
| 1925 | GUM | VC6 | HSNR | | 1 | |
| 1926 | GUM | VC6 | DD3 | | 1 | |
| 1927 | GUM | VC6 | DD2 | | 1 | |
| 1928 | GUM | VC6 | WM | MUE | 10 | |
| 1929 | GUM | VC6 | VC5 | | 1 | |
| 1930 | GUM | VC6 | PVQL | | 1 | |
| 1931 | GUM | VC6 | WM | MVE | 1 | |
| 1932 | GUM | VC6 | VD10 | | 1 | |
| 1933 | GUM | VC6 | DD5 | | 1 | |
| 1934 | GUM | VC6 | VD13 | | 1 | |
| 1935 | GUM | VC6 | AVL | | 1 | |
| 1936 | GUM | VC6 | VD8 | | 1 | |
| 1937 | GUM | VC6 | VD7 | | 1 | |
| 1938 | GUM | VC6 | VD12 | | 1 | |
| 1939 | GUM | VD10 | WM | MVI | 8 | |
| 1940 | GUM | VD11 | WM | MVI | 8 | |
| 1941 | GUM | VD3 | VB2 | | 2 | |
| 1942 | GUM | VD3 | WM | MVI | 25 | |
| 1943 | GUM | VD3 | AS3 | | 2 | |
| 1944 | GUM | VD4 | WM | MVI | 25 | |
| 1945 | GUM | VD5 | WM | MVI | 25 | |
| 1946 | GUM | VD6 | WM | MVI | 25 | |
| 1947 | GUM | VD7 | WM | MVI | 25 | |
| 1948 | GUM | VD8 | WM | MVI | 25 | |
| 1949 | GUM | VD9 | WM | MVI | 25 | |
| 1950 | HUI | PVT | WN | NONE5 | 1 | |
| 1951 | HLI | PVPL | AVDR | | 2 | |
| 1952 | HLI | PVPL | DD2 | | 1 | |
| 1953 | HLI | PVPL | PVCR | | 4 | |
| 1954 | HLI | PVPL | AVBR | | 6 | |
| 1955 | HLI | PVPL | AVER | | 1 | |
| 1956 | HLI | PVPL | DVC | | 2 | |
| 1957 | HLI | PVPL | AVHR | | 2 | |
| 1958 | HLI | PVPL | AVAL | | 2 | |

152

| Entry Number | Type | Pre | Post | Ref | Den | Uniq |
|---|---|---|---|---|---|---|
| 1959 | HLI | PVPL | AVL | | 4 | |
| 1960 | HLI | PVPL | ADAL | | 1 | |
| 1961 | HLI | PVPL | WN | HDC | 3 | |
| 1962 | HLI | PVPL | AVBL | | 5 | |
| 1963 | HLI | PVPL | AQR | | 1 | |
| 1964 | HLI | PVPL | RIGL | | 2 | |
| 1965 | HLI | PVPL | AVAR | | 2 | |
| 1966 | HLI | PVPL | PVQR | | 1 | |
| 1967 | HUM | AS11 | DA3 | | 1 | |
| 1968 | HUM | AS11 | VD10 | | 2 | |
| 1969 | HUM | AS11 | DA9 | | 1 | |
| 1970 | HUM | AS11 | WM | MDE | 15 | |
| 1971 | HUM | AS11 | DA8 | | 1 | |
| 1972 | HUM | AS11 | VD11 | | 16 | |
| 1973 | HUM | DA8 | WM | MDE | 29 | |
| 1974 | HUM | DA8 | VD11 | | 23 | |
| 1975 | HUM | DA8 | DA9 | | 1 | |
| 1976 | HUM | DA8 | DD6 | | 1 | |
| 1977 | HUM | DA8 | VD12 | | 1 | |
| 1978 | HUM | DA9 | DD6 | | 1 | |
| 1979 | HUM | DA9 | VD11 | | 23 | |
| 1980 | HUM | DA9 | WM | MDE | 29 | |
| 1981 | HUM | DA9 | VD12 | | 1 | |
| 1982 | HUM | DD6 | WM | MDI | 29 | |
| 1983 | HUM | DD6 | DA8 | | 2 | |
| 1984 | HUM | DD6 | VD11 | | 1 | |
| 1985 | HUM | DD6 | VD12 | | 1 | |
| 1986 | HUM | PDA | WM | NMJ | 1 | |
| 1987 | HUM | PDA | DD6 | | 1 | |
| 1988 | HUM | PDA | PVNR | | 1 | |
| 1989 | HUM | PDA | DA9 | | 1 | |
| 1990 | HUM | PDB | WM | NMJ | 1 | |
| 1991 | HUM | PDB | DD6 | | 1 | |
| 1992 | HUM | VA12 | VD12 | | 2 | |
| 1993 | HUM | VA12 | PVCR | | 3 | |
| 1994 | HUM | VA12 | DB7 | | 4 | |
| 1995 | HUM | VA12 | AS11 | | 2 | |
| 1996 | HUM | VA12 | VD13 | | 11 | |
| 1997 | HUM | VA12 | WM | MVE | 14 | |
| 1998 | HUM | VA12 | DD6 | | 2 | |

| Entry Number | Type | Pre | Post | Ref | Den | Uniq |
|---|---|---|---|---|---|---|
| 1999 | HUM | VA12 | DA8 | | 3 | |
| 2000 | HUM | VA12 | DA9 | | 5 | |
| 2001 | HUM | VA12 | PVCL | | 2 | |
| 2002 | HUM | VA12 | LUAL | | 2 | |
| 2003 | HUM | VA12 | VA11 | | 1 | |
| 2004 | HUM | VD12 | WM | MVI | 8 | |
| 2005 | HUM | VD12 | VA12 | | 1 | |
| 2006 | HUM | VD13 | VA12 | | 1 | |
| 2007 | HUM | VD13 | WM | MVI | 8 | |
| 2008 | HRI | PVPR | AVBL | | 3 | |
| 2009 | HRI | PVPR | DVC | | 2 | |
| 2010 | HRI | PVPR | DD2 | | 1 | |
| 2011 | HRI | PVPR | AVAR | | 2 | |
| 2012 | HRI | PVPR | AVL | | 4 | |
| 2013 | HRI | PVPR | RIAR | | 2 | |
| 2014 | HRI | PVPR | AVEL | | 1 | |
| 2015 | HRI | PVPR | AVAL | | 1 | |
| 2016 | HRI | PVPR | AVBR | | 7 | |
| 2017 | HRI | PVPR | WN | HDC | 3 | |
| 2018 | HRI | PVPR | AVHL | | 3 | |
| 2019 | HRI | PVPR | PVCR | | 7 | |
| 2020 | HRI | PVPR | RIGR | | 1 | |
| 2021 | HRI | PVPR | RIMR | | 1 | |
| 2022 | HRI | PVPR | PVCL | | 5 | |
| 2023 | HRI | PVPR | PVQL | | 1 | |
| 2024 | HRI | PVPR | ADFR | | 1 | |
| 2025 | HRI | PVPR | AQR | | 1 | |
| 2026 | JUI | DVA | DB2 | | 1 | |
| 2027 | JUI | DVA | AVBL | | 1 | |
| 2028 | JUI | DVA | PVCL | | 5 | |
| 2029 | JUI | DVA | PVR | | 2 | |
| 2030 | JUI | DVA | DB4 | | 1 | |
| 2031 | JUI | DVA | VB11 | | 2 | |
| 2032 | JUI | DVA | DB5 | | 1 | |
| 2033 | JUI | DVA | SAAVL | | 1 | |
| 2034 | JUI | DVA | SMBDL | | 3 | |
| 2035 | JUI | DVA | SMBVL | | 3 | |
| 2036 | JUI | DVA | SAAVR | | 1 | |
| 2037 | JUI | DVA | SAADR | | 1 | |
| 2038 | JUI | DVA | DB7 | | 2 | |

154

| Entry Number | Type | Pre | Post | Ref | Den | Uniq |
|---|---|---|---|---|---|---|
| 2039 | JUI | DVA | AVAL | | 3 | |
| 2040 | JUI | DVA | PDER | | 2 | |
| 2041 | JUI | DVA | RIR | | 1 | |
| 2042 | JUI | DVA | AIZL | | 3 | |
| 2043 | JUI | DVA | DB3 | | 3 | |
| 2044 | JUI | DVA | RIAL | | 1 | |
| 2045 | JUI | DVA | RIMR | | 1 | |
| 2046 | JUI | DVA | SMBVR | | 2 | |
| 2047 | JUI | DVA | AQR | | 3 | |
| 2048 | JUI | DVA | AUAR | | 1 | |
| 2049 | JUI | DVA | VA12 | | 1 | |
| 2050 | JUI | DVA | AUAL | | 1 | |
| 2051 | JUI | DVA | AVER | | 6 | |
| 2052 | JUI | DVA | SMBDR | | 2 | |
| 2053 | JUI | DVA | VB1 | | 1 | |
| 2054 | JUI | DVA | AVEL | | 8 | |
| 2055 | JUI | DVA | RIAR | | 2 | |
| 2056 | JUI | DVC | RIGR | | 5 | |
| 2057 | JUI | DVC | AVKR | | 1 | |
| 2058 | JUI | DVC | RMFL | | 2 | |
| 2059 | JUI | DVC | RIBL | | 1 | |
| 2060 | JUI | DVC | AVAR | | 7 | |
| 2061 | JUI | DVC | RIGL | | 5 | |
| 2062 | JUI | DVC | AIBL | | 1 | |
| 2063 | JUI | DVC | AIBR | | 3 | |
| 2064 | JUI | DVC | AVKL | | 2 | |
| 2065 | JUI | DVC | AVAL | | 5 | |
| 2066 | JUI | DVC | DVB | | 1 | |
| 2067 | JUI | DVC | AVBL | | 1 | |
| 2068 | JUI | DVC | RMFR | | 4 | |
| 2069 | JUM | DVB | RIGR | | 3 | |
| 2070 | JUM | DVB | AVL | | 2 | |
| 2071 | JUM | DVB | DD6 | | 3 | |
| 2072 | JUM | DVB | RMEV | | 1 | |
| 2073 | JUM | DVB | DD6 | | 1 | |
| 2074 | JUM | DVB | RIGL | | 2 | |
| 2075 | JUM | DVB | WN | HDC | 5 | |
| 2076 | JUM | DVB | AVKL | | 7 | |
| 2077 | JUM | DVB | DVC | | 5 | |
| 2078 | JUM | DVB | AVKR | | 7 | |

| Entry Number | Type | Pre | Post | Ref | Den | Uniq |
|---|---|---|---|---|---|---|
| 2079 | JUM | DVB | AIBL | | 2 | |
| 2080 | JUM | DVB | WM | NMJ | 5 | |
| 2081 | JUM | DVB | AIBR | | 5 | |
| 2082 | JUM | DVB | RMFR | | 4 | |
| 2083 | JUM | DVB | DA8 | | 2 | |
| 2084 | JUM | DVB | RIH | | 1 | |
| 2085 | JUM | DVB | RMFL | | 2 | |
| 2086 | JUM | DVB | SMBDR | | 1 | |
| 2087 | JUM | DVB | PDA | | 1 | |
| 2088 | JUM | DVB | PVPR | | 2 | |
| 2089 | KLI | LUAL | AVJL | | 1 | |
| 2090 | KLI | LUAL | PHBL | | 1 | |
| 2091 | KLI | LUAL | PVNL | | 1 | |
| 2092 | KLI | LUAL | AVAL | | 5 | |
| 2093 | KLI | LUAL | AVAR | | 4 | |
| 2094 | KLI | LUAL | AVDR | | 2 | |
| 2095 | KLI | LUAL | PVWL | | 1 | |
| 2096 | KLI | LUAL | AVDL | | 4 | |
| 2097 | KLI | PVCL | LUAR | | 1 | |
| 2098 | KLI | PVCL | AVJL | | 3 | |
| 2099 | KLI | PVCL | AVER | | 1 | |
| 2100 | KLI | PVCL | PVR | | 1 | |
| 2101 | KLI | PVCL | PVWL | | 1 | |
| 2102 | KLI | PVCL | RIS | | 2 | |
| 2103 | KLI | PVCL | DB5 | | 3 | |
| 2104 | KLI | PVCL | VB5 | | 1 | |
| 2105 | KLI | PVCL | DB7 | | 4 | |
| 2106 | KLI | PVCL | DB4 | | 7 | |
| 2107 | KLI | PVCL | DVA | | 4 | |
| 2108 | KLI | PVCL | VB6 | | 5 | |
| 2109 | KLI | PVCL | VB11 | | 1 | |
| 2110 | KLI | PVCL | AVAR | | 9 | |
| 2111 | KLI | PVCL | PDEL | | 1 | |
| 2112 | KLI | PVCL | AVEL | | 2 | |
| 2113 | KLI | PVCL | PVCR | | 3 | |
| 2114 | KLI | PVCL | AVL | | 1 | |
| 2115 | KLI | PVCL | VB4 | | 4 | |
| 2116 | KLI | PVCL | AS1 | | 1 | |
| 2117 | KLI | PVCL | VB3 | | 1 | |
| 2118 | KLI | PVCL | DB2 | | 4 | |

| Entry Number | Type | Pre | Post | Ref | Den | Uniq |
|---|---|---|---|---|---|---|
| 2119 | KLI | PVCL | DA2 | | 1 | |
| 2120 | KLI | PVCL | VB8 | | 2 | |
| 2121 | KLI | PVCL | AVAL | | 7 | |
| 2122 | KLI | PVCL | AVBR | | 16 | |
| 2123 | KLI | PVCL | AVDL | | 3 | |
| 2124 | KLI | PVCL | AS2 | | 2 | |
| 2125 | KLI | PVCL | DB3 | | 5 | |
| 2126 | KLI | PVCL | SIBVL | | 2 | |
| 2127 | KLI | PVCL | AVBL | | 4 | |
| 2128 | KLI | PVCL | RID | | 5 | |
| 2129 | KLI | PVCL | AVDR | | 6 | |
| 2130 | KLI | PVCL | VB7 | | 3 | |
| 2131 | KLI | PVNL | DD1 | | 2 | |
| 2132 | KLI | PVNL | AVER | | 3 | |
| 2133 | KLI | PVNL | DD2 | | 1 | |
| 2134 | KLI | PVNL | AVG | | 1 | |
| 2135 | KLI | PVNL | VD3 | | 2 | |
| 2136 | KLI | PVNL | AVJL | | 3 | |
| 2137 | KLI | PVNL | WN | HDC | 2 | |
| 2138 | KLI | PVNL | AVJR | | 7 | |
| 2139 | KLI | PVNL | RIFR | | 1 | |
| 2140 | KLI | PVNL | AVL | | 4 | |
| 2141 | KLI | PVNL | WM | NMJ | 7 | |
| 2142 | KLI | PVNL | BDUR | | 3 | |
| 2143 | KLI | PVNL | AVDR | | 6 | |
| 2144 | KLI | PVNL | PVT | | 2 | |
| 2145 | KLI | PVNL | PQR | | 3 | |
| 2146 | KLI | PVNL | VD4 | | 4 | |
| 2147 | KLI | PVNL | VC3 | | 1 | |
| 2148 | KLI | PVNL | VC2 | | 2 | |
| 2149 | KLI | PVNL | PVNR | | 4 | |
| 2150 | KLI | PVNL | AVBR | | 3 | |
| 2151 | KLI | PVNL | AVAR | | 6 | |
| 2152 | KLI | PVNL | VD12 | | 1 | |
| 2153 | KLI | PVNL | AVFR | | 1 | |
| 2154 | KLI | PVNL | PVWL | | 3 | |
| 2155 | KLI | PVNL | PVCR | | 2 | |
| 2156 | KLI | PVQL | VD1 | | 1 | |
| 2157 | KLI | PVQL | RMGL | | 1 | |
| 2158 | KLI | PVQL | AVL | | 1 | |

| Entry Number | Type | Pre | Post | Ref | Den | Uniq |
|---|---|---|---|---|---|---|
| 2159 | KLI | PVQL | DVB | | 1 | |
| 2160 | KLI | PVQL | ASKL | | 4 | |
| 2161 | KLI | PVQL | AIAL | | 6 | |
| 2162 | KLI | PVQL | DD1 | | 1 | |
| 2163 | KLI | PVQL | DVC | | 1 | |
| 2164 | KLI | PVQL | ASJL | | 1 | |
| 2165 | KLI | PVQL | WE | CEPshVL | 1 | |
| 2166 | KLI | PVQL | HSNL | | 3 | |
| 2167 | KLI | PVWL | VA12 | | 1 | |
| 2168 | KLI | PVWL | PVWR | | 1 | |
| 2169 | KLI | PVWL | PVCL | | 1 | |
| 2170 | KLI | PVWL | AVAL | | 1 | |
| 2171 | KLI | PVWL | AVDL | | 1 | |
| 2172 | KLI | PVWL | AVJR | | 1 | |
| 2173 | KLI | PVWL | PVT | | 3 | |
| 2174 | KLS | ALNL | SAAVL | | 3 | |
| 2175 | KLS | ALNL | SMBDR | | 2 | |
| 2176 | KLS | ALNL | SMDVL | | 1 | |
| 2177 | KLS | PHAL | AVG | | 5 | |
| 2178 | KLS | PHAL | AVHR | | 1 | |
| 2179 | KLS | PHAL | PHBR | | 4 | |
| 2180 | KLS | PHAL | DVA | | 2 | |
| 2181 | KLS | PHAL | PHAR | | 5 | |
| 2182 | KLS | PHAL | DA8 | | 1 | |
| 2183 | KLS | PHAL | AVDR | | 1 | |
| 2184 | KLS | PHAL | PVQL | | 2 | |
| 2185 | KLS | PHAL | PHBL | | 5 | |
| 2186 | KLS | PHAL | AVFL | | 3 | |
| 2187 | KLS | PHAL | AVHL | | 1 | |
| 2188 | KLS | PHBL | AVDL | | 1 | |
| 2189 | KLS | PHBL | VA12 | | 2 | |
| 2190 | KLS | PHBL | AVAR | | 6 | |
| 2191 | KLS | PHBL | PVCL | | 13 | |
| 2192 | KLS | PHBL | PHBR | | 1 | |
| 2193 | KLS | PHBL | AVAL | | 9 | |
| 2194 | KLS | PHCL | LUAL | | 1 | |
| 2195 | KLS | PHCL | PVCL | | 2 | |
| 2196 | KLS | PHCL | AVAL | | 1 | |
| 2197 | KLS | PHCL | VA12 | | 1 | |
| 2198 | KLS | PHCL | DVA | | 5 | |

| Entry Number | Type | Pre | Post | Ref | Den | Uniq |
|---|---|---|---|---|---|---|
| 2199 | KLS | PLML | HSNL | | 2 | |
| 2200 | KLS | PLNL | SMBVL | | 6 | |
| 2201 | KLS | PLNL | SAADL | | 5 | |
| 2202 | KRI | LUAR | AVJR | | 1 | |
| 2203 | KRI | LUAR | AVDL | | 1 | |
| 2204 | KRI | LUAR | PVWL | | 1 | |
| 2205 | KRI | LUAR | PQR | | 1 | |
| 2206 | KRI | LUAR | AVAL | | 3 | |
| 2207 | KRI | LUAR | PVR | | 1 | |
| 2208 | KRI | LUAR | AVAR | | 7 | |
| 2209 | KRI | LUAR | PVCR | | 3 | |
| 2210 | KRI | LUAR | AVDR | | 3 | |
| 2211 | KRI | PVCR | AS1 | | 1 | |
| 2212 | KRI | PVCR | FLPL | | 1 | |
| 2213 | KRI | PVCR | LUAL | | 1 | |
| 2214 | KRI | PVCR | VB4 | | 4 | |
| 2215 | KRI | PVCR | PDER | | 1 | |
| 2216 | KRI | PVCR | AQR | | 1 | |
| 2217 | KRI | PVCR | PVCL | | 1 | |
| 2218 | KRI | PVCR | VB5 | | 1 | |
| 2219 | KRI | PVCR | PVR | | 1 | |
| 2220 | KRI | PVCR | AVAR | | 10 | |
| 2221 | KRI | PVCR | PVWR | | 1 | |
| 2222 | KRI | PVCR | AVBR | | 6 | |
| 2223 | KRI | PVCR | RID | | 4 | |
| 2224 | KRI | PVCR | DB3 | | 5 | |
| 2225 | KRI | PVCR | SIBVR | | 1 | |
| 2226 | KRI | PVCR | VB7 | | 3 | |
| 2227 | KRI | PVCR | VB11 | | 1 | |
| 2228 | KRI | PVCR | DVA | | 4 | |
| 2229 | KRI | PVCR | VB3 | | 1 | |
| 2230 | KRI | PVCR | VB8 | | 2 | |
| 2231 | KRI | PVCR | AVER | | 1 | |
| 2232 | KRI | PVCR | AS2 | | 2 | |
| 2233 | KRI | PVCR | AVAL | | 8 | |
| 2234 | KRI | PVCR | AVDL | | 6 | |
| 2235 | KRI | PVCR | DB2 | | 4 | |
| 2236 | KRI | PVCR | DB5 | | 3 | |
| 2237 | KRI | PVCR | DB7 | | 4 | |
| 2238 | KRI | PVCR | DA2 | | 1 | |

| Entry Number | Type | Pre | Post | Ref | Den | Uniq |
|---|---|---|---|---|---|---|
| 2239 | KRI | PVCR | VB6 | | 5 | |
| 2240 | KRI | PVCR | AVL | | 1 | |
| 2241 | KRI | PVCR | DB4 | | 7 | |
| 2242 | KRI | PVCR | AVBL | | 9 | |
| 2243 | KRI | PVCR | AVEL | | 3 | |
| 2244 | KRI | PVCR | AVDR | | 1 | |
| 2245 | KRI | PVNR | PVCR | | 2 | |
| 2246 | KRI | PVNR | PVT | | 2 | |
| 2247 | KRI | PVNR | AVG | | 1 | |
| 2248 | KRI | PVNR | AVDR | | 6 | |
| 2249 | KRI | PVNR | PVWR | | 3 | |
| 2250 | KRI | PVNR | AVAR | | 6 | |
| 2251 | KRI | PVNR | AVJL | | 2 | |
| 2252 | KRI | PVNR | AVJR | | 7 | |
| 2253 | KRI | PVNR | AVBR | | 3 | |
| 2254 | KRI | PVNR | VC3 | | 1 | |
| 2255 | KRI | PVNR | AVER | | 3 | |
| 2256 | KRI | PVNR | VC2 | | 2 | |
| 2257 | KRI | PVNR | RIFR | | 1 | |
| 2258 | KRI | PVNR | VD3 | | 2 | |
| 2259 | KRI | PVNR | BDUR | | 3 | |
| 2260 | KRI | PVNR | DD1 | | 2 | |
| 2261 | KRI | PVNR | VD4 | | 4 | |
| 2262 | KRI | PVNR | DD2 | | 1 | |
| 2263 | KRI | PVNR | PVNL | | 5 | |
| 2264 | KRI | PVNR | WN | HDC | 2 | |
| 2265 | KRI | PVNR | BDUL | | 2 | |
| 2266 | KRI | PVNR | WM | NMJ | 7 | |
| 2267 | KRI | PVNR | AVL | | 4 | |
| 2268 | KRI | PVNR | AVFR | | 1 | |
| 2269 | KRI | PVNR | VD12 | | 1 | |
| 2270 | KRI | PVNR | PQR | | 3 | |
| 2271 | KRI | PVQR | VD1 | | 1 | |
| 2272 | KRI | PVQR | RIFR | | 1 | |
| 2273 | KRI | PVQR | AVL | | 1 | |
| 2274 | KRI | PVQR | AVFR | | 1 | |
| 2275 | KRI | PVQR | DVB | | 1 | |
| 2276 | KRI | PVQR | WE | CEPshVR | 1 | |
| 2277 | KRI | PVQR | HSNR | | 1 | |
| 2278 | KRI | PVQR | DD1 | | 1 | |

| Entry Number | Type | Pre | Post | Ref | Den | Uniq |
|---|---|---|---|---|---|---|
| 2279 | KRI | PVQR | DVC | | 1 | |
| 2280 | KRI | PVQR | AIAR | | 7 | |
| 2281 | KRI | PVQR | ASKR | | 4 | |
| 2282 | KRI | PVWR | AVAR | | 1 | |
| 2283 | KRI | PVWR | PVCR | | 1 | |
| 2284 | KRI | PVWR | PVWL | | 1 | |
| 2285 | KRI | PVWR | AVDR | | 1 | |
| 2286 | KRI | PVWR | VA12 | | 1 | |
| 2287 | KRI | PVWR | PVT | | 3 | |
| 2288 | KRI | PVWR | AVJL | | 1 | |
| 2289 | KRS | ALNR | SAAVR | | 3 | |
| 2290 | KRS | ALNR | SMBDL | | 1 | |
| 2291 | KRS | ALNR | RMHR | | 1 | |
| 2292 | KRS | ALNR | SMDVL | | 1 | |
| 2293 | KRS | ALNR | SMDDR | | 1 | |
| 2294 | KRS | PHAR | PHBL | | 1 | |
| 2295 | KRS | PHAR | AVHR | | 1 | |
| 2296 | KRS | PHAR | PHAL | | 6 | |
| 2297 | KRS | PHAR | DVA | | 2 | |
| 2298 | KRS | PHAR | PVQL | | 2 | |
| 2299 | KRS | PHAR | AVG | | 3 | |
| 2300 | KRS | PHAR | PHBR | | 5 | |
| 2301 | KRS | PHBR | AVDR | | 1 | |
| 2302 | KRS | PHBR | AVAL | | 7 | |
| 2303 | KRS | PHBR | PHBL | | 1 | |
| 2304 | KRS | PHBR | AVDL | | 1 | |
| 2305 | KRS | PHBR | PVCL | | 6 | |
| 2306 | KRS | PHBR | VA12 | | 1 | |
| 2307 | KRS | PHBR | AVAR | | 7 | |
| 2308 | KRS | PHBR | PVCR | | 3 | |
| 2309 | KRS | PHCR | VA12 | | 1 | |
| 2310 | KRS | PHCR | AVHR | | 1 | |
| 2311 | KRS | PHCR | PVCR | | 8 | |
| 2312 | KRS | PHCR | DA9 | | 6 | |
| 2313 | KRS | PHCR | DVA | | 7 | |
| 2314 | KRS | PHCR | LUAR | | 1 | |
| 2315 | KRS | PHCR | PHCL | | 1 | |
| 2316 | KRS | PLMR | PDER | | 5 | |
| 2317 | KRS | PLMR | HSNR | | 2 | |
| 2318 | KRS | PLMR | PVCR | | 1 | |

| Entry Number | Type | Pre | Post | Ref | Den | Uniq |
|---|---|---|---|---|---|---|
| 2319 | KRS | PLMR | AVDL | | 4 | |
| 2320 | KRS | PLMR | AVAL | | 5 | |
| 2321 | KRS | PLMR | DVA | | 5 | |
| 2322 | KRS | PLNR | SMBVR | | 5 | |
| 2323 | KRS | PLNR | SAADR | | 3 | |
| 2324 | KUS | PQR | AVDR | | 23 | |
| 2325 | KUS | PQR | AVAR | | 7 | |
| 2326 | KUS | PQR | AVG | | 1 | |
| 2327 | KUS | PQR | PVNR | | 1 | |
| 2328 | KUS | PQR | AVAL | | 7 | |
| 2329 | KUS | PQR | LUAL | | 1 | |
| 2330 | KUS | PVR | IL1DL | | 1 | |
| 2331 | KUS | PVR | DA9 | | 1 | |
| 2332 | KUS | PVR | PDER | | 1 | |
| 2333 | KUS | PVR | DVA | | 1 | |
| 2334 | KUS | PVR | AVBR | | 3 | |
| 2335 | KUS | PVR | RIPR | | 3 | |
| 2336 | KUS | PVR | AVKR | | 1 | |
| 2337 | KUS | PVR | RIPL | | 4 | |
| 2338 | KUS | PVR | DB2 | | 1 | |
| 2339 | KUS | PVR | AVBL | | 4 | |
| 2340 | KUS | PVR | AVJL | | 2 | |
| 2341 | KUS | PVR | DB3 | | 1 | |
| 2342 | KUS | PVR | IL1VL | | 1 | |
| 2343 | KUS | PVR | SABD | | 1 | |
| 2344 | KUS | PVR | AVJR | | 2 | |
| 2345 | KUS | PVR | IL1DR | | 1 | |
| 2346 | KUS | PVR | IL1VR | | 1 | |
| 2347 | KUS | PVR | PVCR | | 1 | |
| 2348 | PLIS | I1L | | | | |
| 2349 | PRIS | I1R | | | | |
| 2350 | PLIS | I2L | | | | |
| 2351 | PRIS | I2R | | | | |
| 2352 | PUIS | I3 | | | | |
| 2353 | PUIS | I4 | | | | |
| 2354 | PUIS | I5 | | | | |
| 2355 | PUIS | I6 | | | | |
| 2356 | PUM | M1 | WM | PHARYN1 | 1 | |
| 2357 | PLM | M2L | WM | PHARYN1 | 1 | |
| 2358 | PRM | M2R | WM | PHARYN1 | 1 | |

| Entry Number | Type | Pre | Post | Ref | Den | Uniq |
|---|---|---|---|---|---|---|
| 2359 | PLSM | M3L | WM | PHARYN2 | 1 | |
| 2360 | PRSM | M3R | WM | PHARYN2 | 1 | |
| 2361 | PUM | M4 | WM | PHARYN1 | 1 | |
| 2362 | PUM | M5 | WM | PHARYN1 | 1 | |
| 2363 | PLI | MCL | WM | PHARYN3 | 1 | |
| 2364 | PRI | MCR | WM | PHARYN3 | 1 | |
| 2365 | PUIM | MI | WM | PHARYN4 | 1 | |
| 2366 | PLM | NSML | WN | PHARYN5 | 1 | |
| 2367 | PRM | NSMR | WN | PHARYN5 | 1 | |
| 2368 | ALS | WE | BAGL | | 1 | |
| 2369 | ALS | WE | CEPVL | | 1 | |
| 2370 | ALS | WE | IL2DL | | 1 | |
| 2371 | ALS | WE | IL2L | | 1 | |
| 2372 | ALS | WE | IL2VL | | 1 | |
| 2373 | ALS | WE | OLLL | | 1 | |
| 2374 | ALS | WE | OLQDL | | 1 | |
| 2375 | ALS | WE | OLQVL | | 1 | |
| 2376 | ALS | WE | URBL | | 1 | |
| 2377 | ALS | WE | URYDL | | 1 | |
| 2378 | ALS | WE | URYVL | | 1 | |
| 2379 | ARS | WE | BAGR | | 1 | |
| 2380 | ARS | WE | CEPVR | | 1 | |
| 2381 | ARS | WE | IL2DR | | 1 | |
| 2382 | ARS | WE | IL2R | | 1 | |
| 2383 | ARS | WE | IL2VR | | 1 | |
| 2384 | ARS | WE | OLLR | | 1 | |
| 2385 | ARS | WE | OLQDR | | 1 | |
| 2386 | ARS | WE | OLQVR | | 1 | |
| 2387 | ARS | WE | URBR | | 1 | |
| 2388 | ARS | WE | URYDR | | 1 | |
| 2389 | ARS | WE | URYVR | | 1 | |
| 2390 | ALMS | WE | IL1DL | | 1 | |
| 2391 | ALMS | WE | IL1L | | 1 | |
| 2392 | ALMS | WE | IL1VL | | 1 | |
| 2393 | ALMS | WE | URADL | | 1 | |
| 2394 | ALMS | WE | URAVL | | 1 | |
| 2395 | ARMS | WE | IL1DR | | 1 | |
| 2396 | ARMS | WE | IL1R | | 1 | |
| 2397 | ARMS | WE | IL1VR | | 1 | |
| 2398 | ARMS | WE | URADR | | 1 | |

| Entry Number | Type | Pre | Post | Ref | Den | Uniq |
|---|---|---|---|---|---|---|
| 2399 | ARMS | WE | URAVR | | 1 | |
| 2400 | BLS | WE | CEPDL | | 1 | |
| 2401 | BLS | WE | CEPVL | | 1 | |
| 2402 | BLSI | WE | URXL | | 1 | |
| 2403 | BRS | WE | CEPDR | | 1 | |
| 2404 | BRSI | WE | URXR | | 1 | |
| 2405 | CLS | WE | ADFL | | 1 | |
| 2406 | CLS | WE | ADLL | | 1 | |
| 2407 | CLS | WE | AFDL | | 1 | |
| 2408 | CLS | WE | ASEL | | 1 | |
| 2409 | CLS | WE | ASGL | | 1 | |
| 2410 | CLS | WE | ASHL | | 1 | |
| 2411 | CLS | WE | ASIL | | 1 | |
| 2412 | CLS | WE | ASJL | | 1 | |
| 2413 | CLS | WE | ASKL | | 1 | |
| 2414 | CLS | WE | AUAL | | 1 | |
| 2415 | CLS | WE | AWAL | | 1 | |
| 2416 | CLS | WE | AWBL | | 1 | |
| 2417 | CLS | WE | AWCL | | 1 | |
| 2418 | CRS | WE | ADFR | | 1 | |
| 2419 | CRS | WE | ADLR | | 1 | |
| 2420 | CRS | WE | AFDR | | 1 | |
| 2421 | CRS | WE | ASER | | 1 | |
| 2422 | CRS | WE | ASGR | | 1 | |
| 2423 | CRS | WE | ASHR | | 1 | |
| 2424 | CRS | WE | ASIR | | 1 | |
| 2425 | CRS | WE | ASJR | | 1 | |
| 2426 | CRS | WE | ASKR | | 1 | |
| 2427 | CRS | WE | AWAR | | 1 | |
| 2428 | CRS | WE | AWBR | | 1 | |
| 2429 | CRS | WE | AWCR | | 1 | |
| 2430 | ELS | WE | ADEL | | 1 | |
| 2431 | ELS | WE | FLPL | | 1 | |
| 2432 | ERS | WE | ADER | | 1 | |
| 2433 | ERS | WE | FLPR | | 1 | |
| 2434 | EUS | WE | AQR | | 1 | |
| 2435 | EUS | WE | AVG | | 1 | |
| 2436 | FLS | WE | ALML | | 1 | |
| 2437 | FLS | WE | PDEL | | 1 | |
| 2438 | FRS | WE | ALMR | | 1 | |

164

| Entry Number | Type | Pre | Post | Ref | Den | Uniq |
|---|---|---|---|---|---|---|
| 2439 | FRS | WE | PDER | | 1 | |
| 2440 | FUS | WE | AVM | | 1 | |
| 2441 | FUS | WE | PVM | | 1 | |
| 2442 | KLS | WE | ALNL | | 1 | |
| 2443 | KLS | WE | PHAL | | 1 | |
| 2444 | KLS | WE | PHBL | | 1 | |
| 2445 | KLS | WE | PHCL | | 1 | |
| 2446 | KLS | WE | PLML | | 1 | |
| 2447 | KLS | WE | PLNL | | 1 | |
| 2448 | KRS | WE | ALNR | | 1 | |
| 2449 | KRS | WE | PHAR | | 1 | |
| 2450 | KRS | WE | PHBR | | 1 | |
| 2451 | KRS | WE | PHCR | | 1 | |
| 2452 | KRS | WE | PLMR | | 1 | |
| 2453 | KRS | WE | PLNR | | 1 | |
| 2454 | KUS | WE | PQR | | 1 | |
| 2455 | KUS | WE | PVR | | 1 | |
| 2456 | PLIS | WE | I1L | | 1 | |
| 2457 | PRIS | WE | I1R | | 1 | |
| 2458 | PLIS | WE | I2L | | 1 | |
| 2459 | PRIS | WE | I2R | | 1 | |
| 2460 | PUIS | WE | I3 | | 1 | |
| 2461 | PUIS | WE | I4 | | 1 | |
| 2462 | PUIS | WE | I5 | | 1 | |
| 2463 | PUIS | WE | I6 | | 1 | |

--- End of File ---

## F. PARTLIST FILE SET

### F.1. *PARTLIST.TXT*

#### F.1.a. Description

The PARTLIST data file has four (4) columns.

1) The **Parts List** column lists the cells of *C. elegans*, in both hermaphrodite and male, and in both embryonic and post-embryonic stages.

2) The **Cell Lineage** column contains codes that trace the cellular ancestry of each cell in the Parts List column from its origin. These codes are explained in the reference given below, and modifications made on these codes are explained in the Derivation section of this text.

3) The **Short Description** column contains a 40-character description of the worm part listed in the Parts List column. Conventions used in these descriptions are explained in the Derivation section of this text.

4) The **Number of Cells** column gives the number of cells per given worm part listed in the Parts List column.

#### F.1.b. Derivation

Data were derived from:

Wood, W.B. The Nematode *Caenorhabditis elegans* (Monograph 17), edited by W. Wood and the Community of C. elegans Researchers. Cold Spring Harbor: Cold Spring Harbor Laboratory, 1988.

The framework of the PARTLIST data file was mainly derived from APPENDIX 1 - Parts List (pp. 415 to 431) and from APPENDIX 3 - Part A: Lineage Charts (pp. 457 to 478) of the reference above. A segment of the Parts List as seen in literature is reproduced in Figure 1.3, p. 4, and a portion of the Lineage Charts as referenced in literature is presented in Figure 1.4, p. 4.

The zygote ($P_0$) becomes the anterior cell AB and the posterior cell $P_1$ in the first division. $P_1$ divides to yield the anterior cell EMS and the posterior cell $P_2$. EMS divides to yield the anterior cell MS and the posterior cell E, while $P_2$ divides and yields the anterior cell C and the posterior cell $P_3$. $P_3$ divides to yield the anterior cell D and the posterior cell $P_4$. AB, MS, E, C, D, and $P_4$ are the six embryonic founder cells. The embryonic founder cells undergo successive divisions yielding unnamed intermediate cells and some cells that seem to die naturally. At the end of the embryonic phase, the divisions of the embryonic founder cells have produced the embryonic, terminally differentiated cells and the postembryonic blast cells. The unpaired postembryonic blast cells are: B (male), F (male), G1, G2, K, K', M, U (male), W, Y, and Z1 to Z4. The paired left (L) and right (R) postembryonic blast cells are: H1, H2, P1 to P12, Q,

T, and V1 to V6 (e.g., H1L and H1R). The postembryonic blast cells undergo successive divisions yielding unnamed intermediate cells and some cells that seem to die naturally. At the end of the postembryonic phase, the divisions of the postembryonic blast cells have produced the postembryonic, terminally differentiated cells. The general scheme may be outlined in the following manner:

  I. Zygote (P0)

 II. Divisions of the zygote yield named intermediate cells: EMS, $P_1$, $P_2$, $P_3$

III. Six embryonic founder cells result from divisions of zygote: AB, MS, E, C, D, and $P_4$

IV. Divisions of the six embryonic founder cells yield unnamed intermediate cells and some cells that seem to die naturally

 V. Embryonic, terminally differentiated cells and postembryonic blast cells result from divisions of embryonic founder cells. The unpaired postembryonic blast cells are: B (male), F (male), G1, G2, K, K', M, U (male), W, Y, and Z1 to Z4. The paired left (L) and right (R) postembryonic blast cells are: H1, H2, P1 to P12, Q, T, and V1 to V6 (e.g., H1L and H1R)

VI. Divisions of the postembryonic blast cells yield unnamed intermediate cells and some cells that seem to die naturally

VII. Postembryonic, terminally differentiated cells result from divisions of post-embryonic blast cells.

To render the subscripts of P0 and P04 amenable to electronic spreadsheet manipulation, and to distinguish cells $P_1$ to $P_4$ in levels II and III of the scheme above from cells P1 to P12 in level V, $P_0$ is written as P0 (P zero) and $P_4$ is written as P04 (P zero four) in the succeeding text and in the .PRN file.

The only other general change in notation is that K' is listed as K prime.

*F.1.b.i.* The Parts List column lists all cells in levels I, III, V, and VII of the scheme above. In literature, the Parts List contains cells with the "or" notation represented by the slash ("/"). For example,

| Functional Name (or Parts List) | Lineage Name (or Cell Lineage) |
| --- | --- |
| AVFL/R | P1 aaaa |
| AVFL/R | W aaa |
| P1/2 | AB plapaapp |
| P1/2 | AB prapaapp. |

This means that AVFL or AVFR can arise from either P1 aaaa or W aaa, and that P1 or P2 can arise from either AB plapaapp or AB prapaap. In these cases, each cell in the

Parts List column was made to correspond with only one lineage. The example above thus becomes

| Parts List | Cell Lineage |
|---|---|
| AVFL | P1 aaaa |
| AVFR | W aaa |
| P1 | AB plapaapp |
| P2 | AB prapaapp. |

*F.1.b.ii.* Except the zygote (P0), each Cell Lineage column entry in literature has two segments separated by a space. The first segment is either the notation for the zygote, an embryonic founder cell, or a postembryonic blast cell. The second segment reflects the spatial direction of the successive divisions of the cell in the first segment (seen in levels II, IV, and VI of the scheme above). Thus, except level II intermediate cells which are named, the geometric positions of otherwise unnamed intermediate daughter cells during successive divisions are seen and preserved in the alphabetical string of the second segment. The string codes are: a = anterior, p = posterior, l = left, r = right, d = dorsal, and v = ventral. The terminal cell in the string is the named cell in the Parts List column. Examples of cells and their cell lineages are: AB = P0 a (P0 divides anteroposteriorly and AB is its anterior daughter cell), ADAL = AB plapaaaapp, and PVM = QL paa.

The Cell Lineage column was expanded to include the left (L) and right (R) cells of paired postembryonic blast cells individually (see level V in the general scheme above). In literature, each "L&R" pair was one entry. For example,

| Functional Name (or Parts List) | Lineage Name (or Cell Lineage) |
|---|---|
| ADEso | H2 aa    L&R. |

Upon expansion of the lineage, modifications to the corresponding entries in the Parts List column were also made. The example above thus becomes

| Parts List | Cell Lineage |
|---|---|
| ADEsoL | H2L aa |
| ADEsoR | H2R aa. |

The Cell Lineage column also contains entries with the "or" notation, represented by the slash ("/"). For example,

| Functional Name (or Parts List) | Lineage Name (or Cell Lineage) |
|---|---|
| SVPL | B a(l/r)aalda |
| SPVR | B a(l/r)aarda. |

This means that B alaalda or B araalda are the alternate pathways that give rise to SPVL. While the same scheme is true for SPVR, the cell lineages of SPVL and SPVR are still unique despite the presence of alternate pathways. This notation, therefore, was preserved in the .PRN file.

Explicit "or" statements retained in the Cell Lineage column result from the capacity of postembryonic blast cells Z1 and Z4 to give rise to the same set of terminally differentiated cells by two interchangeable pathways. For example,

| Parts List | Cell Lineage |
|---|---|
| gon herm anch | Z1 ppp (5L) |
| | or |
| | Z4 aaa (5R) |
| gon male link | Z1 paa |
| | or |
| | Z4 aaa. |

The two interchangeable pathways of Z1 and Z4 that produce the same set of terminally differentiated cells in the *C. elegans* hermaphrodite are labelled "(5L)" and "(5R)". These labels are appended to the appropriate cell lineages for the purpose of identification. Since no such labels exist for the male, the labels "(ZA)" and "(ZB)" were chosen arbitrarily to serve the same function. The example above thus becomes

| Parts List | Cell Lineage |
|---|---|
| gon herm anch | Z1 ppp (5L) |
| | or |
| gon herm anch | Z4 aaa (5R) |
| gon male link | Z1 paa (ZA) |
| | or |
| gon male link | Z4 aaa (ZB). |

Multiplicity data (i.e., the number of cells per worm part) are also provided in the Cell Lineage column. These are discussed in *F.1.b.iv.* below.

The entry "In a and In p" (n = a number designation) in the Cell Lineage column refers to postembryonic divisions of 14 intestinal nuclei without cell division. Since the terminal divisions of embryonic founder cell E accounts for 20 intestinal cells already, there is no increase in the number of cells (Number of Cells = 0).

The entry "Pn" (n = a number designation) in the Cell Lineage column of the reference was fully expanded in all of its occurrences in the Cell Lineage column of the .PRN file. Corresponding modifications in the Parts List and Number of Cells columns were also made. Expansion was based on the Lineage Charts (APPENDIX 3) of the reference above.

For parsing, note that no synonyms are included in the Cell Lineage column, although they are suggested in the large cell lineage charts in the reference above (APPENDIX 3). Synonyms may arise because a, d, l and p, v, r represent two groups of letters which are interchangeable to some extent. Synonyms which may be present could lead to computational inconsistencies in higher generations, e.g., when d is equivalent to l, E pdaa = E plaa. Other sources of equivalence are alternative progenitors, from which the cell lines of destroyed normal progenitors arise. Known equivalence groups in *C. elegans* are in Table 1, page 143 of the reference above.

Ten cell generations in general contribute to cells of the embryo, with about 20% loss from a complete binary tree. On the average, an additional 5 to 6 generations are necessary to include all postembryonic lineage strings. This suggests that most of the possible 65k cells from 16 binary divisions of the zygote do not appear in the adult because of death of progenitors in the course of development.

*F.1.b.iii.* The Short Description column is again derived from APPENDIX 1 - Parts List (pp. 415 to 429). Description was limited to forty (40) characters or less per line, so that the first three (3) columns of the PARTLIST data file would show completely on the computer screen. To preserve data while limiting a Short Description entry to 40 characters, numbers were included within each description. Description of the part following the number applies to all parts with the same number in the Short Description column. For example,

| Functional Name | Lineage Name | Brief Description |
|---|---|---|
| ADEL | AB plapaaaapp | Anterior deirids, sensory receptors |
| ADER | AB prapaaaapa | in lateral alae, contain dopamine |

becomes

| Parts List | Cell Lineage | Short Description |
|---|---|---|
| ADEL | AB plapaaaap | Anterior deirid 1 sensory receptor in |
| ADER | AB prapaaaapa | Anterior deirid 1 lateral alae-dopamine |

where all descriptions following the number apply to all parts with the same number. In this example, "sensory receptor in lateral alae-dopamine" would be the description for both anterior deirids containing the number 1. A hyphen indicates the beginning of the next descriptive element (as in the example above, "-dopamine"). The numbers can also be viewed as delimiters for descriptions common to a group. This method limits loss of information when rows are sorted.

*F.1.b.iv.* The Number of Cells column is derived from the multiplicity data provided in the Cell Lineage column of the reference. For example,

| Parts List | Cell Lineage | |
|---|---|---|
| int | E | X20 |

becomes

| Parts List | Cell Lineage | Number of Cells |
|---|---|---|
| int | E | 20. |

Note that the multiplicity value was moved to the Number of Cells column. In expanding the "L&R" postembryonic blast cell pairs discussed above, each cell was given a value of one (1).

### F.1.c. Comments

The format of APPENDIX 1 - Parts List of the reference above is shown in Figure 1.3, p. 4. The format of APPENDIX 3 - Part A: Lineage Charts of the reference above is shown in Figure 1.4, p. 4.

All commas in the .PRN file in Disk 1 were taken out in order to facilitate worksheet file exportation to and importation from text files.

IMPORTANT: Two (2) files with the .PRN extension were created for the PARTLIST file set, so that corresponding information can be seen side by side on the computer screen during text file manipulation. PARTLIS1.PRN has three (3) columns, namely, Parts List, Cell Lineage, and Short Description. PARTLIS2.PRN also has three columns, namely, Parts List, Short Description, and Number of Cells.

--- End of File ---

**F.2.** *PARTLIS1.PRN*

| Parts List | Cell Lineage | Short Description |
| --- | --- | --- |
| AB | P0 a | Embryonic founder cell |
| ADAL | AB plapaaaapp | Ring interneuron |
| ADAR | AB prapaaaapp | Ring interneuron |
| ADEL | AB plapaaaapa | Anterior deirid 1 sensory receptor in |
| ADER | AB prapaaaapa | Anterior deirid 1 lateral alae-dopamine |
| ADEshL | AB arppaaaa | Anterior deirid sheath cell |
| ADEshR | AB arpppaaa | Anterior deirid sheath cell |
| ADEsoL | H2L aa | Anterior deirid socket |
| ADEsoR | H2R aa | Anterior deirid socket |
| ADFL | AB alpppppaa | Amphid neuron 2 ciliated-chemosensory- |
| ADFR | AB praaappaa | Amphid neuron 2 commissure fm vent.gang |
| ADLL | AB alppppaad | Amphid neuron 3 ciliated-chemosensory- |
| ADLR | AB praaapaad | Amphid neuron 3 direct to ring |
| AFDL | AB alpppapav | Amphid finger cell 4 associated with |
| AFDR | AB praaaapav | Amphid finger cell 4 amphid sheath |
| AIAL | AB plppaappa | Amphid interneuron |
| AIAR | AB prppaappa | Amphid interneuron |
| AIBL | AB plaapappa | Amphid interneuron |
| AIBR | AB praapappa | Amphid interneuron |
| AIML | AB plpaapppa | Ring interneuron |
| AIMR | AB prpaapppa | Ring interneuron |
| AINL | AB alaaaalal | Ring interneuron |
| AINR | AB alaapaaar | Ring interneuron |
| AIYL | AB plpapaaap | Amphid interneuron |
| AIYR | AB prpapaaap | Amphid interneuron |
| AIZL | AB plapaaapav | Amphid interneuron |
| AIZR | AB prapaaapav | Amphid interneuron |
| ALA | AB alapppaaa | Neuron to excretory canal / dorsal cord |
| ALML | AB arppaappa | Touch receptor neuron 5 anterolateral |
| ALMR | AB arpppappa | Touch receptor neuron 5 microtubules |
| ALNL | AB plapappppap | Neuron assoc. with ALM 6 send processes |
| ALNR | AB prapappppap | Neuron assoc. with ALM 6 to tailspike |
| AMshL | AB plaapaapp | Amphid sheath cell |
| AMshR | AB praapaapp | Amphid sheath cell |
| AMsoL | AB plpaapapa | Amphid socket cell |
| AMsoR | AB prpaapapa | Amphid socket cell |
| AQR | QR ap | Neuron ciliate to ring |
| AS1 | P1 apa | Ventral cord motorneuron 7 innervate |

| Parts List | Cell Lineage | Short Description |
|---|---|---|
| AS2 | P2 apa | Ventral cord motorneuron 7 dorsal mus.- |
| AS3 | P3 apa | Ventral cord motorneuron 7 no ventral |
| AS4 | P4 apa | Ventral cord motorneuron 7 counterpart- |
| AS5 | P5 apa | Ventral cord motorneuron 7 cholinergic? |
| AS6 | P6 apa | Ventral cord motorneuron 7 like VAn but |
| AS7 | P7 apa | Ventral cord motorneuron 7 with added |
| AS8 | P8 apa | Ventral cord motorneuron 7 AVB input |
| AS9 | P9 apa | Ventral cord motorneuron 7 |
| AS10 | P10 apa | Ventral cord motorneuron 7 |
| AS11 | P11 apa | Ventral cord motorneuron 7 |
| ASEL | AB alpppppppaa | Amphid neuron 8 ciliated-chemosensory- |
| ASER | AB praaapppaa | Amphid neuron 8 to ring by commissure |
| ASGL | AB plaapapap | Amphid neuron 8 from ventral ganglion- |
| ASGR | AB praapapap | Amphid neuron 8 diverse connections in |
| ASHL | AB plpaappaa | Amphid neuron 8 ring neuropil |
| ASHR | AB prpaappaa | Amphid neuron 8 |
| ASIL | AB plaapappa | Amphid neuron 8 |
| ASIR | AB praapappa | Amphid neuron 8 |
| ASJL | AB alpppppppa | Amphid neuron 8 |
| ASJR | AB praaapppa | Amphid neuron 8 |
| ASKL | AB alpppappa | Amphid neuron 8 |
| ASKR | AB praaaappa | Amphid neuron 8 |
| AUAL | AB alppppppppp | Neuron runs with amphid neuron 9 lack |
| AUAR | AB praaappppp | Neuron runs with amphid neuron 9 cilia |
| AVAL | AB alppaaapa | Ventral cord interneuron 10 to VA & |
| AVAR | AB alaappapa | Ventral cord interneuron 10 DA & AS |
| AVBL | AB plpaapaap | Ventral cord interneuron 11 to VB & |
| AVBR | AB prpaapaap | Ventral cord interneuron 11 DB & AS |
| AVDL | AB alaaapalr | Ventral cord interneuron 12 to VA & |
| AVDR | AB alaaapprl | Ventral cord interneuron 12 DA & AS |
| AVEL | AB alpppaaaa | Ventral cord interneuron 13 like AVD in |
| AVER | AB praaaaaaa | Ventral cord interneuron 13 ant. cord |
| AVFL | P1 aaaa | Interneuron in ventral cord 14 & ring- |
| AVFR | W aaa | Interneuron in ventral cord 14 edited |
| AVG | AB prpapppap | Ventral cord interneuron to tailspike |
| AVHL | AB alapaaaaa | Neuron 15 presynaptic in ring & |
| AVHR | AB alappapaa | Neuron 15 postsynaptic in ventral cord |
| AVJL | AB alapapppa | Neuron 16 postsynaptic in ventral cord |
| AVJR | AB alapppppa | Neuron 16 & presynaptic in ring |
| AVKL | AB plpapapap | Ring & ventral cord interneuron |

| Parts List | Cell Lineage | Short Description |
|---|---|---|
| AVKR | AB prpapapap | Ring & ventral cord interneuron |
| AVL | AB prpappaap | Ring & vent.cord interneuron/motoneuron |
| AVM | QR paa | Microtubule cell touch receptor |
| AWAL | AB plaapapaa | Amphid wing neuron 17 ciliated-sensory- |
| AWAR | AB praapapaa | Amphid wing neuron 17 assoc. with |
| AWBL | AB alpppppap | Amphid wing neuron 17 amphid sheath |
| AWBR | AB praaappap | Amphid wing neuron 17 |
| AWCL | AB plpaaaaap | Amphid wing neuron 17 |
| AWCR | AB prpaaaaap | Amphid wing neuron 17 |
| B | AB prpppappa | Rectal cell -postemb.blast cell in male |
| BAGL | AB alppappap | Neuron 18 ciliated in head-not part of |
| BAGR | AB arappppap | Neuron 18 sensillium-assoc. with ILso |
| BDUL | AB arppaappp | Neuron 19 along excretory canal & also |
| BDUR | AB arpppappp | Neuron 19 in nerve ring-dark vesicles |
| C | PO ppa | Embryonic founder cell |
| CA1 | P3 aapa | Male cell - not reconstructed |
| CA2 | P4 aapa | Male cell - not reconstructed |
| CA3 | P5 aapa | Male cell - not reconstructed |
| CA4 | P6 aapa | Male neuron to dorsal muscles |
| CA5 | P7 aapa | Male neuron to dorsal muscles |
| CA6 | P8 aapa | Male neuron to dorsal muscles |
| CA7 | P9 aapa | Male neuron to dorsal muscles |
| CA8 | P10 aapa | Male neuron? 20 in ventral cord-neuron |
| CA9 | P11 aapa | Male neuron? 20 like but lacks synapses |
| CANL | AB alapaaapa | Neuron 21 along excretory canal-no |
| CANR | AB alappappa | Neuron 21 synapses-essential for life |
| CEMDL | AB plaaaaaap | Male cephalic neuron 22 die in hermaph. |
| CEMDR | AB arpapaaap | Male cephalic neuron 22 open to outside |
| CEMVL | AB plpaapapp | Male cephalic neuron 22 sex chemotaxis? |
| CEMVR | AB prpaapapp | Male cephalic neuron 22 |
| CEPDL | AB plaaaappa | Neuron cephalic sensillum dopamine |
| CEPDR | AB arpapaappa | Neuron cephalic sensillum dopamine |
| CEPVL | AB plpaappppa | Neuron cephalic sensillum dopamine |
| CEPVR | AB prpaappppa | Neuron cephalic sensillum dopamine |
| CEPshDL | AB arpaaapp | Cephalic sheath cell 23 sheet-like |
| CEPshDR | AB arpaaapap | Cephalic sheath cell 23 processes |
| CEPshVL | AB plpaaapap | Cephalic sheath cell 23 envelop ring |
| CEPshVR | AB prpaaapap | Cephalic sheath cell 23 neuropil & |
| CEPsoDL | AB alapapppp | Cephalic socket cell 23 part of ventral |
| CEPsoDR | AB alapppppp | Cephalic socket cell 23 ganglion |

| Parts List | Cell Lineage | Short Description |
| --- | --- | --- |
| CEPsoVL | AB alppaappp | Cephalic socket cell 23 |
| CEPsoVR | AB alaapappp | Cephalic socket cell 23 |
| CP0 | P2 aap | Male ventral cord - not reconstructed |
| CP1 | P3 aapp | Male ventral cord - not reconstructed |
| CP2 | P4 aapp | Male ventral cord - not reconstructed |
| CP3 | P5 aapp | Male ventral cord - not reconstructed |
| CP4 | P6 aapp | Male motorneuron in ventral cord |
| CP5 | P7 aapp | Male motorneuron in ventral cord |
| CP6 | P8 aapp | Male motorneuron in ventral cord |
| CP7 | P9 aapp | Male motorneuron in ventral cord |
| CP8 | P10 aapp | Male interneuron to preanal ganglion |
| CP9 | P11 aapp | Male interneuron to preanal ganglion |
| D | P0 pppa | Embryonic founder cell |
| DA1 | AB prppapaap | Ventral cord motorneuron 24 to dorsal |
| DA2 | AB plppapapa | Ventral cord motorneuron 24 muscles- |
| DA3 | AB prppapapa | Ventral cord motorneuron 24 cholinergic |
| DA4 | AB plppapapp | Ventral cord motorneuron 24 |
| DA5 | AB prppapapp | Ventral cord motorneuron 24 |
| DA6 | AB plpppaaap | Ventral cord motorneuron 24 |
| DA7 | AB prpppaaap | Ventral cord motorneuron 24 |
| DA8 | AB prpapappp | Ventral cord motorneuron 24 |
| DA9 | AB plpppaaaa | Ventral cord motorneuron 24 |
| DB1 | AB plpaaaapp | Ventral cord motorneuron 25 to dorsal |
| DB2 | AB arappappa | Ventral cord motorneuron 25 muscles- |
| DB3 | AB prpaaaapp | Ventral cord motorneuron 25 cholinergic |
| DB4 | AB prpappapp | Ventral cord motorneuron 25 |
| DB5 | AB plpapappp | Ventral cord motorneuron 25 |
| DB6 | AB plppaappp | Ventral cord motorneuron 25 |
| DB7 | AB prppaappp | Ventral cord motorneuron 25 |
| DD1 | AB plppappap | Ventral cord motorneuron 26 reciprocal |
| DD2 | AB prppappap | Ventral cord motorneuron 26 inhibitor- |
| DD3 | AB plppapppa | Ventral cord motorneuron 26 change |
| DD4 | AB prppapppa | Ventral cord motorneuron 26 pattern of |
| DD5 | AB plppapppp | Ventral cord motorneuron 26 synapses- |
| DD6 | AB prppapppp | Ventral cord motorneuron 26 GABA |
| DVA | AB prppppapp | Ring interneuron -large vesicle in ring |
| DVB | K p | Ring interneuron -to rectal muscles |
| DVC | C aapaa | Ring interneuron |
| DVE | B ppap | Male specific neuron 27 cell body in |
| DVF | B pppppa | Male specific neuron 27 preanal gang.- |

| Parts List | Cell Lineage | Short Description |
|---|---|---|
| DX1 | F lvda | Male specific neuron 27 penetrate |
| DX2 | F rvda | Male specific neuron 27 basement memb. |
| DX3 | U laa | Male specific neuron 27 & contact |
| DX4 | U raa | Male specific neuron 27 muscles |
| E | P0 pap | Embryonic founder cell |
| EF1 | F lvdp | Male specific neuron 28 in preanal gang |
| EF2 | F rvdp | Male specific neuron 28 -input fm ray |
| EF3 | U lap | Male specific neuron 28 neurons |
| EF4 | U rap | Male specific neuron 28 |
| F | AB plppppapp | Rectal cell -postemb.blast cell in male |
| FLPL | AB plapaaapad | Neuron 29 ciliated in head-not part of |
| FLPR | AB prapaaapad | Neuron 29 sensillium-assoc. with ILso |
| G1 | AB prpaaaapa | Postemb.blast cell 30 excretory socket- |
| G2 | AB plapaapa | Postemb.blast cell 30 G1(emb)G2(L1)G2.p |
| GLRDL | MS aaaaaal | Glial cell 30 sheet bet. pharynx & ring |
| GLRDR | MS aaaaaar | Glial cell 30 neuropil-no chemical |
| GLRL | MS apaaaad | Glial cell 30 synapses-GJs with muscles |
| GLRR | MS ppaaaad | Glial cell 30 & RME-processes to head |
| GLRVL | MS apaaaav | Glial cell 30 |
| GLRVR | MS ppaaaav | Glial cell 30 |
| H0L | AB plaaappa | Seam hypodermal cell |
| H0R | AB arpapppa | Seam hypodermal cell |
| H1L | AB plaaappp | Seam hypodermal cell postemb. blast |
| H1R | AB arpapppp | Seam hypodermal cell postemb. blast |
| H2L | AB arppaaap | Seam hypodermal cell postemb. blast |
| H2R | AB arpppaap | Seam hypodermal cell postemb. blast |
| HOA | P10 pppa | Neuron male hook sensillum |
| HOB | P10 ppap | Neuron male hook sensillum |
| HOsh | P10 ppppp | Sheath male hook sensillum |
| HOso | P10 ppaa | Socket male hook sensillum |
| HSNL | AB plapppappa | Hermaph.motoneuron vulval mus.serotonin |
| HSNR | AB prapppappa | Hermaph.motoneuron vulval mus.serotonin |
| I1L | AB alpapppaa | Pharyngeal interneuron -sensory fm RIP |
| I1R | AB arapappaa | Pharyngeal interneuron -sensory fm RIP |
| I2L | AB alpappaapa | Pharyngeal interneuron -ant. sensory |
| I2R | AB arapapaapa | Pharyngeal interneuron -ant. sensory |
| I3 | MS aaaaapaa | Pharyngeal interneuron -ant. sensory |
| I4 | MS aaaapaa | Pharyngeal interneuron |
| I5 | AB arapapapp | Pharyngeal interneuron -post. sensory |
| I6 | MS paaapaa | Pharyngeal interneuron -post. sensory |

| Parts List | Cell Lineage | Short Description |
|---|---|---|
| IL1DL | AB alapappaaa | Neuron inner labial sensorimotor |
| IL1DR | AB alappppaaa | Neuron inner labial sensorimotor |
| IL1L | AB alapaappaa | Neuron inner labial sensorimotor |
| IL1R | AB alaappppaa | Neuron inner labial sensorimotor |
| IL1VL | AB alppapppaa | Neuron inner labial sensorimotor |
| IL1VR | AB arapppppaa | Neuron inner labial sensorimotor |
| IL2DL | AB alapappap | Neuron inner labial sensory open out |
| IL2DR | AB alappppap | Neuron inner labial sensory open out |
| IL2L | AB alapaappp | Neuron inner labial sensory open out |
| IL2R | AB alaappppp | Neuron inner labial sensory open out |
| IL2VL | AB alppapppp | Neuron inner labial sensory open out |
| IL2VR | AB arapppppp | Neuron inner labial sensory open out |
| ILshDL | AB alaaaparr | Inner labial sheath cell |
| ILshDR | AB alaaapll | Inner labial sheath cell |
| ILshL | AB alaaaalpp | Inner labial sheath cell |
| ILshR | AB alaapaapp | Inner labial sheath cell |
| ILshVL | AB alppapaap | Inner labial sheath cell |
| ILshVR | AB arapppaap | Inner labial sheath cell |
| ILsoDL | AB plaapaaap | Inner labial socket cell -tip of labium |
| ILsoDR | AB praapaaap | Inner labial socket cell -tip of labium |
| ILsoL | AB alaaapall | Inner labial socket cell -tip of labium |
| ILsoR | AB alaaapprr | Inner labial socket cell -tip of labium |
| ILsoVL | AB alppapapp | Inner labial socket cell -tip of labium |
| ILsoVR | AB arapppapp | Inner labial socket cell -tip of labium |
| K | AB plpapppaa | Rectal cell postemb. blast cell |
| K prime | AB plpapppap | Rectal cell |
| LUAL | AB plpppaapap | Interneuron posterior ventral cord |
| LUAR | AB prpppaapap | Interneuron posterior ventral cord |
| M | MS apaapp | Postembryonic mesoblast |
| M1 | MS paapaaa | Pharyngeal motorneuron |
| M2L | AB araapappa | Pharyngeal motorneuron |
| M2R | AB araappppa | Pharyngeal motorneuron |
| M3L | AB araapappp | Pharyngeal sensory-motorneuron |
| M3P | AB araappppp | Pharyngeal sensory-motorneuron |
| M4 | MS paaaaaa | Pharyngeal motorneuron |
| M5 | MS paaapap | Pharyngeal motorneuron |
| MCL | AB alpaaappp | Pharyngeal neuron to marginal cell |
| MCR | AB arapaappp | Pharyngeal neuron to marginal cell |
| MI | AB araappaaa | Pharyngeal motorneuron-interneuron |
| MS | P0 paa | Embryonic founder cell |

| Parts List | Cell Lineage | Short Description |
| --- | --- | --- |
| NSML | AB araapapaav | Pharyngeal motoneuron 31 neurosecretory |
| NSMR | AB araapppaav | Pharyngeal motoneuron 31 -serotonin |
| OLLL | AB alppppapaa | Neuron outer labial sensillum |
| OLLR | AB praaapapaa | Neuron outer labial sensillum |
| OLLshL | AB alpppaapd | Lateral outer labial sheath cell |
| OLLshR | AB praaaaapd | Lateral outer labial sheath cell |
| OLLsoL | AB alapaaapp | Lateral outer labial socket cell |
| OLLsoR | AB alappappp | Lateral outer labial socket cell |
| OLQDL | AB alapapapaa | Neuron outer labial sensilla |
| OLQDR | AB alapppapaa | Neuron outer labial sensilla |
| OLQVL | AB plpaaappaa | Neuron outer labial sensilla |
| OLQVR | AB prpaaappaa | Neuron outer labial sensilla |
| OLQshDL | AB arpaaaapa | Quadrant outer labial sheath cell |
| OLQshDR | AB arpaaapaa | Quadrant outer labial sheath cell |
| OLQshVL | AB alpppaaap | Quadrant outer labial sheath cell |
| OLQshVR | AB praaaaaap | Quadrant outer labial sheath cell |
| OLQsoDL | AB arpaaaaal | Quadrant outer labial socket cell |
| OLQsoDR | AB arpaaaaar | Quadrant outer labial socket cell |
| OLQsoVL | AB alppaaapp | Quadrant outer labial socket cell |
| OLQsoVR | AB alaappapp | Quadrant outer labial socket cell |
| P0 |  | The single-cell zygote |
| P04 | P0 pppp | Embryonic founder cell-germ line |
| P1 | AB plapaapp | Postemb. blast cell 32 for ventral |
| P2 | AB prapaapp | Postemb. blast cell 32 cord & ventral |
| P3 | AB plappaaa | Postemb. blast cell 32 hypodermis & |
| P4 | AB prappaaa | Postemb. blast cell 32 vulva & male |
| P5 | AB plappaap | Postemb. blast cell 32 preanal gang. - |
| P6 | AB prappaap | Postemb. blast cell 32 ventral hypoder- |
| P7 | AB plappapp | Postemb. blast cell 32 mis in L1 |
| P8 | AB prappapp | Postemb. blast cell 32 |
| P9 | AB plapapap | Postemb. blast cell 32 |
| P10 | AB prapapap | Postemb. blast cell 32 |
| P11 | AB plapappa | Postemb. blast cell 32 |
| P12 | AB prapappa | Postemb. blast cell 32 |
| PCAL | Y plppd | Sensory neuron male postcloac. sensil. |
| PCAR | Y prppd | Sensory neuron male postcloac. sensil. |
| PCBL | Y plpa | Neuron in sheath male postcloac.sensil. |
| PCBR | Y prpa | Neuron in sheath male postcloac.sensil. |
| PCCL | B arpaaa | Sensory neuron male postcloac. sensil. |
| PCCR | B alpaaa | Sensory neuron male postcloac. sensil. |

| Parts List | Cell Lineage | Short Description |
|---|---|---|
| PChL | Y plaa | Hypodermal cell male postcloac. sensil. |
| PChR | Y praa | Hypodermal cell male postcloac. sensil. |
| PCshL | Y plppv | Sheath cell male postcloac. sensil. |
| PCshR | Y prppv | Sheath cell male postcloac. sensil. |
| PCsoL | Y plap | Socket cell male postcloac. sensil. |
| PCsoR | Y prap | Socket cell male postcloac. sensil. |
| PDA herm | Y | Motorneuron preanal ganglion herm. |
| PDA male | Y a | Motorneuron preanal ganglion male |
| PDB | P12 apa | Interneuron-motorneuron post.vent.cord |
| PDC | P11 papa | Interneuron preanal ganglion male |
| PDEL | V5L paaa | Neuron dopaminergic postdeirid |
| PDER | V5R paaa | Neuron dopaminergic postdeirid |
| PDEshL | V5L papp | Sheath sensillum postdeirid dopamine |
| PDEshR | V5R papp | Sheath sensillum postdeirid dopamine |
| PDEsoL | V5L papa | Socket sensillum postdeirid dopamine |
| PDEsoR | V5R papa | Socket sensillum postdeirid dopamine |
| PGA | P11 papp | Interneuron male preanal ganglion |
| PHAL | AB plpppaapp | Phasmid neuron ciliated chemosensory |
| PHAR | AB prpppaapp | Phasmid neuron ciliated chemosensory |
| PHBL | AB plapppappp | Phasmid neuron ciliated chemosensory |
| PHBR | AB prapppappp | Phasmid neuron ciliated chemosensory |
| PHCL | TL pppaa | Neuron male tail spike |
| PHCR | TR pppaa | Neuron male tail spike |
| PHshL | AB plpppapaa | Phasmid sheath cell |
| PHshR | AB prpppapaa | Phasmid sheath cell |
| PHso1L | TL paa | Phasmid socket cell |
| PHso1R | TR paa | Phasmid socket cell |
| PHso2L | TL pap | Phasmid socket cell |
| PHso2R | TR pap | Phasmid socket cell |
| PLML | AB plapappppaa | Neuron 33 posterolateral microtubule- |
| PLMR | AB prapappppaa | Neuron 33 touch receptor |
| PLNL | TL pppap | Interneuron with PLM to tailspike |
| PLNR | TR pppap | Interneuron with PLM to tailspike |
| PQR | QL ap | Neuron cilium preanal ganglion |
| PVCL | AB plpppaapaa | Vent.cord interneuron 34 in lumbar gang |
| PVCR | AB prpppaapaa | Vent.cord interneuron 34 -to VB & DB |
| PVDL | V5L paapa | Neuron near excretory canal |
| PVDR | V5R paapa | Neuron near excretory canal |
| PVM | QL paa | Post.vent. microtubule touch receptor |
| PVNL | TL appp | Interneuron to ring and vulva |

| Parts List | Cell Lineage | Short Description |
|---|---|---|
| PVNR | TR appp | Interneuron to ring and vulva |
| PVPL | AB plppppaaa | Interneuron along ventral cord to ring |
| PVPR | AB prppppaaa | Interneuron along ventral cord to ring |
| PVQL | AB plapppaaa | Interneuron along ventral cord to ring |
| PVQR | AB prapppaaa | Interneuron 35 along ventral cord to |
| PVR | C aappa | Interneuron 35 ring & tailspike |
| PVT | AB plpappppa | Interneuron ventral cord |
| PVV | P11 paaa | Motorneuron male ventral cord |
| PVWL | TL ppa | Interneuron posterior ventral cord |
| PVWR | TR ppa | Interneuron posterior ventral cord |
| PVX | P12 aap | Male interneuron 36 to ring |
| PVY | P11 paap | Male interneuron 36 & ventral cord |
| PVZ | P10 pppppa | Male motorneuron ventral cord |
| QL | AB plapapaaa | Postemb.neuroblast migrates anteriorly |
| QR | AB prapapaaa | Postemb.neuroblast migrates posteriorly |
| R1AL | V5L pppppaaa | Neuron male sensory ray type A |
| R1AR | V5R pppppaaa | Neuron male sensory ray type A |
| R1BL | V5L pppppapa | Neuron male sensory ray type B |
| R1BR | V5R pppppapa | Neuron male sensory ray type B |
| R1stL | V5L pppppapp | Structural cell male sensory ray |
| R1stR | V5R pppppapp | Structural cell male sensory ray |
| R2AL | V6L papapaaa | Neuron male sensory ray |
| R2AR | V6R papapaaa | Neuron male sensory ray |
| R2BL | V6L papapapa | Neuron male sensory ray |
| R2BR | V6R papapapa | Neuron male sensory ray |
| R2stL | V6L papapapp | Structural cell male sensory ray |
| R2stR | V6R papapapp | Structural cell male sensory ray |
| R3AL | V6L papppaaa | Neuron male sensory ray |
| R3AR | V6R papppaaa | Neuron male sensory ray |
| R3BL | V6L papppapa | Neuron male sensory ray |
| R3BR | V6R papppapa | Neuron male sensory ray |
| R3stL | V6L papppapp | Structural call male sensory ray |
| R3stR | V6R papppapp | Structural cell male sensory ray |
| R4AL | V6L pppapaaa | Neuron male sensory ray type A |
| R4AR | V6R pppapaaa | Neuron male sensory ray type A |
| R4BL | V6L pppapapa | Neuron male sensory ray type B |
| R4BR | V6R pppapapa | Neuron male sensory ray type B |
| R4stL | V6L pppapapp | Structural cell male sensory ray |
| R4stR | V6R pppapapp | Structural cell male sensory ray |
| R5AL | V6L pppppaaa | Neuron male sensory type A dopamine |

| Parts List | Cell Lineage | Short Description |
| --- | --- | --- |
| R5AR | V6R pppppaaa | Neuron male sensory type A dopamine |
| R5BL | V6L pppppapa | Neuron male sensory type B |
| R5BR | V6R pppppapa | Neuron male sensory type B |
| R5stL | V6L pppppapp | Structural cell male sensory ray |
| R5stR | V6R pppppapp | Structural cell male sensory ray |
| R6AL | V6L ppppaaaa | Neuron male sensory ray type A |
| R6AR | V6R ppppaaaa | Neuron male sensory ray type A |
| R6BL | V6L ppppaapa | Neuron male sensory ray |
| R6BR | V6R ppppaapa | Neuron male sensory ray |
| R6stL | V6L ppppaapp | Structural call male sensory ray |
| R6stR | V6R ppppaapp | Structural cell male sensory ray |
| R7AL | TL apappaaa | Neuron male sensory type A dopamine |
| R7AR | TR apappaaa | Neuron male sensory type A dopamine |
| R7BL | TL apappapa | Neuron male sensory ray type B |
| R7BR | TR apappapa | Neuron male sensory ray type B |
| R7stL | TL apappapp | Structural cell male sensory ray |
| R7stR | TR apappapp | Structural cell male sensory ray |
| R8AL | TL appaaaaa | Neuron male sensory ray |
| R8AR | TR appaaaaa | Neuron male sensory ray |
| R8BL | TL appaaapa | Neuron male sensory ray |
| R8BR | TR appaaapa | Neuron male sensory ray |
| R8stL | TL appaaapp | Structural cell male sensory ray |
| R8stR | TR appaaapp | Structrual cell male sensory ray |
| R9AL | TL appapaaa | Neuron male sensory type A dopamine |
| R9AR | TR appapaaa | Neuron male sensory type A dopamine |
| R9BL | TL appapapa | Neuron male sensory ray |
| R9BR | TR appapapa | Neuron male sensory ray |
| R9stL | TL appapapp | Structural cell male sensory ray |
| R9stR | TR appapapp | Structural cell male sensory ray |
| RIAL | AB alapaapaa | Interneuron ring to RMD and SMD |
| RIAR | AB alaapppaa | Interneuron ring to RMD and SMD |
| RIBL | AB plpaappap | Interneuron ring |
| RIBR | AB prpaappap | Interneuron ring |
| RICL | AB plppaaaapp | Interneuron ring |
| RICR | AB prppaaaapp | Interneuron ring |
| RID | AB alappaapa | Motorneuron dorsal cord and muscles |
| RIFL | AB plppapaaap | Ring interneuron |
| RIFR | AB prppapaaap | Ring interneuron |
| RIGL | AB plppappaa | Ring interneuron |
| RIGR | AB prppappaa | Ring interneuron |

| Parts List | Cell Lineage | Short Description |
|---|---|---|
| RIH | AB prpappaaa | Ring interneuron |
| RIML | AB plppaapap | Ring motorneuron/interneuron |
| RIMR | AB prppaapap | Ring motorneuron/interneuron |
| RIPL | AB alpapaaaa | Ring/pharynx interneuron 37 direct con. |
| RIPR | AB arappaaaa | Ring/pharynx interneuron 37 no synapses |
| RIR | AB prpapppaa | Ring interneuron |
| RIS | AB prpappapa | Ring interneuron |
| RIVL | AB plpaapaaa | Ring interneuron |
| RIVR | AB prpaapaaa | Ring interneuron |
| RMDDL | AB alpapapaa | Ring motorneuron/interneuron 38 many |
| RMDDR | AB arappapaa | Ring motorneuron/interneuron 38 synap- |
| RMDL | AB alpppapad | Ring motorneuron/interneuron 38 ses-re- |
| RMDR | AB praaaapad | Ring motorneuron/interneuron 38 cipro- |
| RMDVL | AB alppapaaa | Ring motorneuron/interneuron 38 cal in- |
| RMDVR | AB arapppaaa | Ring motorneuron/interneuron 38 hibitor? |
| RMED | AB alapppaap | Ring motorneuron GABA |
| RMEL | AB alaaaarlp | Ring motorneuron GABA |
| RMER | AB alaaaarrp | Ring motorneuron GABA |
| RMEV | AB plpappaaa | Ring motorneuron GABA |
| RMFL | G2 al | Ring motorneuron/interneuron |
| RMFR | G2 ar | Ring motorneuron/interneuron |
| RMGL | AB plapaaapp | Ring motorneuron/interneuron |
| RMGR | AB prapaaapp | Ring motorneuron/interneuron |
| RMHL | G1 l | Ring motorneuron/interneuron |
| RMHR | G1 r | Ring motorneuron/interneuron |
| SAADL | AB alppapapa | Ring interneuron anteriorly directed |
| SAADR | AB arapppapa | Ring interneuron anteriorly directed |
| SAAVL | AB plpaaaaaa | Ring interneuron anteriorly directed |
| SAAVR | AB prpaaaaaa | Ring interneuron anteriorly directed |
| SABD | AB plppapaap | Ring interneuron 39 ant. directed- |
| SABVL | AB plppapaaaa | Ring interneuron 39 motor in L1 |
| SABVR | AB prppapaaaa | Ring interneuron 39 |
| SDQL | QL pap | Interneuron posterior lateral to ring |
| SDQR | QR pap | Interneuron anterior lateral to ring |
| SIADL | AB plpapaapa | Interneuron posteriorly directed |
| SIADR | AB prpapaapa | Interneuron posteriorly directed |
| SIAVL | AB plpapappa | Interneuron posteriorly directed |
| SIAVR | AB prpapappa | Interneuron posteriorly directed |
| SIBDL | AB plppaaaaa | Interneuron |
| SIBDR | AB prppaaaaa | Interneuron |

| Parts List | Cell Lineage | Short Description |
|---|---|---|
| SIBVL | AB plpapaapp | Interneuron |
| SIBVR | AB prpapaapp | Interneuron |
| SMBDL | AB alpapapapp | Ring motorneuron/interneuron posterior |
| SMBDR | AB arappapapp | Ring motorneuron/interneuron posterior |
| SMBVL | AB alpapappp | Ring motorneuron/interneuron posterior |
| SMBVR | AB arappappp | Ring motorneuron/interneuron posterior |
| SMDDL | AB plpapaaaa | Ring motorneuron/interneuron posterior |
| SMDDR | AB prpapaaaa | Ring motorneuron/interneuron posterior |
| SMDVL | AB alppappaa | Ring motorneuron/interneuron posterior |
| SMDVR | AB arappppaa | Ring motorneuron/interneuron posterior |
| SPCL | B alpaap | Sensory/motorneuron male 40 to spicule |
| SPCR | B arpaap | Sensory/motorneuron male 40 protractor |
| SPDL | B alpapaa | Sensory neuron male 41 copulatory spi- |
| SPDR | B arpapaa | Sensory neuron male 41 cule-open to out |
| SPVL | B a(l/r)aalda | Sensory neuron male 41 at spicule tip |
| SPVR | B a(l/r)aarda | Sensory neuron male 41 |
| SPshDL | B alpapap | Sheath cell male copulatory spicule |
| SPshDR | B arpapap | Sheath cell male copulatory spicule |
| SPshVL | B a(l/r)aaldp | Sheath cell male copulatory spicule |
| SPshVR | B a(l/r)aardp | Sheath cell male copulatory spicule |
| SPso1L | B a(l/r)pppl | Socket cell male copulatory spicule |
| SPso1R | B a(l/r)pppr | Socket cell male copulatory spicule |
| SPso2L | B a(l/r)aald | Socket cell male copulatory spicule |
| SPso2R | B a(l/r)aard | Socket cell male copulatory spicule |
| SPso3L | B a(l/r)aalv | Socket cell male copulatory spicule |
| SPso3R | B a(l/r)aarv | Socket cell male copulatory spicule |
| SPso4L | B alpapp | Socket cell male copulatory spicule |
| SPso4R | B arpapp | Socket cell male copulatory spicule |
| TL | AB plappppp | Tail seam hypoderm.42 cell postemb.- |
| TR | AB prappppp | Tail seam hypoderm.42 phasmid socket L1 |
| U | AB plppppapa | Rectal cell postemb blast cell male |
| URADL | AB plaaaaaaa | Ring motorneuron 43 non-ciliated head- |
| URADR | AB arpapaaaa | Ring motorneuron 43 with OLQ in embryo |
| URAVL | AB plpaaapaa | Ring motorneuron 43 |
| URAVR | AB prpaaapaa | Ring motorneuron 43 |
| URBL | AB plaapaapa | Neuron 44 presyn. in ring-nonciliated |
| URBR | AB praapaapa | Neuron 44 head-with OLL in embryo |
| URXL | AB plaaaaappp | Ring interneuron 45 non-ciliated head- |
| URXR | AB arpapaappp | Ring interneuron 45 with CEPD in embryo |
| URYDL | AB alapapapp | Neuron 46 presyn. in ring-nonciliated |

| Parts List | Cell Lineage | Short Description |
|---|---|---|
| URYDR | AB alapppapp | Neuron 46 head-with OLQ in embryo |
| URYVL | AB plpaaappp | Neuron 46 |
| URYVR | AB prpaaappp | Neuron 46 |
| V1L | AB arppapaa | Seam hypodermal cell postemb blast |
| V1R | AB arppppaa | Seam hypodermal cell postemb blast |
| V2L | AB arppapap | Seam hypodermal cell postemb blast |
| V2R | AB arppppap | Seam hypodermal cell postemb blast |
| V3L | AB plappapa | Seam hypodermal cell postemb blast |
| V3R | AB prappapa | Seam hypodermal cell postemb blast |
| V4L | AB arppappa | Seam hypodermal cell postemb blast |
| V4R | AB arpppppa | Seam hypodermal cell postemb blast |
| V5L | AB plapapaap | Seam hypodermal cell postemb blast |
| V5R | AB prapapaap | Seam hypodermal cell postemb blast |
| V6L | AB arppappp | Seam hypodermal cell postemb blast |
| V6R | AB arpppppp | Seam hypodermal cell postemb blast |
| VA1 | W pa | Ventral cord motorneuron ventral mus. |
| VA2 | P2 aaaa | Ventral cord motorneuron ventral mus. |
| VA3 | P3 aaaa | Ventral cord motorneuron ventral mus. |
| VA4 | P4 aaaa | Ventral cord motorneuron ventral mus. |
| VA5 | P5 aaaa | Ventral cord motorneuron ventral mus. |
| VA6 | P6 aaaa | Ventral cord motorneuron ventral mus. |
| VA7 | P7 aaaa | Ventral cord motorneuron ventral mus. |
| VA8 | P8 aaaa | Ventral cord motorneuron ventral mus. |
| VA9 | P9 aaaa | Ventral cord motorneuron ventral mus. |
| VA10 | P10 aaaa | Ventral cord motorneuron ventral mus. |
| VA11 | P11 aaaa | Ventral cord motorneuron ventral mus. |
| VA12 | P12 aaaa | Ventral cord moto/interneuron preanal g |
| VB1 | P1 aaap | Ventral cord motoneuron vent.mus. + ring |
| VB2 | W aap | Ventral cord motorneuron ventral mus. |
| VB3 | P2 aaap | Ventral cord motorneuron ventral mus. |
| VB4 | P3 aaap | Ventral cord motorneuron ventral mus. |
| VB5 | P4 aaap | Ventral cord motorneuron ventral mus. |
| VB6 | P5 aaap | Ventral cord motorneuron ventral mus. |
| VB7 | P6 aaap | Ventral cord motorneuron ventral mus. |
| VB8 | P7 aaap | Ventral cord motorneuron ventral mus. |
| VB9 | P8 aaap | Ventral cord motorneuron ventral mus. |
| VB10 | P9 aaap | Ventral cord motorneuron ventral mus. |
| VB11 | P10 aaap | Ventral cord motorneuron ventral mus. |
| VC1 | P3 aap | Motorneuron vulval & body mus. hermaph. |
| VC2 | P4 aap | Motorneuron vulval & body mus. hermaph. |

| Parts List | Cell Lineage | Short Description |
|---|---|---|
| VC3 | P5 aap | Motorneuron vulval & body mus. hermaph. |
| VC4 | P6 aap | Motorneuron vulval & body mus. hermaph. |
| VC5 | P7 aap | Motorneuron vulval & body mus. hermaph. |
| VC6 | P8 aap | Motorneuron vulval & body mus. hermaph. |
| VD1 | W pp | Ventral cord motorneuron 47 vent. body |
| VD2 | P1 app | Ventral cord motorneuron 47 mus.-recip- |
| VD3 | P2 app | Ventral cord motorneuron 47 rocal inhi- |
| VD4 | P3 app | Ventral cord motorneuron 47 bitor?-GABA |
| VD5 | P4 app | Ventral cord motorneuron 47 |
| VD6 | P5 app | Ventral cord motorneuron 47 |
| VD7 | P6 app | Ventral cord motorneuron 47 |
| VD8 | P7 app | Ventral cord motorneuron 47 |
| VD9 | P8 app | Ventral cord motorneuron 47 |
| VD10 | P9 app | Ventral cord motorneuron 47 |
| VD11 | P10 app | Ventral cord motorneuron 47 |
| VD12 | P11 app | Ventral cord motorneuron 47 |
| VD13 | P12 app | Ventral cord motorneuron 47 |
| W | AB prapaapa | Postemb neuroblast analogous to Pn.a |
| XXXL | AB plaaapaa | Embryonic head hypodermal cell |
| XXXR | AB arpappaa | Embryonic head hypodermal cell |
| Y | AB prpppaaaa | Rectal cell PDA/hermaph.-blast/male |
| Z1 | MS pppaap | Somatic gonad precursor cell |
| Z2 | P04 p | Germ line precursor cell |
| Z3 | P04 a | Germ line precursor cell |
| Z4 | MS appaap | Somatic gonad precursor cell |
| arc ant | AB | Cell 48 between pharynx and hypodermis- |
| arc post | AB | Cell 48 anterior part of buccal cavity |
| cc herm DL | M dlpa | Postemb coelomocyte hermaphrodite |
| cc herm DR | M drpa | Postemb coelomocyte hermaphrodite |
| cc male D | M dlpappp | Postemb coelomocyte male single |
| ccAL | MS apapaaa | Embryonic coelomocyte |
| ccAR | MS ppapaaa | Embryonic coelomocyte |
| ccPL | MS apapaap | Embryonic coelomocyte |
| ccPR | MS ppapaap | Embryonic coelomocyte |
| e1D | AB araaaapap | Pharyngeal epithelial cell |
| e1VL | AB araaaaaaa | Pharyngeal epithelial cell |
| e1VR | AB araaaaapa | Pharyngeal epithelial cell |
| e2DL | AB alpaapaap | Pharyngeal epithelial cell |
| e2DR | AB araaapaap | Pharyngeal epithelial cell |
| e2V | AB alpappapa | Pharyngeal epithelial cell |

| Parts List | Cell Lineage | Short Description |
|---|---|---|
| e3D | AB araapaaaa | Pharyngeal epithelial cell |
| e3VL | AB alpaaaaaa | Pharyngeal epithelial cell |
| e3VR | AB arapaaaaa | Pharyngeal epithelial cell |
| exc cell | AB plpappaap | H-shaped excretory cell |
| exc duct | AB plpaaaapa | Excretory duct |
| exc gl L | AB plpapapaa | Excretory gland fused open to exc duct |
| exc gl R | AB prpapapaa | Excretory gland fused open to exc duct |
| exc socket | G2 p | Excretory socket at duct to hypodermis |
| g1AL | MS aapaapaa | Pharyngeal gland cell |
| g1AR | MS papaapaa | Pharyngeal gland cell |
| g1P | MS aaaaapap | Pharyngeal gland cell |
| g2L | MS aapapaa | Pharyngeal gland cell |
| g2R | MS papapaa | Pharyngeal gland cell |
| gon herm anch | Z1 ppp (5L) or | Anchor cell induces vulva (5L) |
| gon herm anch | Z4 aaa (5R) | Anchor cell induces vulva (5R) |
| gon herm dtc A | Z1 aa | Ant. distal tip cell 49 inhibit meiosis |
| gon herm dtc P | Z4 pp | Post. distal tip cell 49 -leads gonad |
| gon herm dish A | Z1 apa | Ant. epithelial distal sheath no mus. |
| gon herm dish A | Z1 paaa | Ant. epithelial distal sheath no mus. |
| gon herm dish P | Z4 pap | Post. epithelial distal sheath no mus. |
| gon herm dish P | Z4 appp | Post. epithelial distal sheath no mus. |
| gon herm prsh A | Z1 appa | Ant. epithelial proximal sheath w/ mus. |
| gon herm prsh A | Z1 paapa | Ant. epithelial proximal sheath w/ mus. |
| gon herm prsh P | Z4 paap | Post. epithelial proximal sheath w/ mus |
| gon herm prsh P | Z4 appap | Post. epithelial proximal sheath w/ mus |
| gon herm spth A | Z1 appp | Anterior spermatheca |
| gon herm spth A | Z1 paapp | Anterior spermatheca |
| gon herm spth A | Z1 papaa | Anterior spermatheca |
| gon herm spth A | Z4 apaaa | Anterior spermatheca |
| gon herm spth P | Z4 paaa | Posterior spermatheca |
| gon herm spth P | Z4 appaa | Posterior spermatheca |
| gon herm spth P | Z4 apapp | Posterior spermatheca |
| gon herm spth P | Z1 pappp | Posterior spermatheca |
| gon herm sujn A | Z1 papa | Anterior spermatheca uterine jct |
| gon herm sujn A | Z1 ppaaa | Anterior spermatheca uterine jct |
| gon herm sujn A | Z4 apaa | Anterior spermatheca uterine jct |
| gon herm sujn P | Z1 papp | Posterior spermatheca uterine jct |
| gon herm sujn P | Z4 apap | Posterior spermatheca uterine jct |
| gon herm sujn P | Z4 aappp | Posterior spermatheca uterine jct |

| Parts List | Cell Lineage | Short Description |
| --- | --- | --- |
| gon herm dut | Z1 pap | Dorsal uterus |
| gon herm dut | Z4 apa | Dorsal uterus |
| gon herm vut | Z1 ppa (5L) | Ventral uterus (5L) |
| gon herm vut | Z4 aaa (5L) | Ventral uterus (5L) |
| gon herm vut | Z4 aap (5L) | Ventral uterus (5L) |
| | or | |
| gon herm vut | Z1 ppa (5R) | Ventral uterus (5R) |
| gon herm vut | Z1 ppp (5R) | Ventral uterus (5R) |
| gon herm vut | Z4 aap (5R) | Ventral uterus (5R) |
| gon male dtc | Z1 a | Distal tip cell inhibit meiosis |
| gon male dtc | Z4 p | Distal tip cell inhibit meiosis |
| gon male link | Z1 paa (ZA) | Linker cell initiates union w/ cloaca (ZA) |
| | or | |
| gon male link | Z4 aaa (ZB) | Linker cell initiates union w/ cloaca (ZB) |
| gon male sves | Z1 pap | Seminal vesicle |
| gon male sves | Z1 pp | Seminal vesicle |
| gon male sves | Z4 (p/a)aa | Seminal vesicle |
| gon male sves | Z4 aap | Seminal vesicle |
| gon male sves | Z4 ap | Seminal vesicle |
| gon male vdef | Z1 pap (ZA) | Vas deferens (ZA) |
| gon male vdef | Z4 aa (ZA) | Vas deferens (ZA) |
| | or | |
| gon male vdef | Z1 pa (ZB) | Vas deferens (ZB) |
| gon male vdef | Z4 aap (ZB) | Vas deferens (ZB) |
| hmc | MS appaaa | Head mesodermal cell |
| hyp1 | AB | Hypodermal syncytium in head |
| hyp2 | AB | Hypodermal syncitium in head |
| hyp3 | AB | Hypodermal syncitium in head |
| hyp4 | AB | Hypodermal syncitium in head |
| hyp5 | AB | Hypodermal syncitium in head |
| hyp6 | AB | Hypodermal syncitium in head |
| hyp7 | AB | Embryonic hypodermal syncitium |
| hyp7 | C | Embryonic hypodermal symcitium |
| hyp7L | H1L | Postemb hypodermal syncitium |
| hyp7R | H1R | Postemb hypodermal syncitium |
| hyp7L | H2L | Postemb hypodermal syncitium |
| hyp7R | H2R | Postemb hypodermal syncitium |
| hyp7 | P1 p | Postemb hypodermal syncitium |
| hyp7 | P2 p | Postemb hypodermal syncitium |
| hyp7 | P9 p | Postemb hypodermal syncitium |

| Parts List | Cell Lineage | Short Description |
|---|---|---|
| hyp7L | TL | Postemb hypodermal syncitium |
| hyp7R | TR | Postemb hypodermal syncitium |
| hyp7L | V1L | Postemb hypodermal syncitium |
| hyp7R | V1R | Postemb hypodermal syncitium |
| hyp7L | V2L | Postemb hypodermal syncitium |
| hyp7R | V2R | Postemb hypodermal syncitium |
| hyp7L | V3L | Postemb hypodermal syncitium |
| hyp7R | V3R | Postemb hypodermal syncitium |
| hyp7L | V4L | Postemb hypodermal syncitium |
| hyp7R | V4R | Postemb hypodermal syncytium |
| hyp7L | V5L | Postemb hypodermal syncitium |
| hyp7R | V5R | Postemb hypodermal syncitium |
| hyp7L | V6L | Postemb hypodermal syncitium |
| hyp7R | V6R | Postemb hypodermal syncitium |
| hyp7 herm | P3 pa | Postemb hypodermal syncitium |
| hyp7 herm | P3 pp | Postemb hypodermal syncitium |
| hyp7 herm | P4 pa | Postemb hypodermal syncitium |
| hyp7 herm | P4 pp | Postemb hypodermal syncitium |
| hyp7 herm | P8 pa | Postemb hypodermal syncitium |
| hyp7 herm | P8 pp | Postemb hypodermal syncitium |
| hyp7 herm | P10 p | Postemb hypodermal syncitium |
| hyp7 herm | P11 p | Postemb hypodermal syncitium |
| hyp7 herm L | TL | Postemb hypodermal syncitium |
| hyp7 herm R | TR | Postemb hypodermal syncitium |
| hyp7 herm L | V5L | Postemb hypodermal syncitium |
| hyp7 herm R | V5R | Postemb hypodermal syncitium |
| hyp7 herm L | V6L | Postemb hypodermal syncitium |
| hyp7 herm R | V6R | Postemb hypodermal syncitium |
| hyp7 male | P3 p | Postemb hypodermal syncitium |
| hyp7 male | P4 p | Postemb hypodermal syncitium |
| hyp7 male | P5 p | Postemb hypodermal syncitium |
| hyp7 male | P6 p | Postemb hypodermal syncitium |
| hyp7 male | P7 p | Postemb hypodermal syncitium |
| hyp7 male | P8 p | Postemb hypodermal syncitium |
| hyp7 male | P10 paaa | Postemb hypodermal syncitium |
| hyp7 male | P10 paap | Postemb hypodermal syncitium |
| hyp7 male | P10 papa | Postemb hypodermal syncitium |
| hyp7 male L | TL | Postemb hypodermal syncitium |
| hyp7 male R | TR | Postemb hypodermal syncitium |
| hyp7 male L | V5L | Postemb hypodermal syncitium |

| Parts List | Cell Lineage | Short Description |
|---|---|---|
| hyp7 male R | V5R | Postemb hypodermal syncitium |
| hyp7 male L | V6L | Postemb hypodermal syncitium |
| hyp7 male R | V6R | Postemb hypodermal syncitium |
| hyp8 | AB plpppapap | Tail ventral hypodermis |
| hyp9 | AB prpppapap | Tail ventral hypodermis |
| hyp10 | AB | Tail ventral hypodermis |
| hyp11 | C pappa | Tail dorsal hypodermis |
| hyp P12 | P12 pa | Preanal hypodermis |
| hyp hook | P10 papp | Hypodermis with hook sensillum male |
| hyp hook | P11 | Hypodermis with hook sensillum male |
| int | E | Embryonic intestinal cell |
| int | In a & In p | Postemb 14 intest. nuclei no cell div. |
| linker killer | U lp | 1 cell 50 sometimes fused with U l/ra- |
| linker killer | U rp | 1 cell 50 phagocytoses male linker cell |
| m1DL | AB araapaaap | Pharyngeal muscle cell |
| m1DR | AB araappaap | Pharyngeal muscle cell |
| m1L | AB araaaaaap | Pharyngeal muscle cell |
| m1R | AB araaaaapp | Pharyngeal muscle cell |
| m1VL | AB alpaaaapa | Pharyngeal muscle cell |
| m1VR | AB arapaaapa | Pharyngeal muscle cell |
| m2DL | AB araapaapa | Pharyngeal muscle cell |
| m2DR | AB araappapa | Pharyngeal muscle cell |
| m2L | AB alpaaapaa | Pharyngeal muscle cell |
| m2R | AB arapaapaa | Pharyngeal muscle cell |
| m2VL | AB alpaaaaap | Pharyngeal muscle cell |
| m2VR | AB arapaaaap | Pharyngeal muscle cell |
| m3DL | MS aaapaaa | Pharyngeal muscle cell |
| m3DR | MS paaaapa | Pharyngeal muscle cell |
| m3L | AB alpaapapp | Pharyngeal muscle cell |
| m3R | AB arapaappa | Pharyngeal muscle cell |
| m3VL | AB alpappppp | Pharyngeal muscle cell |
| m3VR | AB arapapppp | Pharyngeal muscle cell |
| m4DL | MS aaaaapp | Pharyngeal muscle cell |
| m4DR | MS paaaapp | Pharyngeal muscle cell |
| m4L | MS aaapaap | Pharyngeal muscle cell |
| m4R | AB araaapapp | Pharyngeal muscle cell |
| m4VL | MS aapaaaa | Pharyngeal muscle cell |
| m4VR | MS papaaaa | Pharyngeal muscle cell |
| m5DL | MS aaaapap | Pharyngeal muscle cell |
| m5DR | MS paaappa | Pharyngeal muscle cell |

| Parts List | Cell Lineage | Short Description |
|------------|--------------|-------------------|
| m5L | AB araapapap | Pharyngeal muscle cell |
| m5R | AB araapppap | Pharyngeal muscle cell |
| m5VL | MS aapaaap | Pharyngeal muscle cell |
| m5VR | MS papaaap | Pharyngeal muscle cell |
| m6D | MS paaappp | Pharyngeal muscle cell |
| m6VL | MS aapappa | Pharyngeal muscle cell |
| m6VR | MS papappa | Pharyngeal muscle cell |
| m7D | MS aaaappp | Pharyngeal muscle cell |
| m7VL | MS aapaapp | Pharyngeal muscle cell |
| m7VR | MS papaapp | Pharyngeal muscle cell |
| m8 | MS aaapapp | Pharyngeal muscle cell |
| mc1DL | AB alpaapapa | Pharyngeal marginal cell |
| mc1DR | AB araaapapa | Pharyngeal marginal cell |
| mc1V | AB alpappppa | Pharyngeal marginal cell |
| mc2DL | AB araapaapp | Pharyngeal marginal cell |
| mc2DR | AB araappapp | Pharyngeal marginal cell |
| mc2V | AB arapapppa | Pharyngeal marginal cell |
| mc3DL | MS aaapapa | Pharyngeal marginal cell |
| mc3DR | MS paapapa | Pharyngeal marginal cell |
| mc3V | AB alpappapp | Pharyngeal marginal cell |
| mu anal | AB plpppppap | Anal depressor muscle |
| mu bod | AB prpppppaa | Embryonic body wall muscle |
| mu bod | C | Embryonic body wall muscle |
| mu bod | D | Embryonic body wall muscle |
| mu bod | MS | Embryonic body wall muscle |
| mu bod | M | Postemb body wall muscle |
| mu male diag | M | Diagonal muscle body wall |
| mu male gub | M | Male gubernacular muscle |
| mu int L | AB plpppppaa | Intestinal muscle to int. & body wall |
| mu int R | MS ppaapp | Intestinal muscle to int. & body wall |
| mu male long | M | Longitudinal muscle body wall |
| mu male obl | M | Oblique muscle body wall |
| mu sph | AB prpppppap | Sphincter intestino-rectal valve |
| mu male spic | M | Muscle copulatory spicule |
| mu herm ut | M | Uterine muscle |
| mu herm vul | M | Vulval muscle |
| proct | B | Male proctodeum 51 union of int. & vas |
| proct | F | Male proctodeum 51 def.-w/ copul. spic. |
| rect D | AB plpapppp | Rectal epithelial cell 52 near intesti- |
| rect VL | AB plppppaap | Rectal epithelial cell 52 no-rectal val |

| Parts List | Cell Lineage | Short Description |
|---|---|---|
| rect VR | AB prpppaap | Rectal epithelial cell 52 -ve-microvil. |
| rect hyp | K a | Rectal hypodermis |
| rect hyp | K prime | Rectal hypodermis |
| seL | H1L | Seam hypodermal cell |
| seR | H1R | Seam hypodermal cell |
| seL | H2L | Seam hypodermal cell |
| seR | H2R | Seam hypodermal cell |
| seL | V1L | Seam hypodermal cell |
| seR | V1R | Seam hypodermal cell |
| seL | V2L | Seam hypodermal cell |
| seR | V2R | Seam hypodermal cell |
| seL | V3L | Seam hypodermal cell |
| seR | V3R | Seam hypodermal cell |
| seL | V4L | Seam hypodermal cell |
| seR | V4R | Seam hypodermal cell |
| se herm L | TL | Seam hypodermal cell |
| se herm R | TR | Seam hypodermal cell |
| se herm L | V5L | Seam hypodermal cell |
| se herm R | V5R | Seam hypodermal cell |
| se herm L | V6L | Seam hypodermal cell |
| se herm R | V6R | Seam hypodermal cell |
| se male L | V5L | Seam hypodermal cell |
| se male R | V5R | Seam hypodermal cell |
| setL | V5L | Tail seam hypodermal cell male |
| setR | V5R | Tail seam hypodermal cell male |
| setL | V6L | Tail seam hypodermal cell male |
| setR | V6R | Tail seam hypodermal cell male |
| spike/death | AB plppppppa | Temporary cell for tail spike |
| spike/death | AB prppppppa | Temporary cell for tail spike |
| virL | AB prpapppp | Intestino-rectal valve |
| virR | AB prpappppa | Intestino rectal valve |
| vpi1 | MS paapapp | Pharyngo-intestinal valve |
| vpi2DL | MS aapapp | Pharyngo-intestinal valve |
| vpi2DR | MS papappp | Pharyngo-intestinal valve |
| vpi2V | MS aappaa | Pharyngo-intestinal valve |
| vpi3D | MS aaappp | Pharyngo-intestinal valve |
| vpi3V | MS aappap | Pharyngo-intestinal valve |
| vulva | P5 | Hermaphrodite vulva |
| vulva | P6 | Hermaphrodite vulva |
| vulva | P7 | Hermaphrodite vulva |

--- End of File ---

**F.3.** *PARTLIS2.PRN*

| Parts List | Short Description | Number of Cells |
|---|---|---|
| AB | Embryonic founder cell | 1 |
| ADAL | Ring interneuron | 1 |
| ADAR | Ring interneuron | 1 |
| ADEL | Anterior deirid 1 sensory receptor in | 1 |
| ADER | Anterior deirid 1 lateral alae-dopamine | 1 |
| ADEshL | Anterior deirid sheath cell | 1 |
| ADEshR | Anterior deirid sheath cell | 1 |
| ADEsoL | Anterior deirid socket | 1 |
| ADEsoR | Anterior deirid socket | 1 |
| ADFL | Amphid neuron 2 ciliated-chemosensory- | 1 |
| ADFR | Amphid neuron 2 commissure fm vent.gang | 1 |
| ADLL | Amphid neuron 3 ciliated-chemosensory- | 1 |
| ADLR | Amphid neuron 3 direct to ring | 1 |
| AFDL | Amphid finger cell 4 associated with | 1 |
| AFDR | Amphid finger cell 4 amphid sheath | 1 |
| AIAL | Amphid interneuron | 1 |
| AIAR | Amphid interneuron | 1 |
| AIBL | Amphid interneuron | 1 |
| AIBR | Amphid interneuron | 1 |
| AIML | Ring interneuron | 1 |
| AIMR | Ring interneuron | 1 |
| AINL | Ring interneuron | 1 |
| AINR | Ring interneuron | 1 |
| AIYL | Amphid interneuron | 1 |
| AIYR | Amphid interneuron | 1 |
| AIZL | Amphid interneuron | 1 |
| AIZR | Amphid interneuron | 1 |
| ALA | Neuron to excretory canal / dorsal cord | 1 |
| ALML | Touch receptor neuron 5 anterolateral | 1 |
| ALMR | Touch receptor neuron 5 microtubules | 1 |
| ALNL | Neuron assoc. with ALM 6 send processes | 1 |
| ALNR | Neuron assoc. with ALM 6 to tailspike | 1 |
| AMshL | Amphid sheath cell | 1 |
| AMshR | Amphid sheath cell | 1 |
| AMsoL | Amphid socket cell | 1 |
| AMsoR | Amphid socket cell | 1 |
| AQR | Neuron ciliate to ring | 1 |
| AS1 | Ventral cord motorneuron 7 innervate | 1 |

| Parts List | Short Description | Number of Cells |
|---|---|---|
| AS2 | Ventral cord motorneuron 7 dorsal mus.- | 1 |
| AS3 | Ventral cord motorneuron 7 no ventral | 1 |
| AS4 | Ventral cord motorneuron 7 counterpart- | 1 |
| AS5 | Ventral cord motorneuron 7 cholinergic? | 1 |
| AS6 | Ventral cord motorneuron 7 like VAn but | 1 |
| AS7 | Ventral cord motorneuron 7 with added | 1 |
| AS8 | Ventral cord motorneuron 7 AVB input | 1 |
| AS9 | Ventral cord motorneuron 7 | 1 |
| AS10 | Ventral cord motorneuron 7 | 1 |
| AS11 | Ventral cord motorneuron 7 | 1 |
| ASEL | Amphid neuron 8 ciliated-chemosensory- | 1 |
| ASER | Amphid neuron 8 to ring by commissure | 1 |
| ASGL | Amphid neuron 8 from ventral ganglion- | 1 |
| ASGR | Amphid neuron 8 diverse connections in | 1 |
| ASHL | Amphid neuron 8 ring neuropil | 1 |
| ASHR | Amphid neuron 8 | 1 |
| ASIL | Amphid neuron 8 | 1 |
| ASIR | Amphid neuron 8 | 1 |
| ASJL | Amphid neuron 8 | 1 |
| ASJR | Amphid neuron 8 | 1 |
| ASKL | Amphid neuron 8 | 1 |
| ASKR | Amphid neuron 8 | 1 |
| AUAL | Neuron runs with amphid neuron 9 lack | 1 |
| AUAR | Neuron runs with amphid neuron 9 cilia | 1 |
| AVAL | Ventral cord interneuron 10 to VA & | 1 |
| AVAR | Ventral cord interneuron 10 DA & AS | 1 |
| AVBL | Ventral cord interneuron 11 to VB & | 1 |
| AVBR | Ventral cord interneuron 11 DB & AS | 1 |
| AVDL | Ventral cord interneuron 12 to VA & | 1 |
| AVDR | Ventral cord interneuron 12 DA & AS | 1 |
| AVEL | Ventral cord interneuron 13 like AVD in | 1 |
| AVER | Ventral cord interneuron 13 ant. cord | 1 |
| AVFL | Interneuron in ventral cord 14 & ring- | 1 |
| AVFR | Interneuron in ventral cord 14 edited | 1 |
| AVG | Ventral cord interneuron to tailspike | 1 |
| AVHL | Neuron 15 presynaptic in ring & | 1 |
| AVHR | Neuron 15 postsynaptic in ventral cord | 1 |
| AVJL | Neuron 16 postsynaptic in ventral cord | 1 |
| AVJR | Neuron 16 & presynaptic in ring | 1 |
| AVKL | Ring & ventral cord interneuron | 1 |

| Parts List | Short Description | Number of Cells |
|---|---|---|
| AVKR | Ring & ventral cord interneuron | 1 |
| AVL | Ring & vent.cord interneuron/motoneuron | 1 |
| AVM | Microtubule cell touch receptor | 1 |
| AWAL | Amphid wing neuron 17 ciliated-sensory- | 1 |
| AWAR | Amphid wing neuron 17 assoc. with | 1 |
| AWBL | Amphid wing neuron 17 amphid sheath | 1 |
| AWBR | Amphid wing neuron 17 | 1 |
| AWCL | Amphid wing neuron 17 | 1 |
| AWCR | Amphid wing neuron 17 | 1 |
| B | Rectal cell -postemb.blast cell in male | 1 |
| BAGL | Neuron 18 ciliated in head-not part of | 1 |
| BAGR | Neuron 18 sensillium-assoc. with ILso | 1 |
| BDUL | Neuron 19 along excretory canal & also | 1 |
| BDUR | Neuron 19 in nerve ring-dark vesicles | 1 |
| C | Embryonic founder cell | 1 |
| CA1 | Male cell - not reconstructed | 1 |
| CA2 | Male cell - not reconstructed | 1 |
| CA3 | Male cell - not reconstructed | 1 |
| CA4 | Male neuron to dorsal muscles | 1 |
| CA5 | Male neuron to dorsal muscles | 1 |
| CA6 | Male neuron to dorsal muscles | 1 |
| CA7 | Male neuron to dorsal muscles | 1 |
| CA8 | Male neuron? 20 in ventral cord-neuron | 1 |
| CA9 | Male neuron? 20 like but lacks synapses | 1 |
| CANL | Neuron 21 along excretory canal-no | 1 |
| CANR | Neuron 21 synapses-essential for life | 1 |
| CEMDL | Male cephalic neuron 22 die in hermaph. | 1 |
| CEMDR | Male cephalic neuron 22 open to outside | 1 |
| CEMVL | Male cephalic neuron 22 sex chemotaxis? | 1 |
| CEMVR | Male cephalic neuron 22 | 1 |
| CEPDL | Neuron cephalic sensillum dopamine | 1 |
| CEPDR | Neuron cephalic sensillum dopamine | 1 |
| CEPVL | Neuron cephalic sensillum dopamine | 1 |
| CEPVR | Neuron cephalic sensillum dopamine | 1 |
| CEPshDL | Cephalic sheath cell 23 sheet-like | 1 |
| CEPshDR | Cephalic sheath cell 23 processes | 1 |
| CEPshVL | Cephalic sheath cell 23 envelop ring | 1 |
| CEPshVR | Cephalic sheath cell 23 neuropil & | 1 |
| CEPsoDL | Cephalic socket cell 23 part of ventral | 1 |
| CEPsoDR | Cephalic socket cell 23 ganglion | 1 |

| Parts List | Short Description | Number of Cells |
|---|---|---|
| CEPsoVL | Cephalic socket cell 23 | 1 |
| CEPsoVR | Cephalic socket cell 23 | 1 |
| CP0 | Male ventral cord - not reconstructed | 1 |
| CP1 | Male ventral cord - not reconstructed | 1 |
| CP2 | Male ventral cord - not reconstructed | 1 |
| CP3 | Male ventral cord - not reconstructed | 1 |
| CP4 | Male motorneuron in ventral cord | 1 |
| CP5 | Male motorneuron in ventral cord | 1 |
| CP6 | Male motorneuron in ventral cord | 1 |
| CP7 | Male motorneuron in ventral cord | 1 |
| CP8 | Male interneuron to preanal ganglion | 1 |
| CP9 | Male interneuron to preanal ganglion | 1 |
| D | Embryonic founder cell | 1 |
| DA1 | Ventral cord motorneuron 24 to dorsal | 1 |
| DA2 | Ventral cord motorneuron 24 muscles- | 1 |
| DA3 | Ventral cord motorneuron 24 cholinergic | 1 |
| DA4 | Ventral cord motorneuron 24 | 1 |
| DA5 | Ventral cord motorneuron 24 | 1 |
| DA6 | Ventral cord motorneuron 24 | 1 |
| DA7 | Ventral cord motorneuron 24 | 1 |
| DA8 | Ventral cord motorneuron 24 | 1 |
| DA9 | Ventral cord motorneuron 24 | 1 |
| DB1 | Ventral cord motorneuron 25 to dorsal | 1 |
| DB2 | Ventral cord motorneuron 25 muscles- | 1 |
| DB3 | Ventral cord motorneuron 25 cholinergic | 1 |
| DB4 | Ventral cord motorneuron 25 | 1 |
| DB5 | Ventral cord motorneuron 25 | 1 |
| DB6 | Ventral cord motorneuron 25 | 1 |
| DB7 | Ventral cord motorneuron 25 | 1 |
| DD1 | Ventral cord motorneuron 26 reciprocal | 1 |
| DD2 | Ventral cord motorneuron 26 inhibitor- | 1 |
| DD3 | Ventral cord motorneuron 26 change | 1 |
| DD4 | Ventral cord motorneuron 26 pattern of | 1 |
| DD5 | Ventral cord motorneuron 26 synapses- | 1 |
| DD6 | Ventral cord motorneuron 26 GABA | 1 |
| DVA | Ring interneuron -large vesicle in ring | 1 |
| DVB | Ring interneuron -to rectal muscles | 1 |
| DVC | Ring interneuron | 1 |
| DVE | Male specific neuron 27 cell body in | 1 |
| DVF | Male specific neuron 27 preanal gang.- | 1 |

| Parts List | Short Description | Number of Cells |
|---|---|---|
| DX1 | Male specific neuron 27 penetrate | 1 |
| DX2 | Male specific neuron 27 basement memb. | 1 |
| DX3 | Male specific neuron 27 & contact | 1 |
| DX4 | Male specific neuron 27 muscles | 1 |
| E | Embryonic founder cell | 1 |
| EF1 | Male specific neuron 28 in preanal gang | 1 |
| EF2 | Male specific neuron 28 -input fm ray | 1 |
| EF3 | Male specific neuron 28 neurons | 1 |
| EF4 | Male specific neuron 28 | 1 |
| F | Rectal cell -postemb.blast cell in male | 1 |
| FLPL | Neuron 29 ciliated in head-not part of | 1 |
| FLPR | Neuron 29 sensillium-assoc. with ILso | 1 |
| G1 | Postemb.blast cell 30 excretory socket- | 1 |
| G2 | Postemb.blast cell 30 G1(emb)G2(L1)G2.p | 1 |
| GLRDL | Glial cell 30 sheet bet. pharynx & ring | 1 |
| GLRDR | Glial cell 30 neuropil-no chemical | 1 |
| GLRL | Glial cell 30 synapses-GJs with muscles | 1 |
| GLRR | Glial cell 30 & RME-processes to head | 1 |
| GLRVL | Glial cell 30 | 1 |
| GLRVR | Glial cell 30 | 1 |
| H0L | Seam hypodermal cell | 1 |
| H0R | Seam hypodermal cell | 1 |
| H1L | Seam hypodermal cell postemb. blast | 1 |
| H1R | Seam hypodermal cell postemb. blast | 1 |
| H2L | Seam hypodermal cell postemb. blast | 1 |
| H2R | Seam hypodermal cell postemb. blast | 1 |
| HOA | Neuron male hook sensillum | 1 |
| HOB | Neuron male hook sensillum | 1 |
| HOsh | Sheath male hook sensillum | 1 |
| HOso | Socket male hook sensillum | 1 |
| HSNL | Hermaph.motoneuron vulval mus.serotonin | 1 |
| HSNR | Hermaph.motoneuron vulval mus.serotonin | 1 |
| I1L | Pharyngeal interneuron -sensory fm RIP | 1 |
| I1R | Pharyngeal interneuron -sensory fm RIP | 1 |
| I2L | Pharyngeal interneuron -ant. sensory | 1 |
| I2R | Pharyngeal interneuron -ant. sensory | 1 |
| I3 | Pharyngeal interneuron -ant. sensory | 1 |
| I4 | Pharyngeal interneuron | 1 |
| I5 | Pharyngeal interneuron -post. sensory | 1 |
| I6 | Pharyngeal interneuron -post. sensory | 1 |

| Parts List | Short Description | Number of Cells |
|---|---|---|
| IL1DL | Neuron inner labial sensorimotor | 1 |
| IL1DR | Neuron inner labial sensorimotor | 1 |
| IL1L | Neuron inner labial sensorimotor | 1 |
| IL1R | Neuron inner labial sensorimotor | 1 |
| IL1VL | Neuron inner labial sensorimotor | 1 |
| IL1VR | Neuron inner labial sensorimotor | 1 |
| IL2DL | Neuron inner labial sensory open out | 1 |
| IL2DR | Neuron inner labial sensory open out | 1 |
| IL2L | Neuron inner labial sensory open out | 1 |
| IL2R | Neuron inner labial sensory open out | 1 |
| IL2VL | Neuron inner labial sensory open out | 1 |
| IL2VR | Neuron inner labial sensory open out | 1 |
| ILshDL | Inner labial sheath cell | 1 |
| ILshDR | Inner labial sheath cell | 1 |
| ILshL | Inner labial sheath cell | 1 |
| ILshR | Inner labial sheath cell | 1 |
| ILshVL | Inner labial sheath cell | 1 |
| ILshVR | Inner labial sheath cell | 1 |
| ILsoDL | Inner labial socket cell -tip of labium | 1 |
| ILsoDR | Inner labial socket cell -tip of labium | 1 |
| ILsoL | Inner labial socket cell -tip of labium | 1 |
| ILsoR | Inner labial socket cell -tip of labium | 1 |
| ILsoVL | Inner labial socket cell -tip of labium | 1 |
| ILsoVR | Inner labial socket cell -tip of labium | 1 |
| K | Rectal cell postemb. blast cell | 1 |
| K prime | Rectal cell | 1 |
| LUAL | Interneuron posterior ventral cord | 1 |
| LUAR | Interneuron posterior ventral cord | 1 |
| M | Postembryonic mesoblast | 1 |
| M1 | Pharyngeal motorneuron | 1 |
| M2L | Pharyngeal motorneuron | 1 |
| M2R | Pharyngeal motorneuron | 1 |
| M3L | Pharyngeal sensory-motorneuron | 1 |
| M3R | Pharyngeal sensory-motorneuron | 1 |
| M4 | Pharyngeal motorneuron | 1 |
| M5 | Pharyngeal motorneuron | 1 |
| MCL | Pharyngeal neuron to marginal cell | 1 |
| MCR | Pharyngeal neuron to marginal cell | 1 |
| MI | Pharyngeal motorneuron-interneuron | 1 |
| MS | Embryonic founder cell | 1 |

| Parts List | Short Description | Number of Cells |
|---|---|---|
| NSML | Pharyngeal motoneuron 31 neurosecretory | 1 |
| NSMR | Pharyngeal motoneuron 31 -serotonin | 1 |
| OLLL | Neuron outer labial sensillum | 1 |
| OLLR | Neuron outer labial sensillum | 1 |
| OLLshL | Lateral outer labial sheath cell | 1 |
| OLLshR | Lateral outer labial sheath cell | 1 |
| OLLsoL | Lateral outer labial socket cell | 1 |
| OLLsoR | Lateral outer labial socket cell | 1 |
| OLQDL | Neuron outer labial sensilla | 1 |
| OLQDR | Neuron outer labial sensilla | 1 |
| OLQVL | Neuron outer labial sensilla | 1 |
| OLQVR | Neuron outer labial sensilla | 1 |
| OLQshDL | Quadrant outer labial sheath cell | 1 |
| OLQshDR | Quadrant outer labial sheath cell | 1 |
| OLQshVL | Quadrant outer labial sheath cell | 1 |
| OLQshVR | Quadrant outer labial sheath cell | 1 |
| OLQsoDL | Quadrant outer labial socket cell | 1 |
| OLQsoDR | Quadrant outer labial socket cell | 1 |
| OLQsoVL | Quadrant outer labial socket cell | 1 |
| OLQsoVR | Quadrant outer labial socket cell | 1 |
| P0 | The single-cell zygote | 1 |
| P04 | Embryonic founder cell-germ line | 1 |
| P1 | Postemb. blast cell 32 for ventral | 1 |
| P2 | Postemb. blast cell 32 cord & ventral | 1 |
| P3 | Postemb. blast cell 32 hypodermis & | 1 |
| P4 | Postemb. blast cell 32 vulva & male | 1 |
| P5 | Postemb. blast cell 32 preanal gang. - | 1 |
| P6 | Postemb. blast cell 32 ventral hypoder- | 1 |
| P7 | Postemb. blast cell 32 mis in L1 | 1 |
| P8 | Postemb. blast cell 32 | 1 |
| P9 | Postemb. blast cell 32 | 1 |
| P10 | Postemb. blast cell 32 | 1 |
| P11 | Postemb. blast cell 32 | 1 |
| P12 | Postemb. blast cell 32 | 1 |
| PCAL | Sensory neuron male postcloac. sensil. | 1 |
| PCAR | Sensory neuron male postcloac. sensil. | 1 |
| PCBL | Neuron in sheath male postcloac.sensil. | 1 |
| PCBR | Neuron in sheath male postcloac.sensil. | 1 |
| PCCL | Sensory neuron male postcloac. sensil. | 1 |
| PCCR | Sensory neuron male postcloac. sensil. | 1 |

| Parts List | Short Description | Number of Cells |
|---|---|---|
| PChL | Hypodermal cell male postcloac. sensil. | 1 |
| PChR | Hypodermal cell male postcloac. sensil. | 1 |
| PCshL | Sheath cell male postcloac. sensil. | 1 |
| PCshR | Sheath cell male postcloac. sensil. | 1 |
| PCsoL | Socket cell male postcloac. sensil. | 1 |
| PCsoR | Socket cell male postcloac. sensil. | 1 |
| PDA herm | Motorneuron preanal ganglion herm. | 1 |
| PDA male | Motorneuron preanal ganglion male | 1 |
| PDB | Interneuron-motorneuron post.vent.cord | 1 |
| PDC | Interneuron preanal ganglion male | 1 |
| PDEL | Neuron dopaminergic postdeirid | 1 |
| PDER | Neuron dopaminergic postdeirid | 1 |
| PDEshL | Sheath sensillum postdeirid dopamine | 1 |
| PDEshR | Sheath sensillum postdeirid dopamine | 1 |
| PDEsoL | Socket sensillum postdeirid dopamine | 1 |
| PDEsoR | Socket sensillum postdeirid dopamine | 1 |
| PGA | Interneuron male preanal ganglion | 1 |
| PHAL | Phasmid neuron ciliated chemosensory | 1 |
| PHAR | Phasmid neuron ciliated chemosensory | 1 |
| PHBL | Phasmid neuron ciliated chemosensory | 1 |
| PHBR | Phasmid neuron ciliated chemosensory | 1 |
| PHCL | Neuron male tail spike | 1 |
| PHCR | Neuron male tail spike | 1 |
| PHshL | Phasmid sheath cell | 1 |
| PHshR | Phasmid sheath cell | 1 |
| PHso1L | Phasmid socket cell | 1 |
| PHso1R | Phasmid socket cell | 1 |
| PHso2L | Phasmid socket cell | 1 |
| PHso2R | Phasmid socket cell | 1 |
| PLML | Neuron 33 posterolateral microtubule- | 1 |
| PLMR | Neuron 33 touch receptor | 1 |
| PLNL | Interneuron with PLM to tailspike | 1 |
| PLNR | Interneuron with PLM to tailspike | 1 |
| PQR | Neuron cilium preanal ganglion | 1 |
| PVCL | Vent.cord interneuron 34 in lumbar gang | 1 |
| PVCR | Vent.cord interneuron 34 -to VB & DB | 1 |
| PVDL | Neuron near excretory canal | 1 |
| PVDR | Neuron near excretory canal | 1 |
| PVM | Post.vent. microtubule touch receptor | 1 |
| PVNL | Interneuron to ring and vulva | 1 |

| Parts List | Short Description | Number of Cells |
|---|---|---|
| PVNR | Interneuron to ring and vulva | 1 |
| PVPL | Interneuron along ventral cord to ring | 1 |
| PVPR | Interneuron along ventral cord to ring | 1 |
| PVQL | Interneuron along ventral cord to ring | 1 |
| PVQR | Interneuron 35 along ventral cord to | 1 |
| PVR | Interneuron 35 ring & tailspike | 1 |
| PVT | Interneuron ventral cord | 1 |
| PVV | Motorneuron male ventral cord | 1 |
| PVWL | Interneuron posterior ventral cord | 1 |
| PVWR | Interneuron posterior ventral cord | 1 |
| PVX | Male interneuron 36 to ring | 1 |
| PVY | Male interneuron 36 & ventral cord | 1 |
| PVZ | Male motorneuron ventral cord | 1 |
| QL | Postemb.neuroblast migrates anteriorly | 1 |
| QR | Postemb.neuroblast migrates posteriorly | 1 |
| R1AL | Neuron male sensory ray type A | 1 |
| R1AR | Neuron male sensory ray type A | 1 |
| R1BL | Neuron male sensory ray type B | 1 |
| R1BR | Neuron male sensory ray type B | 1 |
| R1stL | Structural cell male sensory ray | 1 |
| R1stR | Structural cell male sensory ray | 1 |
| R2AL | Neuron male sensory ray | 1 |
| R2AR | Neuron male sensory ray | 1 |
| R2BL | Neuron male sensory ray | 1 |
| R2BR | Neuron male sensory ray | 1 |
| R2stL | Structural cell male sensory ray | 1 |
| R2stR | Structural cell male sensory ray | 1 |
| R3AL | Neuron male sensory ray | 1 |
| R3AR | Neuron male sensory ray | 1 |
| R3BL | Neuron male sensory ray | 1 |
| R3BR | Neuron male sensory ray | 1 |
| R3stL | Structural call male sensory ray | 1 |
| R3stR | Structural cell male sensory ray | 1 |
| R4AL | Neuron male sensory ray type A | 1 |
| R4AR | Neuron male sensory ray type A | 1 |
| R4BL | Neuron male sensory ray type B | 1 |
| R4BR | Neuron male sensory ray type B | 1 |
| R4stL | Structural cell male sensory ray | 1 |
| R4stR | Structural cell male sensory ray | 1 |
| R5AL | Neuron male sensory type A dopamine | 1 |

| Parts List | Short Description | Number of Cells |
|---|---|---|
| R5AR | Neuron male sensory type A dopamine | 1 |
| R5BL | Neuron male sensory type B | 1 |
| R5BR | Neuron male sensory type B | 1 |
| R5stL | Structural cell male sensory ray | 1 |
| R5stR | Structural cell male sensory ray | 1 |
| R6AL | Neuron male sensory ray type A | 1 |
| R6AR | Neuron male sensory ray type A | 1 |
| R6BL | Neuron male sensory ray | 1 |
| R6BR | Neuron male sensory ray | 1 |
| R6stL | Structural call male sensory ray | 1 |
| R6stR | Structural cell male sensory ray | 1 |
| R7AL | Neuron male sensory type A dopamine | 1 |
| R7AR | Neuron male sensory type A dopamine | 1 |
| R7BL | Neuron male sensory ray type B | 1 |
| R7BR | Neuron male sensory ray type B | 1 |
| R7stL | Structural cell male sensory ray | 1 |
| R7stR | Structural cell male sensory ray | 1 |
| R8AL | Neuron male sensory ray | 1 |
| R8AR | Neuron male sensory ray | 1 |
| R8BL | Neuron male sensory ray | 1 |
| R8BR | Neuron male sensory ray | 1 |
| R8stL | Structural cell male sensory ray | 1 |
| R8stR | Structrual cell male sensory ray | 1 |
| R9AL | Neuron male sensory type A dopamine | 1 |
| R9AR | Neuron male sensory type A dopamine | 1 |
| R9BL | Neuron male sensory ray | 1 |
| R9BR | Neuron male sensory ray | 1 |
| R9stL | Structural cell male sensory ray | 1 |
| R9stR | Structural cell male sensory ray | 1 |
| RIAL | Interneuron ring to RMD and SMD | 1 |
| RIAR | Interneuron ring to RMD and SMD | 1 |
| RIBL | Interneuron ring | 1 |
| RIBR | Interneuron ring | 1 |
| RICL | Interneuron ring | 1 |
| RICR | Interneuron ring | 1 |
| RID | Motorneuron dorsal cord and muscles | 1 |
| RIFL | Ring interneuron | 1 |
| RIFR | Ring interneuron | 1 |
| RIGL | Ring interneuron | 1 |
| RIGR | Ring interneuron | 1 |

| Parts List | Short Description | Number of Cells |
|---|---|---|
| RIH | Ring interneuron | 1 |
| RIML | Ring motorneuron/interneuron | 1 |
| RIMR | Ring motorneuron/interneuron | 1 |
| RIPL | Ring/pharynx interneuron 37 direct con. | 1 |
| RIPR | Ring/pharynx interneuron 37 no synapses | 1 |
| RIR | Ring interneuron | 1 |
| RIS | Ring interneuron | 1 |
| RIVL | Ring interneuron | 1 |
| RIVR | Ring interneuron | 1 |
| RMDDL | Ring motorneuron/interneuron 38 many | 1 |
| RMDDR | Ring motorneuron/interneuron 38 synap- | 1 |
| RMDL | Ring motorneuron/interneuron 38 ses-re- | 1 |
| RMDR | Ring motorneuron/interneuron 38 cipro- | 1 |
| RMDVL | Ring motorneuron/interneuron 38 cal in- | 1 |
| RMDVR | Ring motorneuron/interneuron 38 hibitor? | 1 |
| RMED | Ring motorneuron GABA | 1 |
| RMEL | Ring motorneuron GABA | 1 |
| RMER | Ring motorneuron GABA | 1 |
| RMEV | Ring motorneuron GABA | 1 |
| RMFL | Ring motorneuron/interneuron | 1 |
| RMFR | Ring motorneuron/interneuron | 1 |
| RMGL | Ring motorneuron/interneuron | 1 |
| RMGR | Ring motorneuron/interneuron | 1 |
| RMHL | Ring motorneuron/interneuron | 1 |
| RMHR | Ring motorneuron/interneuron | 1 |
| SAADL | Ring interneuron anteriorly directed | 1 |
| SAADR | Ring interneuron anteriorly directed | 1 |
| SAAVL | Ring interneuron anteriorly directed | 1 |
| SAAVR | Ring interneuron anteriorly directed | 1 |
| SABD | Ring interneuron 39 ant. directed- | 1 |
| SABVL | Ring interneuron 39 motor in L1 | 1 |
| SABVR | Ring interneuron 39 | 1 |
| SDQL | Interneuron posterior lateral to ring | 1 |
| SDQR | Interneuron anterior lateral to ring | 1 |
| SIADL | Interneuron posteriorly directed | 1 |
| SIADR | Interneuron posteriorly directed | 1 |
| SIAVL | Interneuron posteriorly directed | 1 |
| SIAVR | Interneuron posteriorly directed | 1 |
| SIBDL | Interneuron | 1 |
| SIBDR | Interneuron | 1 |

| Parts List | Short Description | Number of Cells |
|---|---|---|
| SIBVL | Interneuron | 1 |
| SIBVR | Interneuron | 1 |
| SMBDL | Ring motorneuron/interneuron posterior | 1 |
| SMBDR | Ring motorneuron/interneuron posterior | 1 |
| SMBVL | Ring motorneuron/interneuron posterior | 1 |
| SMBVR | Ring motorneuron/interneuron posterior | 1 |
| SMDDL | Ring motorneuron/interneuron posterior | 1 |
| SMDDR | Ring motorneuron/interneuron posterior | 1 |
| SMDVL | Ring motorneuron/interneuron posterior | 1 |
| SMDVR | Ring motorneuron/interneuron posterior | 1 |
| SPCL | Sensory/motorneuron male 40 to spicule | 1 |
| SPCR | Sensory/motorneuron male 40 protractor | 1 |
| SPDL | Sensory neuron male 41 copulatory spi- | 1 |
| SPDR | Sensory neuron male 41 cule-open to out | 1 |
| SPVL | Sensory neuron male 41 at spicule tip | 1 |
| SPVR | Sensory neuron male 41 | 1 |
| SPshDL | Sheath cell male copulatory spicule | 1 |
| SPshDR | Sheath cell male copulatory spicule | 1 |
| SPshVL | Sheath cell male copulatory spicule | 1 |
| SPshVR | Sheath cell male copulatory spicule | 1 |
| SPso1L | Socket cell male copulatory spicule | 1 |
| SPso1R | Socket cell male copulatory spicule | 1 |
| SPso2L | Socket cell male copulatory spicule | 1 |
| SPso2R | Socket cell male copulatory spicule | 1 |
| SPso3L | Socket cell male copulatory spicule | 1 |
| SPso3R | Socket cell male copulatory spicule | 1 |
| SPso4L | Socket cell male copulatory spicule | 1 |
| SPso4R | Socket cell male copulatory spicule | 1 |
| TL | Tail seam hypoderm.42 cell postemb.- | 1 |
| TR | Tail seam hypoderm.42 phasmid socket L1 | 1 |
| U | Rectal cell postemb blast cell male | 1 |
| URADL | Ring motorneuron 43 non-ciliated head- | 1 |
| URADR | Ring motorneuron 43 with OLQ in embryo | 1 |
| URAVL | Ring motorneuron 43 | 1 |
| URAVR | Ring motorneuron 43 | 1 |
| URBL | Neuron 44 presyn. in ring-nonciliated | 1 |
| URBR | Neuron 44 head-with OLL in embryo | 1 |
| URXL | Ring interneuron 45 non-ciliated head- | 1 |
| URXR | Ring interneuron 45 with CEPD in embryo | 1 |
| URYDL | Neuron 46 presyn. in ring-nonciliated | 1 |

| Parts List | Short Description | Number of Cells |
|---|---|---|
| URYDR | Neuron 46 head-with OLQ in embryo | 1 |
| URYVL | Neuron 46 | 1 |
| URYVR | Neuron 46 | 1 |
| V1L | Seam hypodermal cell postemb blast | 1 |
| V1R | Seam hypodermal cell postemb blast | 1 |
| V2L | Seam hypodermal cell postemb blast | 1 |
| V2R | Seam hypodermal cell postemb blast | 1 |
| V3L | Seam hypodermal cell postemb blast | 1 |
| V3R | Seam hypodermal cell postemb blast | 1 |
| V4L | Seam hypodermal cell postemb blast | 1 |
| V4R | Seam hypodermal cell postemb blast | 1 |
| V5L | Seam hypodermal cell postemb blast | 1 |
| V5R | Seam hypodermal cell postemb blast | 1 |
| V6L | Seam hypodermal cell postemb blast | 1 |
| V6R | Seam hypodermal cell postemb blast | 1 |
| VA1 | Ventral cord motorneuron ventral mus. | 1 |
| VA2 | Ventral cord motorneuron ventral mus. | 1 |
| VA3 | Ventral cord motorneuron ventral mus. | 1 |
| VA4 | Ventral cord motorneuron ventral mus. | 1 |
| VA5 | Ventral cord motorneuron ventral mus. | 1 |
| VA6 | Ventral cord motorneuron ventral mus. | 1 |
| VA7 | Ventral cord motorneuron ventral mus. | 1 |
| VA8 | Ventral cord motorneuron ventral mus. | 1 |
| VA9 | Ventral cord motorneuron ventral mus. | 1 |
| VA10 | Ventral cord motorneuron ventral mus. | 1 |
| VA11 | Ventral cord motorneuron ventral mus. | 1 |
| VA12 | Ventral cord moto/interneuron preanal g | 1 |
| VB1 | Ventral cord motoneuron vent.mus. + ring | 1 |
| VB2 | Ventral cord motorneuron ventral mus. | 1 |
| VB3 | Ventral cord motorneuron ventral mus. | 1 |
| VB4 | Ventral cord motorneuron ventral mus. | 1 |
| VB5 | Ventral cord motorneuron ventral mus. | 1 |
| VB6 | Ventral cord motorneuron ventral mus. | 1 |
| VB7 | Ventral cord motorneuron ventral mus. | 1 |
| VB8 | Ventral cord motorneuron ventral mus. | 1 |
| VB9 | Ventral cord motorneuron ventral mus. | 1 |
| VB10 | Ventral cord motorneuron ventral mus. | 1 |
| VB11 | Ventral cord motorneuron ventral mus. | 1 |
| VC1 | Motorneuron vulval & body mus. hermaph. | 1 |
| VC2 | Motorneuron vulval & body mus. hermaph. | 1 |

| Parts List | Short Description | Number of Cells |
|---|---|---|
| VC3 | Motorneuron vulval & body mus. hermaph. | 1 |
| VC4 | Motorneuron vulval & body mus. hermaph. | 1 |
| VC5 | Motorneuron vulval & body mus. hermaph. | 1 |
| VC6 | Motorneuron vulval & body mus. hermaph. | 1 |
| VD1 | Ventral cord motorneuron 47 vent. body | 1 |
| VD2 | Ventral cord motorneuron 47 mus.-recip- | 1 |
| VD3 | Ventral cord motorneuron 47 rocal inhi- | 1 |
| VD4 | Ventral cord motorneuron 47 bitor?-GABA | 1 |
| VD5 | Ventral cord motorneuron 47 | 1 |
| VD6 | Ventral cord motorneuron 47 | 1 |
| VD7 | Ventral cord motorneuron 47 | 1 |
| VD8 | Ventral cord motorneuron 47 | 1 |
| VD9 | Ventral cord motorneuron 47 | 1 |
| VD10 | Ventral cord motorneuron 47 | 1 |
| VD11 | Ventral cord motorneuron 47 | 1 |
| VD12 | Ventral cord motorneuron 47 | 1 |
| VD13 | Ventral cord motorneuron 47 | 1 |
| W | Postemb neuroblast analogous to Pn.a | 1 |
| XXXL | Embryonic head hypodermal cell | 1 |
| XXXR | Embryonic head hypodermal cell | 1 |
| Y | Rectal cell PDA/hermaph.-blast/male | 1 |
| Z1 | Somatic gonad precursor cell | 1 |
| Z2 | Germ line precursor cell | 1 |
| Z3 | Germ line precursor cell | 1 |
| Z4 | Somatic gonad precursor cell | 1 |
| arc ant | Cell 48 between pharynx and hypodermis- | 3 |
| arc post | Cell 48 anterior part of buccal cavity | 3 |
| cc herm DL | Postemb coelomocyte hermaphrodite | 1 |
| cc herm DR | Postemb coelomocyte hermaphrodite | 1 |
| cc male D | Postemb coelomocyte male single | 1 |
| ccAL | Embryonic coelomocyte | 1 |
| ccAR | Embryonic coelomocyte | 1 |
| ccPL | Embryonic coelomocyte | 1 |
| ccPR | Embryonic coelomocyte | 1 |
| e1D | Pharyngeal epithelial cell | 1 |
| e1VL | Pharyngeal epithelial cell | 1 |
| e1VR | Pharyngeal epithelial cell | 1 |
| e2DL | Pharyngeal epithelial cell | 1 |
| e2DR | Pharyngeal epithelial cell | 1 |
| e2V | Pharyngeal epithelial cell | 1 |

| Parts List | Short Description | Number of Cells |
|---|---|---|
| e3D | Pharyngeal epithelial cell | 1 |
| e3VL | Pharyngeal epithelial cell | 1 |
| e3VR | Pharyngeal epithelial cell | 1 |
| exc cell | H-shaped excretory cell | 1 |
| exc duct | Excretory duct | 1 |
| exc gl L | Excretory gland fused open to exc duct | 1 |
| exc gl R | Excretory gland fused open to exc duct | 1 |
| exc socket | Excretory socket at duct to hypodermis | 1 |
| g1AL | Pharyngeal gland cell | 1 |
| g1AR | Pharyngeal gland cell | 1 |
| g1P | Pharyngeal gland cell | 1 |
| g2L | Pharyngeal gland cell | 1 |
| g2R | Pharyngeal gland cell | 1 |
| gon herm anch | Anchor cell induces vulva (5L) or (by alternate cell lineage) | 1 |
| gon herm anch | Anchor cell induces vulva (5R) | 1 |
| gon herm dtc A | Ant. distal tip cell 49 inhibit meiosis | 1 |
| gon herm dtc P | Post. distal tip cell 49 -leads gonad | 1 |
| gon herm dish A | Ant. epithelial distal sheath no mus. | 1 |
| gon herm dish A | Ant. epithelial distal sheath no mus. | 1 |
| gon herm dish P | Post. epithelial distal sheath no mus. | 1 |
| gon herm dish P | Post. epithelial distal sheath no mus. | 1 |
| gon herm prsh A | Ant. epithelial proximal sheath w/ mus. | 4 |
| gon herm prsh A | Ant. epithelial proximal sheath w/ mus. | 4 |
| gon herm prsh P | Post. epithelial proximal sheath w/ mus | 4 |
| gon herm prsh P | Post. epithelial proximal sheath w/ mus | 4 |
| gon herm spth A | Anterior spermatheca | 9 |
| gon herm spth A | Anterior spermatheca | 9 |
| gon herm spth A | Anterior spermatheca | 3 |
| gon herm spth A | Anterior spermatheca | 3 |
| gon herm spth P | Posterior spermatheca | 9 |
| gon herm spth P | Posterior spermatheca | 9 |
| gon herm spth P | Posterior spermatheca | 3 |
| gon herm spth P | Posterior spermatheca | 3 |
| gon herm sujn A | Anterior spermatheca uterine jct | 2 |
| gon herm sujn A | Anterior spermatheca uterine jct | 2 |
| gon herm sujn A | Anterior spermatheca uterine jct | 2 |
| gon herm sujn P | Posterior spermatheca uterine jct | 2 |
| gon herm sujn P | Posterior spermatheca uterine jct | 2 |
| gon herm sujn P | Posterior spermatheca uterine jct | 2 |

| Parts List | Short Description | Number of Cells |
|---|---|---|
| gon herm dut | Dorsal uterus | 14 |
| gon herm dut | Dorsal uterus | 14 |
| gon herm vut | Ventral uterus (5L) | 10 |
| gon herm vut | Ventral uterus (5L) | 10 |
| gon herm vut | Ventral uterus (5L) or (by alternate cell lineage) | 12 |
| gon herm vut | Ventral uterus (5R) | 12 |
| gon herm vut | Ventral uterus (5R) | 10 |
| gon herm vut | Ventral uterus (5R) | 10 |
| gon male dtc | Distal tip cell inhibit meiosis | 1 |
| gon male dtc | Distal tip cell inhibit meiosis | 1 |
| gon male link | Linker cell initiates union w/ cloaca (ZA) or (by alternate cell lineage) | 1 |
| gon male link | Linker cell initiates union w/ cloaca (ZB) | 1 |
| gon male sves | Seminal vesicle | 1 |
| gon male sves | Seminal vesicle | 10 |
| gon male sves | Seminal vesicle | 1 |
| gon male sves | Seminal vesicle | 1 |
| gon male sves | Seminal vesicle | 10 |
| gon male vdef | Vas deferens (ZA) | 10 |
| gon male vdef | Vas deferens (ZA) or (by alternate cell lineage) | 20 |
| gon male vdef | Vas deferens (ZB) | 20 |
| gon male vdef | Vas deferens (ZB) | 10 |
| hmc | Head mesodermal cell | 1 |
| hyp1 | Hypodermal syncytium in head | 3 |
| hyp2 | Hypodermal syncitium in head | 2 |
| hyp3 | Hypodermal syncitium in head | 2 |
| hyp4 | Hypodermal syncitium in head | 3 |
| hyp5 | Hypodermal syncitium in head | 2 |
| hyp6 | Hypodermal syncitium in head | 6 |
| hyp7 | Embryonic hypodermal syncitium | 11 |
| hyp7 | Embryonic hypodermal symcitium | 12 |
| hyp7L | Postemb hypodermal syncitium | 6 |
| hyp7R | Postemb hypodermal syncitium | 6 |
| hyp7L | Postemb hypodermal syncitium | 8 |
| hyp7R | Postemb hypodermal syncitium | 8 |
| hyp7 | Postemb hypodermal syncitium | 1 |
| hyp7 | Postemb hypodermal syncitium | 1 |
| hyp7 | Postemb hypodermal syncitium | 1 |

| Parts List | Short Description | Number of Cells |
|---|---|---|
| hyp7L | Postemb hypodermal syncitium | 4 |
| hyp7R | Postemb hypodermal syncitium | 4 |
| hyp7L | Postemb hypodermal syncitium | 14 |
| hyp7R | Postemb hypodermal syncitium | 14 |
| hyp7L | Postemb hypodermal syncitium | 14 |
| hyp7R | Postemb hypodermal syncitium | 12 |
| hyp7L | Postemb hypodermal syncitium | 14 |
| hyp7R | Postemb hypodermal syncitium | 14 |
| hyp7L | Postemb hypodermal syncitium | 14 |
| hyp7R | Postemb hypodermal syncytium | 14 |
| hyp7L | Postemb hypodermal syncitium | 4 |
| hyp7R | Postemb hypodermal syncitium | 4 |
| hyp7L | Postemb hypodermal syncitium | 10 |
| hyp7R | Postemb hypodermal syncitium | 10 |
| hyp7 herm | Postemb hypodermal syncitium | 1 |
| hyp7 herm | Postemb hypodermal syncitium | 1 |
| hyp7 herm | Postemb hypodermal syncitium | 1 |
| hyp7 herm | Postemb hypodermal syncitium | 1 |
| hyp7 herm | Postemb hypodermal syncitium | 1 |
| hyp7 herm | Postemb hypodermal syncitium | 1 |
| hyp7 herm | Postemb hypodermal syncitium | 1 |
| hyp7 herm | Postemb hypodermal syncitium | 1 |
| hyp7 herm L | Postemb hypodermal syncitium | 2 |
| hyp7 herm R | Postemb hypodermal syncitium | 2 |
| hyp7 herm L | Postemb hypodermal syncitium | 4 |
| hyp7 herm R | Postemb hypodermal syncitium | 4 |
| hyp7 herm L | Postemb hypodermal syncitium | 4 |
| hyp7 herm R | Postemb hypodermal syncitium | 4 |
| hyp7 male | Postemb hypodermal syncitium | 1 |
| hyp7 male | Postemb hypodermal syncitium | 1 |
| hyp7 male | Postemb hypodermal syncitium | 1 |
| hyp7 male | Postemb hypodermal syncitium | 1 |
| hyp7 male | Postemb hypodermal syncitium | 1 |
| hyp7 male | Postemb hypodermal syncitium | 1 |
| hyp7 male | Postemb hypodermal syncitium | 1 |
| hyp7 male | Postemb hypodermal syncitium | 1 |
| hyp7 male | Postemb hypodermal syncitium | 1 |
| hyp7 male L | Postemb hypodermal syncitium | 8 |
| hyp7 male R | Postemb hypodermal syncitium | 8 |
| hyp7 male L | Postemb hypodermal syncitium | 6 |

| Parts List | Short Description | Number of Cells |
|---|---|---|
| hyp7 male R | Postemb hypodermal syncitium | 6 |
| hyp7 male L | Postemb hypodermal syncitium | 8 |
| hyp7 male R | Postemb hypodermal syncitium | 8 |
| hyp8 | Tail ventral hypodermis | 1 |
| hyp9 | Tail ventral hypodermis | 1 |
| hyp10 | Tail ventral hypodermis | 2 |
| hyp11 | Tail dorsal hypodermis | 1 |
| hyp P12 | Preanal hypodermis | 1 |
| hyp hook | Hypodermis with hook sensillum male | 1 |
| hyp hook | Hypodermis with hook sensillum male | 3 |
| int | Embryonic intestinal cell | 20 |
| int | Postemb 14 intest. nuclei no cell div. | 0 |
| linker killer | 1 cell 50 sometimes fused with U l/ra- | 1 |
| linker killer | 1 cell 50 phagocytoses male linker cell | 1 |
| m1DL | Pharyngeal muscle cell | 1 |
| m1DR | Pharyngeal muscle cell | 1 |
| m1L | Pharyngeal muscle cell | 1 |
| m1R | Pharyngeal muscle cell | 1 |
| m1VL | Pharyngeal muscle cell | 1 |
| m1VR | Pharyngeal muscle cell | 1 |
| m2DL | Pharyngeal muscle cell | 1 |
| m2DR | Pharyngeal muscle cell | 1 |
| m2L | Pharyngeal muscle cell | 1 |
| m2R | Pharyngeal muscle cell | 1 |
| m2VL | Pharyngeal muscle cell | 1 |
| m2VR | Pharyngeal muscle cell | 1 |
| m3DL | Pharyngeal muscle cell | 1 |
| m3DR | Pharyngeal muscle cell | 1 |
| m3L | Pharyngeal muscle cell | 1 |
| m3R | Pharyngeal muscle cell | 1 |
| m3VL | Pharyngeal muscle cell | 1 |
| m3VR | Pharyngeal muscle cell | 1 |
| m4DL | Pharyngeal muscle cell | 1 |
| m4DR | Pharyngeal muscle cell | 1 |
| m4L | Pharyngeal muscle cell | 1 |
| m4R | Pharyngeal muscle cell | 1 |
| m4VL | Pharyngeal muscle cell | 1 |
| m4VR | Pharyngeal muscle cell | 1 |
| m5DL | Pharyngeal muscle cell | 1 |
| m5DR | Pharyngeal muscle cell | 1 |

| Parts List | Short Description | Number of Cells |
|---|---|---|
| m5L | Pharyngeal muscle cell | 1 |
| m5R | Pharyngeal muscle cell | 1 |
| m5VL | Pharyngeal muscle cell | 1 |
| m5VR | Pharyngeal muscle cell | 1 |
| m6D | Pharyngeal muscle cell | 1 |
| m6VL | Pharyngeal muscle cell | 1 |
| m6VR | Pharyngeal muscle cell | 1 |
| m7D | Pharyngeal muscle cell | 1 |
| m7VL | Pharyngeal muscle cell | 1 |
| m7VR | Pharyngeal muscle cell | 1 |
| m8 | Pharyngeal muscle cell | 1 |
| mc1DL | Pharyngeal marginal cell | 1 |
| mc1DR | Pharyngeal marginal cell | 1 |
| mc1V | Pharyngeal marginal cell | 1 |
| mc2DL | Pharyngeal marginal cell | 1 |
| mc2DR | Pharyngeal marginal cell | 1 |
| mc2V | Pharyngeal marginal cell | 1 |
| mc3DL | Pharyngeal marginal cell | 1 |
| mc3DR | Pharyngeal marginal cell | 1 |
| mc3V | Pharyngeal marginal cell | 1 |
| mu anal | Anal depressor muscle | 1 |
| mu bod | Embryonic body wall muscle | 1 |
| mu bod | Embryonic body wall muscle | 32 |
| mu bod | Embryonic body wall muscle | 20 |
| mu bod | Embryonic body wall muscle | 28 |
| mu bod | Postemb body wall muscle | 14 |
| mu male diag | Diagonal muscle body wall | 15 |
| mu male gub | Male gubernacular muscle | 4 |
| mu int L | Intestinal muscle to int. & body wall | 1 |
| mu int R | Intestinal muscle to int. & body wall | 1 |
| mu male long | Longitudinal muscle body wall | 10 |
| mu male obl | Oblique muscle body wall | 4 |
| mu sph | Sphincter intestino-rectal valve | 1 |
| mu male spic | Muscle copulatory spicule | 8 |
| mu herm ut | Uterine muscle | 8 |
| mu herm vul | Vulval muscle | 8 |
| proct | Male proctodeum 51 union of int. & vas | 19 |
| proct | Male proctodeum 51 def.-w/ copul. spic. | 4 |
| rect D | Rectal epithelial cell 52 near intesti- | 1 |
| rect VL | Rectal epithelial cell 52 no-rectal val | 1 |

| Parts List | Short Description | Number of Cells |
|---|---|---|
| rect VR | Rectal epithelial cell 52 -ve-microvil. | 1 |
| rect hyp | Rectal hypodermis | 1 |
| rect hyp | Rectal hypodermis | 1 |
| seL | Seam hypodermal cell | 4 |
| seR | Seam hypodermal cell | 4 |
| seL | Seam hypodermal cell | 2 |
| seR | Seam hypodermal cell | 2 |
| seL | Seam hypodermal cell | 4 |
| seR | Seam hypodermal cell | 4 |
| seL | Seam hypodermal cell | 4 |
| seR | Seam hypodermal cell | 4 |
| seL | Seam hypodermal cell | 4 |
| seR | Seam hypodermal cell | 4 |
| seL | Seam hypodermal cell | 4 |
| seR | Seam hypodermal cell | 4 |
| se herm L | Seam hypodermal cell | 2 |
| se herm R | Seam hypodermal cell | 2 |
| se herm L | Seam hypodermal cell | 2 |
| se herm R | Seam hypodermal cell | 2 |
| se herm L | Seam hypodermal cell | 4 |
| se herm R | Seam hypodermal cell | 4 |
| se male L | Seam hypodermal cell | 2 |
| se male R | Seam hypodermal cell | 2 |
| setL | Tail seam hypodermal cell male | 2 |
| setR | Tail seam hypodermal cell male | 2 |
| setL | Tail seam hypodermal cell male | 8 |
| setR | Tail seam hypodermal cell male | 8 |
| spike/death | Temporary cell for tail spike | 1 |
| spike/death | Temporary cell for tail spike | 1 |
| virL | Intestino-rectal valve | 1 |
| virR | Intestino rectal valve | 1 |
| vpi1 | Pharyngo-intestinal valve | 1 |
| vpi2DL | Pharyngo-intestinal valve | 1 |
| vpi2DR | Pharyngo-intestinal valve | 1 |
| vpi2V | Pharyngo-intestinal valve | 1 |
| vpi3D | Pharyngo-intestinal valve | 1 |
| vpi3V | Pharyngo-intestinal valve | 1 |
| vulva | Hermaphrodite vulva | 7 |
| vulva | Hermaphrodite vulva | 8 |
| vulva | Hermaphrodite vulva | 7 |

--- End of File ---

## G. MOTORCON FILE SET

### G.1. *MOTORCON.TXT*

#### G.1.a. Description

The MOTORCON data file includes a collation of myoneural connections but omits interconnections among motor neurons. The latter are found in the PREPOSGO files. The MOTORCON data file has seven (7) columns.

1) The **Seq** column simply provides a sequential numbering system for the motoneurons in the Name column.

2) The **Name** column contains the names of the motoneurons that have myoneural junctions with muscle segments in the NMJ column.

3) The **Gang** column contains the ganglion or neuron group to which any one motoneuron in the Name column belongs. The ganglionic designation is the same as the first letter of the three to four-letter designations given in the LEGENDGO files. Not all ganglia or neuron groups, however, make connections with muscle segments (i.e., they may have motor connections or function, but have no myoneural junction). The following ganglia or neuron groups have myoneural connections and are used in this column:

A = Anterior Ganglion

C = Lateral Ganglion

D = Ventral Ganglion

E = Retrovesicular Ganglion

G = Ventral Cord Neuron Group

H = Pre-anal Ganglion

Where commissural information is applicable and available, with a motoneuron passing either through a left-sided (L) or a right-sided (R) commissure, a second letter (either L or R) appears in the designation to reflect adequately this information. For example, "GR" in the Gang column means that the corresponding motoneuron in the Name column belongs to the Ventral Cord Neuron Group ("G", the first letter), and this motoneuron passes through a commissure on the right ("R", the second letter) side of the body.

The ventral cord cells ("G", above) are listed in the physical order shown in Figure 4, page 14 of reference (1) below, which appears to be the same as Figure 1, page 340 of reference (3) below.

4) The **Type** column classifies the motoneuron in the Name column as either excitatory (E) or inhibitory (I), where known.

5) The **Comm** column refers to the diagram of commissures for the female *Ascaris* (Zuckerman, vol. 1, reference (2) below, Figure 3, p. 171), where ventral cord segments with repeating commissural patterns take Roman numeral designations, and are in an orderly right and left pattern. A similar commissural tabulation used in the Gang column, above, was developed from line drawings in reference (1) below (White, et. al., Figure 7, p. 18).

6) The **Ascaris** column is an attempt to align the ventral cord neurons of *C. elegans* with the ventral cord neurons of *Ascaris*, using the motoneuron equivalences given reference (2) below (Zuckerman, vol. 1, Table 1, p. 172).

7) The **NMJ** column contains labels of muscle segments that appear to be the best choice to have myoneural connections with the motoneurons in the Name column. Choices were made from specific and detailed reference in the literature (references (1), (2), and (3) below), or by extrapolating general patterns.

The first letter of the muscle segment designation follows the pattern A and B for the head, and C and D for the neck (White, et. al., reference (1) below, Tables 2 & 3, pp. 50 & 51). E through L were used to extend this somatic group labeling convention to the remaining eight body muscle groups, giving a twelve-segment structure to the worm. No effort was made to include uterine or other muscles. The second letter is either D or V, representing the vertical orientation as either "D"orsal or "V"entral. The third letter is either L or R, representing the horizontal orientation as either "L"eft or "R"ight. The fourth letter is either L or M, representing where the oblique orientation is closer, either to Dorsal or Ventral ("M"edial), or to Left or Right ("L"ateral). Thus, "ADLL" means a head muscle segment ("A"), found in the "D"orsal, "L"eft, and "L"ateral part of the worm (see continuation of Figure 1.8, bottom, p. 8). Some entries have only two letters, and the reason for this is in the Comments section of this text file.

This first letter notation in the NMJ column may confuse with the use of a similar letter notation for the ganglia, but no cross over index is required. When the NMJ column entries are sorted on muscle segment order (i.e., A to L), the original order of the motoneurons in the Name column (derived from reference (1) below, Tables 2 & 3, pp. 50 & 51) is permuted together with the corresponding entries in the Gang column. By doing this, it is observed that the order of muscular innervation and ganglionic position do not align. Therefore, this pattern does not coordinate with literature.

### G.1.b. Derivation

Data were derived from:

1) White, J.G.; Southgate, E.; Thomson, J.N.; and Brenner, S. The Structure of the Nervous System of the Nematode Caenorhabditis elegans. Phil. Trans. R. Soc. Lond. 314(B 1165):1-340, 1986.

Specific pages of interest are page 50 (Table 2), page 51 (Table 3), page 18 (Figure 7), and page 14 (Figure 4).

2) Zuckerman, B.M. Nematodes as Biological Models, Behavioral and Developmental Models, vol. 1, edited by B. Zuckerman. New York: Academic Press, 1980.

Specific pages of interest are page 171 (Figure 3), page 172 (Table 1), and page 173 (Figure 5).

3) Wood, W.B. The Nematode Caenorhabditis elegans (Monograph 17), edited by W. Wood and the Community of C. elegans Researchers. Cold Spring Harbor: Cold Spring Harbor Laboratory, 1988.

Specific pages of interest are pages 358 to 363 (D. Analysis of Circuitry: The Motor Nervous System), and page 340 (Figure 1).

By orderly extrapolation from literature, the motor neurons were distributed into segments such that at least one of each cell type VA, VB, VD, DA, DB, or DD was located in each segment. AS acted when necessary as a surrogate for DD. VC was taken wherever it occurred, and without considering its role in uterine innervation, it was assigned a ventral segmental role. The general rule that VA and DA connect to one segment anterior to their physical location, and VB and DB connect one segment posteriorly was followed where possible. Neurons VD and DD are strictly intrasegmental in connection (i.e., they do not overlap innervation with other muscle segments). Connections between motoneurons are shown in the PREPOSGO data file and are not duplicated in this data file.

The connections among motoneurons in the PREPOSGO data file, listed in the stereotypical segmental pattern within a segment (reference (3), page 359, Figure 11), covers the reciprocal inhibitory pattern suggested in the literature between dorsal and ventral musculature. This means that basically, VD motoneurons have synaptic connections with VC, AS, DB, DA, and VB motoneurons, while DD motoneurons have synaptic connections with VA, VB, DA, DB, and AS motoneurons. Ignoring the VC and AS motoneurons, this provides the six fold symmetry of Figure 11, page 359 of reference (3).

### G.1.c. Comments

Several formats of the data referenced in literature are shown in Figure 1.8, middle, bottom right and left, p. 7, and in the continuation of Figure 1.8, bottom, p. 8.

Since the pharyngeal connectivity data are yet incomplete (as stated in the LEGENDGO and PREPOSGO files), myoneural connections of the pharynx are omitted in the MOTORCON data file.

There are numbered gaps in the MOTORCON data file. The only entry in a gap is a number from the Seq column. These gaps were intentional, placed to separate the muscle segments A to L.

There remains ambiguity about the innervation of the anterior head and neck musculature. The rule followed is as stated in reference (1): no ventral cord neuron innervates

214

a head muscle in segments A and B. In literature, neurons of the ventral cord class are also shown in the ventral and retrovesicular ganglia, and appear to overlap the innervation of neck muscles. Entries affected by these ambiguities are therefore conjectural, based upon location, and upon the V or D designation.

Finally the assignment of muscle segments to dorsolateral, dorsomedial, ventrolateral, and ventromedial orientations could not be made on all body muscle segments. Only a dorsal or a ventral connection is hypothetically made for the other muscle segments, on the assumption that four muscles in each half of the body are innervated. The latter is the reason for the two-letter designations (instead of four) in the NMJ column of the body segments.

--- End of File ---

**G.2.** *MOTORCON.PRN*

| Seq | Name | Gang | Type | Comm | Ascaris | NMJ |
|---|---|---|---|---|---|---|
| 1 | RMHR | D | | | | ADLL |
| 2 | IL1L | A | | | | ADLL |
| 3 | RMER | A | | | | ADLL |
| 4 | IL1DL | A | | | | ADLL |
| 5 | RMDDR | D | | | | ADLL |
| 6 | RMDDR | D | | | | ADLM |
| 7 | IL1DL | A | | | | ADLM |
| 8 | RMEV | A | | | | ADLM |
| 9 | URADL | A | | | | ADLM |
| 10 | SMBDR | D | | | | ADLM |
| 11 | RMDL | C | | | | ADRL |
| 12 | RMDDL | D | | | | ADRL |
| 13 | RMHL | D | | | | ADRL |
| 14 | IL1DR | A | | | | ADRL |
| 15 | RMEL | A | | | | ADRL |
| 16 | URADR | A | | | | ADRL |
| 17 | IL1R | A | | | | ADRL |
| 18 | SMBDL | D | | | | ADRL |
| 19 | URADR | A | | | | ADRM |
| 20 | RMEV | A | | | | ADRM |
| 21 | SMBDL | D | | | | ADRM |
| 22 | RMDDL | D | | | | ADRM |
| 23 | IL1DR | A | | | | ADRM |
| 24 | URAVL | A | | | | AVLL |
| 25 | RMER | A | | | | AVLL |
| 26 | SMBVL | D | | | | AVLL |
| 27 | IL1VL | A | | | | AVLL |
| 28 | RMDVR | C | | | | AVLL |
| 29 | RMHR | D | | | | AVLL |
| 30 | RMDDR | D | | | | AVLL |
| 31 | IL1L | A | | | | AVLL |
| 32 | URAVL | A | | | | AVLM |
| 33 | RMED | A | | | | AVLM |
| 34 | RMDVR | C | | | | AVLM |
| 35 | SMBVL | D | | | | AVLM |
| 36 | IL1VL | A | | | | AVLM |
| 37 | SMBVR | D | | | | AVRL |
| 38 | RMDL | C | | | | AVRL |

| Seq | Name | Gang | Type | Comm | Ascaris | NMJ |
|-----|-------|------|------|------|---------|------|
| 39 | RMDVL | C | | | | AVRL |
| 40 | RMEL | A | | | | AVRL |
| 41 | IL1VR | A | | | | AVRL |
| 42 | RMFL | D | | | | AVRL |
| 43 | IL1R | A | | | | AVRL |
| 44 | RMDDL | D | | | | AVRL |
| 45 | URAVR | A | | | | AVRL |
| 46 | RMHL | D | | | | AVRL |
| 47 | URAVR | A | | | | AVRM |
| 48 | SMBVR | D | | | | AVRM |
| 49 | SMDVL | C | | | | AVRM |
| 50 | IL1VR | A | | | | AVRM |
| 51 | RMDVL | C | | | | AVRM |
| 52 | RMER | A | | | | BDLL |
| 53 | RMHR | D | | | | BDLL |
| 54 | RMDL | C | | | | BDLL |
| 55 | RMDDR | D | | | | BDLL |
| 56 | IL1L | A | | | | BDLL |
| 57 | RMDR | C | | | | BDLL |
| 58 | URADL | A | | | | BDLL |
| 59 | SMBDR | D | | | | BDLL |
| 60 | SMDDR | D | | | | BDLM |
| 61 | SMBDR | D | | | | BDLM |
| 62 | SMDDL | D | | | | BDLM |
| 63 | URADL | A | | | | BDLM |
| 64 | RMEV | A | | | | BDLM |
| 65 | RMDDR | D | | | | BDLM |
| 66 | IL1DL | A | | | | BDLM |
| 67 | RMHL | D | | | | BDRL |
| 68 | URADR | A | | | | BDRL |
| 69 | RMEL | A | | | | BDRL |
| 70 | RMDDL | D | | | | BDRL |
| 71 | IL1R | A | | | | BDRL |
| 72 | SMDDL | D | | | | BDRL |
| 73 | SMBDL | D | | | | BDRL |
| 74 | RMFL | D | | | | BDRL |
| 75 | RMDL | C | | | | BDRL |
| 76 | RMEV | A | | | | BDRM |
| 77 | SMBDR | D | | | | BDRM |
| 78 | SMBDL | D | | | | BDRM |

| Seq | Name | Gang | Type | Comm | Ascaris | NMJ |
|-----|------|------|------|------|---------|-----|
| 79 | SMDDL | D | | | | BDRM |
| 80 | RMDVR | C | | | | BVLL |
| 81 | SMBVL | D | | | | BVLL |
| 82 | SMDVR | C | | | | BVLL |
| 83 | SMDVR | C | | | | BVLM |
| 84 | URAVL | A | | | | BVLM |
| 85 | SMBVL | D | | | | BVLM |
| 86 | RMED | A | | | | BVLM |
| 87 | RMEL | A | | | | BVRL |
| 88 | RMFL | D | | | | BVRL |
| 89 | RMDL | C | | | | BVRL |
| 90 | SMDVL | C | | | | BVRL |
| 91 | SMBVR | D | | | | BVRL |
| 92 | URAVR | A | | | | BVRL |
| 93 | IL1R | A | | | | BVRL |
| 94 | SMBVR | D | | | | BVRM |
| 95 | URAVR | A | | | | BVRM |
| 96 | RMED | A | | | | BVRM |
| 97 | RMDL | C | | | | CDLL |
| 98 | RMHR | D | | | | CDLL |
| 99 | SMDDR | D | | | | CDLL |
| 100 | RIMR | C | | | | CDLL |
| 101 | RMDR | C | | | | CDLL |
| 102 | RMGL | E | | | | CDLL |
| 103 | SMBDR | D | | | | CDLM |
| 104 | SMDDL | D | | | | CDLM |
| 105 | RMEV | A | | | | CDLM |
| 106 | SMDDR | D | | | | CDLM |
| 107 | RIML | C | | | | CDRL |
| 108 | RMGR | E | | | | CDRL |
| 109 | SMBDL | D | | | | CDRM |
| 110 | SMDDL | D | | | | CDRM |
| 111 | RMDR | C | | | | CVLL |
| 112 | RIMR | C | | | | CVLL |
| 113 | RIVR | C | | | | CVLL |
| 114 | RMGL | E | | | | CVLL |
| 115 | SMBVL | D | | | | CVLM |
| 116 | RMED | A | | | | CVLM |
| 117 | RIVR | C | | | | CVLM |
| 118 | RIVL | C | | | | CVRL |

| Seq | Name | Gang | Type | Comm | Ascaris | NMJ |
|-----|------|------|------|------|---------|-----|
| 119 | RMGR | E | | | | CVRL |
| 120 | RIML | C | | | | CVRL |
| 121 | RMDL | C | | | | CVRL |
| 122 | RIVR | C | | | | CVRM |
| 123 | RMDVR | C | | | | CVRM |
| 124 | RIVL | C | | | | CVRM |
| 125 | SMBVR | D | | | | CVRM |
| 126 | RMDVL | C | | | | CVRM |
| 127 | RIMR | C | | | | DDLL |
| 128 | RIMR | C | | | | DVLL |
| 129 | RMDVR | C | | | | DVLM |
| 130 | RIVR | C | | | | DVLM |
| 131 | RMGR | E | | | | DVRL |
| 132 | RMDVL | C | | | | DVRM |
| 133 | RIVL | C | | | | DVRM |
| 134 | VB1 | E | E | 0L | V-2 | CV |
| 135 | DB1 | E | E | 0L | DE2 | CV |
| 136 | VD1 | E | I | 0L | VI | CV |
| 137 | VB2 | E | E | IL | V-2 | DV |
| 138 | VA1 | E | E | IL | V-1 | DV |
| 139 | AS1 | E | E | IR | DE1 | DV |
| 140 | DD1 | E | I | IR | DI | DV |
| 141 | DA1 | E | E | IR | DE1 | DV |
| 142 | VD2 | E | I | IR | VI | DV |
| 143 | DB2 | E | E | IL | DE2 | DV |
| 144 | pore | | | | | |
| 145 | VA2 | G | E | IIR | V-1 | CV |
| 146 | VB3 | G | E | IIR | V-2 | CV |
| 147 | AS2 | GR | E | IIR | DE1 | CV |
| 148 | DB3 | GR | E | IIR | DE2 | CD |
| 149 | DA2 | GR | E | IIR | DE1 | CD |
| 150 | VD3 | GR | I | IIR | VI | CV |
| 151 | VA3 | G | E | IIL | V-1 | CV |
| 152 | | | | | | |
| 153 | VB4 | G | E | IIIR | V-2 | DV |
| 154 | VC1 | G | E | IIIR | | DV |
| 155 | DD2 | GR | I | IIIR | DI | DD |
| 156 | AS3 | GR | E | IIIR | DE1 | DV |
| 157 | VD4 | GR | I | IIIR | VI | DV |
| 158 | DA3 | GL | E | IIIR | DE1 | DD |

| Seq | Name | Gang | Type | Comm | Ascaris | NMJ |
|-----|------|------|------|------|---------|-----|
| 159 | VA4 | G | E | IIIL | V-1 | DV |
| 160 | | | | | | |
| 161 | VB5 | G | E | IVR | V-2 | EV |
| 162 | VC2 | G | E | IVR | | EV |
| 163 | DB4 | GL | E | IVR | DE2 | ED |
| 164 | AS4 | GR | E | IVR | DE1 | EV |
| 165 | VD5 | GR | I | IVR | VI | EV |
| 166 | VA5 | G | E | IVR | V-1 | EV |
| 167 | | | | | | |
| 168 | VB6 | G | E | IVL | V-2 | FV |
| 169 | DA4 | GL | E | VR | DE1 | FD |
| 170 | DD3 | GR | I | VR | DI | FD |
| 171 | VC3 | G | E | VR | | FV |
| 172 | AS5 | GR | E | VR | DE1 | FV |
| 173 | VD6 | GR | I | VR | VI | FV |
| 174 | VA6 | G | E | VR | V-1 | FV |
| 175 | | | | | | |
| 176 | VB7 | G | E | VL | V-2 | GV |
| 177 | DB5 | GL | E | | DE2 | GD |
| 178 | AS6 | GR | E | | DE1 | GV |
| 179 | VD7 | GR | I | | VI | GV |
| 180 | VC4 | G | E | | | GV |
| 181 | vagina | | | | | |
| 182 | DA5 | GL | E | | DE1 | GD |
| 183 | VC5 | G | E | | | GV |
| 184 | VA7 | G | E | | V-1 | GV |
| 185 | | | | | | |
| 186 | VB8 | G | E | | V-2 | HV |
| 187 | AS7 | GR | E | | DE1 | HD |
| 188 | DD4 | GR | I | | DI | HD |
| 189 | VD8 | GR | I | | VI | HV |
| 190 | VA8 | G | E | | V-1 | HV |
| 191 | | | | | | |
| 192 | VB9 | G | E | | V-2 | IV |
| 193 | VC6 | G | E | | | IV |
| 194 | DB6 | GR | E | | DE2 | ID |
| 195 | AS8 | GR | E | | DE1 | ID |
| 196 | VD9 | GR | I | | VI | IV |
| 197 | DA6 | GL | E | | DE1 | ID |
| 198 | VA9 | G | E | | V-1 | IV |

220

| Seq | Name | Gang | Type | Comm | Ascaris | NMJ |
|-----|------|------|------|------|---------|-----|
| 199 |      |      |      |      |         |     |
| 200 | VB10 | G    | E    |      | V-2     | JV  |
| 201 | AS9  | GR   | E    |      | DE1     | JD  |
| 202 | DD5  | GR   | I    |      | DI      | JD  |
| 203 | VD10 | G    | I    |      | VI      | JV  |
| 204 | VA10 | G    | E    |      | V-1     | JV  |
| 205 |      |      |      |      |         |     |
| 206 | VB11 | G    | E    |      | V-2     | KV  |
| 207 | DB7  | GR   | E    |      | DE2     | KD  |
| 208 | AS10 | GR   | E    |      | DE1     | KD  |
| 209 | DA7  | GL   | E    |      | DE1     | KD  |
| 210 | VD11 | GR   | I    |      | VI      | KV  |
| 211 | VA11 | G    | E    |      | V-1     | KV  |
| 212 |      |      |      |      |         |     |
| 213 | VD12 | HR   | I    |      | VI      | LV  |
| 214 | DD6  | HR   | I    |      | DI      | LD  |
| 215 | DA8  | HL   | E    |      | DE1     | LD  |
| 216 | AS11 | HR   | E    |      | DE1     | LD  |
| 217 | DA9  | H    | E    |      | DE1     | LD  |
| 218 | VA12 | H    | E    |      | V-1     | LV  |
| 219 | VD13 | HR   | I    |      | VI      | LV  |

--- End of File ---

Chapter IV

## Sample Computer Programs

*The Sixth no sooner had begun*
*About the beast to grope,*
*Than, seizing on the swinging tail*
*That fell within his scope*
*'I see,' quoth he, 'The Elephant*
*Is very like a rope!'*
-The Blind Men and the Elephant, J.G. Saxe

## A. README.1ST

### A.1 *Purpose*

The following computer programs were written to provide examples on manipulation of the data provided in Chapter III. The intended users are in small science, probably students, or in personal speculation, which characterizes interactive display. Curious individuals and investigators are encouraged to execute or modify these programs, or to create their own programs, for the purposes of exploring, examining, or studying the *C. elegans* data through computation. The programs are written in FORTRAN, a computer language commonly used in science and engineering, and in BASIC, to take advantage of the graphics commands and capabilities of this interpreter. The programs written in FORTRAN have been run on personal computers and supercomputers with only minor changes.

### A.2 *Requirements*

An IBM™ or compatible personal computer (from the XT™ model and upwards) with 640 Kb of RAM, a 360 Kb floppy drive, a hard drive, a monitor that accommodates at least 320x320 pixels, and an EGA, VGA, or Hercules card are the minimal hardware requirements to run any of these programs. Larger computers or supercomputers may also be used to modify and execute the sample FORTRAN programs. Access to such computers is not necessary, although some examples are provided on how they may be used.

To run the BASIC programs, a BASIC interpreter is needed. In particular, the BASIC programs in this chapter are written in BASICA (Advanced BASIC), which has extensive graphic capabilities. The BASIC program source codes may be edited or modified by using the interpreter itself. The FORTRAN programs are given in source code. In particular, the FORTRAN programs in this chapter are written in FORTRAN 77. The FORTRAN program source codes may be edited or modified by using any reliable program editor, and compiled by using a suitable FORTRAN compiler. Cray supercomputer users should use the FORTRAN compiler provided for the Cray.

## A.3 *Directions*

As stated subsection A.2. (*Media and Contents*) of Chapter III, Disk 2 is the program diskette. It is a 5.25 inch, 360 Kb floppy diskette which contains seven (7) program files and eleven (11) input data files that are specified by the programs. The program files, the corresponding data files, and their outputs are:

(1) Program:  DISPLAY.BAS
  Input Data Files: PAIRSGJ.DAT, PAIRSR.DAT, PAIRSA.DAT, PAIRSG.DAT, PAIRSGR.DAT, PAIRSLR.DAT, PAIRST.DAT
  Output:  Screen display of adjacency matrix

(2) Program:  MAKEDATA.BAS
  Input Data Files: 1 user-provided look-up table (e.g., UNIQSEQ.REF) and 1 modified version of the PREPOSGO.PRN file presented on Disk 1 (to be modified before input and saved as a separate file, e.g., PREPOSGO.REF)
  Output:  2 data files with user-provided names (e.g., PAIRSG.DAT and LABELG.DAT)

(3) Program:  ELEGRAPH.BAS
  Input Data Files: PAIRSG.DAT, LABELG.DAT
  Output:  Screen display of adjacency matrix; result files named PRSAV.DAT and LABSAV.DAT

(4) Program:  PLOT100.BAS
  Input Data Files: PAIRSG.DAT, LABELG.DAT
  Output:  Hard copy of adjacency matrix at 100 dots per inch resolution; optional result files named PRSAV.DAT and LABSAV.DAT (made by activating a subroutine commented out of the main program)

(5) Program:  MAKELIST.FOR
  Executable File: May be immediately compiled to MAKELIST.EXE
  Input Data File: A modified version of the PREPOSGO.PRN file presented on Disk 1 (to be modified before input and saved as a separate file named PREPOSGO.LIS)
  Output:  Data files named PAIRSG.DAT and LABELG.DAT

(6) Program:  CONNECT.FOR
  Executable File: May be immediately compiled to CONNECT.EXE
  Input Data Files: PAIRSG.DAT, LABELG.DAT
  Output:  Hard copy of results of transitive closure computations

(7) Program:  EIGCRAY.FOR
  Executable File: Submitted using a batch file as a job for the Cray
  Input Data File: PAIRSG.DAT
  Output:  Eigenvalue file named EIGCRAY.CPR.

All program files named with the .BAS extension are written in BASIC, and require a BASIC interpreter to run. All program files named with the .FOR extension are written in FORTRAN 77, and are immediately compilable to directly executable (.EXE) files when no source code modifications are desired. In the list above, the output files are termed *data files* when they are still to serve as input files to the other programs presented, and are called *result files* when they are directly the products for analysis. Names of data and result files may be user-provided and/or specified by the code of the corresponding program. Note that a single data file, e.g., PAIRSG.DAT, may be used by several programs.

IMPORTANT: Before executing any program, make sure that the program to be executed and its required input data files are located in the same directory or subdirectory. An easy way of doing this is to copy the contents of Disk 1 and Disk 2 into one subdirectory on the hard disk. Initially, a BASIC interpreter must be copied into the same subdirectory for the BASIC programs in this chapter to run properly, although later, codes may be modified for programs to execute with their data located in a separate subdirectory. Editors and compilers can work by simply providing the path for access to the subdirectory, together with the file name to be modified or compiled. With enough memory, FORTRAN executable files can be run directly from this subdirectory.

CONVENTION: When the format of a line in a data file is presented, e.g., I4,A7,A4, FORTRAN programming convention is followed. In the example given, entries shorter than A7 or A4 are left-justified within 7 or 4 spaces, respectively, and entries shorter than I4 are right-justified within 4 spaces. Within a line, entries are not comma-delimited, and a line is terminated by a single carriage return. When using electronic spreadsheets to prepare data, column widths and entry justifications should also follow this convention.

## B. The BASIC Programs

### B.1. *DISPLAY.BAS*

#### B.1.a. Description

This program provides color displays of Figures 1.11 to 1.16 in Chapter I, pages 10, 12, 13. Figure 1.11 is the symmetrical matrix representation of the gap junctions between neurons. Figures 1.12 to 1.16, on the other hand, provide the topological deformations of the *C. elegans* vertex adjacency matrix. The standard BASIC colors coded 1 to 15 are equated with connection density, such that a blue matrix element (color code 1) indicates that a connection occurs only once, and a high-intensity white dot (color code 15) indicates that the same connection occurs 15 or more times. The display program and input data have been updated since they were presented in reference [14], and the following differences will be noticed when viewing the matrices:

(1) In the matrix showing ganglionic order of neurons (in either the alphabetical or random order of neurons within each ganglion), lines have been drawn to demarcate the areas of ventral cord connections (adjacent to the square area of the ganglia) and of pharyngeal connections (empty and adjacent to the area of ventral cord connections);

(2) In the matrix showing anatomic laterality, the neurons are arranged in *left-unpaired-right* order, instead of simply left-right order as shown in Chapter I, p. 13.

DISPLAY.BAS uses 7 input data files, listed above. Unlike Figures 1.9 and 1.10 also in Chapter I, to facilitate display, row and column labels are not shown in Figures 1.11 to 1.16. Instead, the program uses input data files where neuron names have been translated to integers. The MAKEDATA.BAS and MAKELIST.FOR programs, discussed below, make these translations from modified versions of the PREPOS-GO.PRN file provided on Disk 1. The output files of these programs also show the structure of the input data files of DISPLAY.BAS.

### B.1.b. Directions

Follow instructions in the paragraph labeled "IMPORTANT" in subsection A.3 above. Run BASICA and load "DISPLAY.BAS". Upon running "DISPLAY.BAS," a menu will appear on the screen. Select the matrix to display from the menu. Press any key to go back to the menu after viewing a matrix display. The option to exit the program is also in the menu.

### B.1.c. Source Code

```
10   REM  **************************************************
20   REM  *            DISPLAY.BAS PROGRAM             *
30   REM  *     Rewritten by Theodore B. Achacoso, M.D.   *
40   REM  *     from an initial program by Duc C. Nguyen   *
50   REM  **************************************************
60   REM *** PREPARE THE SCREEN ***
70   KEY OFF : SCREEN 0 : CLS
80   REM *** THE MAIN MENU ***
90   LOCATE 3,5 : PRINT "(1) Figure 1.11  - Gap Junction Matrix"
100  LOCATE 5,5 : PRINT "(2) Figure 1.12  - Random Order"
110  LOCATE 7,5 : PRINT "(3) Figure 1.13  - Anteroposterior Order"
120  LOCATE 9,5 : PRINT "(4) Figure 1.14  - Ganglionic Order, Alphabetical
     within Ganglion"
130  LOCATE 11,5 : PRINT "(5) Figure 1.14r - Ganglionic Order, Random
     within Ganglion"
140  LOCATE 13,5 : PRINT "(6) Figure 1.15  - Order by Laterality"
150  LOCATE 15,5 : PRINT "(7) Figure 1.16  - Order by Neuron Type"
160  LOCATE 17,5 : PRINT "(8) Exit"
170  LOCATE 20,5 : PRINT "TYPE THE NUMBER OF CHOICE"
180  CH$ = INKEY$ : IF CH$ = "" THEN 180
190  REM *** CHOOSE APPROPRIATE FILE TO OPEN ***
```

```
200  IF CH$ = "1" THEN GOTO 1000
210  IF CH$ = "2" THEN GOTO 2000
220  IF CH$ = "3" THEN GOTO 3000
230  IF CH$ = "4" THEN GOTO 4000
240  IF CH$ = "5" THEN GOTO 4000
250  IF CH$ = "6" THEN GOTO 6000
260  IF CH$ = "7" THEN GOTO 7000
270  IF CH$ = "8" THEN SYSTEM
280  GOTO 180
290  REM *** MAIN CALLING PROGRAM ENDS HERE ***
500  REM *** SUBROUTINE THAT OPENS APPROPRIATE FILE AND
     READS DATA ***
510  SCREEN 9 : CLS
520  OPEN "I",#1,PF$
530  IF EOF(1) THEN 670
540  PRE$ = INPUT$(3,#1)
550  YP = VAL(PRE$)
560  POST$ = INPUT$(3,#1)
570  XP = VAL(POST$)
580  INPUT#1,DS$
590  DENS = VAL(DS$)
600  IF DENS > 15 THEN DENS = 15
610  IF XP > 302 THEN 530
620  IF YP > 302 THEN 530
630  IF CH$ = "1" THEN 700
640  REM *** PLOT THE DATA POINTS WITH COLOR ***
650  LINE (1.1*XP,.9*YP)-(1.1*XP+1,.9*YP),DENS
660  GOTO 530
670  CLOSE #1
680  RETURN
700  REM *** SPECIAL PLOT OF DATA POINTS FOR GAP JUNCTIONS
     ***
710  GOSUB 730
720  GOTO 660
730  LINE (1.3*XP,YP)-(1.3*XP+1,YP),DENS
740  RETURN
800  REM *** GO TO SUBROUTINE THAT DRAWS BORDERS COM-
     MON TO ALL GRAPHS ***
810  GOSUB 900
820  WA$ = INKEY$ : IF WA$ = "" THEN 820
830  GOTO 70
900  REM *** SUBROUTINE THAT DRAWS BORDERS: COMMON TO
     ALL GRAPHS ***
910  LINE (1,1)-((303*1.1),1),7
920  LINE (1,1)-(1,(303*.9)),7
930  LINE (1,(303*.9))-((303*1.1),(303*.9)),7
940  LINE ((303*1.1),1)-((303*1.1),(303*.9)),7
950  RETURN
```

```
1000 REM *** PICK UP APPROPRIATE DATA FILE NAME TO OPEN
     ***
1010 PF$ = "PAIRSGJ.DAT"
1020 REM *** GO TO SUBROUTINE THAT OPENS & READS FILES, &
     PLOTS GRAPHS ***
1030 GOSUB 500
1040 REM *** DRAW THE BORDERS: SPECIFIC TO GRAPH ***
1050 LINE (1,1)-(1.3*264,1),7
1060 LINE (1.3*264,1)-(1.3*264,263),7
1070 LINE (1,263)-(1.3*264,263),7
1080 LINE (1,1)-(1,263),7
1090 GOTO 820
2000 REM *** PICK UP APPROPRIATE DATA FILE NAME TO OPEN
     ***
2010 PF$ = "PAIRSR.DAT"
2020 REM *** GO TO SUBROUTINE THAT OPENS & READS FILES, &
     PLOTS GRAPHS ***
2030 GOSUB 500
2040 REM *** DRAW THE BORDERS: COMMON TO ALL GRAPHS ***
2050 GOTO 800
3000 REM *** PICK UP APPROPRIATE DATA FILE NAME TO OPEN
     ***
3010 PF$ = "PAIRSA.DAT"
3020 REM *** GO TO SUBROUTINE THAT OPENS & READS FILES, &
     PLOTS GRAPHS ***
3030 GOSUB 500
3040 REM *** DRAW THE LINE MARKERS SPECIFIC TO GRAPH ***
3050 LINE (166.65,1.8)-(166.65,271.7),7
3060 LINE (2.2,136.35)-(332.2,136.35),7
3070 LINE (2.2,1.8)-(332.2,271.7),7
3080 REM *** DRAW THE BORDERS: COMMON TO ALL GRAPHS ***
3090 GOTO 800
4000 REM *** PICK UP APPROPRIATE DATA FILE NAME TO OPEN
     ***
4010 IF CH$ = "4" THEN PF$ = "PAIRSG.DAT"
4020 IF CH$ = "5" THEN PF$ = "PAIRSGR.DAT"
4030 REM *** GO TO SUBROUTINE THAT OPENS & READS FILES, &
     PLOTS GRAPHS ***
4040 GOSUB 500
4050 REM *** PLOT BOX MARKERS SPECIFIC TO TWO GRAPHS ***
4060 P1 = 1 : P2 = 38 : GOSUB 4170
4070 P1 = 39 : P2 = 44 : GOSUB 4170
4080 P1 = 45 : P2 = 108 : GOSUB 4170
4090 P1 = 109 : P2 = 140 : GOSUB 4170
4100 P1 = 141 : P2 = 169 : GOSUB 4170
4110 P1 = 170 : P2 = 185 : GOSUB 4170
4120 P1 = 186 : P2 = 197 : GOSUB 4170
4130 P1 = 198 : P2 = 200 : GOSUB 4170
```

```
4140 P1 = 201 : P2 = 224 : GOSUB 4170
4142 LINE (2.2,.9*225)-(1.1*225,.9*225),7
4144 LINE (1.1*225,1.8)-(1.1*225,.9*225),7
4146 LINE (2.2,.9*283)-(1.1*283,.9*283),7
4148 LINE (1.1*283,1.8)-(1.1*283,.9*283),7
4150 REM *** DRAW THE BORDERS: COMMON TO ALL GRAPHS ***
4160 GOTO 800
4170 REM *** SUBROUTINE THAT DRAWS BOXES SPECIFIC TO TWO
     GRAPHS ***
4180 LINE (1.1*P1,.9*P1)-(1.1*P2 + 1,.9*P1),7
4190 LINE (1.1*P2 + 1,.9*P1)-(1.1*P2 + 1,.9*P2),7
4200 LINE (1.1*P1,.9*P2)-(1.1*P2 + 1,.9*P2),7
4210 LINE (1.1*P1,.9*P1)-(1.1*P1,.9*P2),7
4220 RETURN
6000 REM *** PICK UP APPROPRIATE DATA FILE NAME TO OPEN
     ***
6010 PF$ = "PAIRSLR.DAT"
6020 REM *** GO TO SUBROUTINE THAT OPENS & READS FILES, &
     PLOTS GRAPHS ***
6030 GOSUB 500
6050 P1 = 100 : GOSUB 6100
6060 P1 = 204 : GOSUB 6100
6070 REM *** DRAW THE BORDERS: COMMON TO ALL GRAPHS ***
6080 GOTO 800
6090 REM *** SUBROUTINE THAT DRAWS LINES SPECIFIC TO
     GRAPH ***
6100 LINE (1.1*P1,1.8)-(1.1*P1,271.8),7
6110 LINE (2.2,(0.9*P1))-(332.2,(0.9*P1)),7
6120 RETURN
7000 REM *** PICK UP APPROPRIATE DATA FILE NAME TO OPEN
     ***
7010 PF$ = "PAIRST.DAT"
7020 REM *** GO TO SUBROUTINE THAT OPENS & READS FILES, &
     PLOTS GRAPHS ***
7030 GOSUB 500
7040 REM *** PLOT LINE MARKERS SPECIFIC TO GRAPH ***
7050 P1 = 79 : GOSUB 7090
7060 P1 = 176 : GOSUB 7090
7070 REM *** DRAW THE BORDERS: COMMON TO ALL GRAPHS ***
7080 GOTO 800
7090 REM *** SUBROUTINE THAT DRAWS LINES SPECIFIC TO
     GRAPH ***
7100 LINE (1.1*P1,1.8)-(1.1*P1,271.8),7
7110 LINE (2.2,(0.9*P1))-(332.2,(0.9*P1))
7120 RETURN
```

--- End of Program ---

**B.2.** *MAKEDATA.BAS*

**B.2.a. Description**

This program makes the input data files used by the other programs in this chapter. The goal of MAKEDATA.BAS is simply to translate the names of 302 neurons plus 4 non-neural "cells" in the PREPOSGO.PRN file into numbers of particular sequence. Thus, there are two input files: (1) the data file that has to be translated into numbers, and (2) the look-up table for numbers and their sequence. The data file is a modification of the PREPOSGO.PRN file, so that only the three pertinent columns, viz., Pre, Post, and Den, are listed in A7,A7,I4 format. PREPOSGO.REF is the sample data file provided with this program. The look-up table has 306 lines containing three columns, viz., sequence number, unique neuron or non-neural "cell", and cell type (from the Type column of the PREPOSGO.PRN file), listed in I4,A7,A4 format. UNIQSEQ.REF is the sample look-up table provided with this program. The advantage of this two-input file setup is that one needs only to vary the sequence number of the look-up table (or of UNIQSEQ.REF) without remodifying the PREPOSGO.PRN file (or without creating another file like PREPOSGO.REF).

The program generates two output files, both named by the user. The translation of the data file from A7,A7,I4 format into numbers is stored in an ASCII file that has an I3,I3,I4 format. This is the structure of input data like PAIRSG.DAT, which is used by the program DISPLAY.BAS above. The other output file is a reference file that contains the unique neurons and non-neural "cells" in one column, and their corresponding type in another column, in A7,A4 format, respectively. The sequence of the 306 lines in this output file is dictated by the look-up table (e.g., UNIQSEQ.REF). This is the structure of LABELG.DAT, which is used by programs like ELEGRAPH.BAS and PLOT100.BAS below.

The sample data file PREPOSGO.REF and the initial UNIQSEQ.REF file were created using an electronic spreadsdheet. The PREPOSGO.PRN file from Disk 1 was imported into a worksheet, and the file was broken into its respective columns by using the *Parse* function of the spreadsheet. Column headings were added, and column widths and entry justifications were adjusted as specified in Chapter III. Except the Entry Number, Type, Pre, Post, and Den columns, all other columns were deleted. The second data block, from Entry Numbers 2368 to 2463 were also deleted. From this worksheet, the last two columns of UNIQSEQ.REF were created first by using the *Uniq* function of the spreadsheet, the input data block being the Type and Pre columns, the criterion block being a Pre column heading and an empty cell below it, and the output block being a new Pre column (in A7 format) and an immediately adjacent new Type column (in A4 format). Entries WE, WI, WM, and WN were then appended to both columns. Immediately before these two columns, a column with the heading "Seq" in I4 format was added to hold the desired number sequence for each neuron. In UNIQSEQ.REF, Seq is simply 1 to 306, numbered by using the *Fill* function of the spreadsheet. It is in the Seq column that the identification number of a neuron can be changed, thereby changing its sequence in the data file and its presentation in a display program like DISPLAY.BAS. These three new columns, Seq, Pre, Type were then saved (without column headings) as the ASCII file UNIQSEQ.REF in I4,A7,A4 format. PREPOS-

GO.REF was prepared next, by deleting rows with blank Den column entries (Entry Numbers 2348 to 2355), and deleting all other columns except the original Pre, Post, and Den columns. These three columns, Pre, Post, Den were then saved (without column headings) as the ASCII file PREPOSGO.REF in an A7,A7,I4 format. An editor was used to verify that both UNIQSEQ.REF and PREPOSGO.REF were saved in their proper formats, modifying as necessary to conform to the formats.

Subsequently, in order to change the sequence of neurons in the vertex adjacency matrix, changes were made only in the UNIQSEQ.REF file, and the modified UNIQSEQ.REF file was saved under a different file name. For example, to produce the approximate anteroposterior sequence of the neurons, UNIQSEQ.REF was imported into an electronic worksheet, parsed accordingly into Seq, Pre, and Type columns in I4,A7,A4 format, and sorted using the Pre column as primary key, ascending. Seq column was then renumbered 1 to 306 using the *Fill* function of the spreadsheet. The Type column, discussed fully in the LEGENDGO files, pp. 95-98, is important in the production of other input data files. For example, to make PAIRSG.DAT where ganglia are arranged anteroposteriorly and neurons within ganglia are arranged alphabetically, UNIQSEQ.REF was sorted first, using the Type column as primary key so that the ganglia would arrange anteroposteriorly. UNIQSEQ.REF was then resorted alphabetically by the ganglion, using the Type column as the guide to member neurons of each ganglion. In the Seq column, neurons contained in the 9 ganglia were numbered from 1 to 224, and groups G (ventral cord neurons), P (pharyngeal neurons), and W (non-neural "cells") were numbered from 225 to 306. These divisions derived from the Type column made group demarcations in DISPLAY.BAS possible. Also, by using the *Extract* function of the spreadsheet, criterion blocks can be set up using the Type column and wildcards to isolate laterality (e.g., ?L* to extract all left-sided neurons), or to isolate neuron type (e.g., ??S* to extract all neurons with putative sensory function).

### B.2.b. Directions

Run BASICA and load MAKEDATA.BAS. Upon interpretation, the program will ask for file names. In succession, and punctuated by the < ENTER > key, type in UNIQSEQ.REF, PREPOSGO.REF, a new name for the new data sequence, say, PAIRS.DAT, and a new name for the reference file, say, LABEL.DAT. To circumvent string array memory requirements, the program was made disk-intensive, and is slow without compilation. The input data files of DISPLAY.BAS and the PAIRS.DAT output file of MAKEDATA.BAS have the same structure, as discussed above. The data files and their structures, discussed in subsection B.2.a, may be viewed or edited as ASCII files in any file editor. If you wish to display the new data sequence PAIRS.DAT with DISPLAY.BAS, back-up the data files of DISPLAY.BAS, and rename PAIRS.DAT with any of the data file names used by DISPLAY.BAS (e.g., PAIRSR.DAT). Upon executing DISPLAY.BAS, select the appropriate key in the menu that will display the file (e.g., "2" for PAIRSR.DAT).

NOTE: The PAIRSG.DAT file presented in Disk 2 was produced by MAKELIST.FOR, although the PAIRSG.DAT file generated by the

230

MAKEDATA.BAS program (with PREPOSGO.REF) is directly usable by the DIS-PLAY.BAS program as an input file. The difference is explained in the note on page 246.

### B.2.c. Source Code

```
10    KEY OFF : CLS : SCREEN 0
20    REM **********************************************
30    REM *               MAKEDATA.BAS                 *
40    REM *     Written by Theodore B. Achacoso, M.D.   *
50    REM **********************************************
60    INPUT "NAME OF UNIQ FILE FROM .PRN FILE
      <UNIQSEQ.REF>: ",UNIQ$
70    INPUT "NAME OF DATA FILE FROM .PRN FILE
      <PREPOGO.REF>: ",REDATA$
80    INPUT "NAME OF NEW DATA FILE TO BE MADE
      <PAIRS?.DAT>: ",PAIRS$
90    INPUT "NAME OF NEW FILE TO CHECK DATA ORDER
      <LABEL?.DAT>: ",LABEL$
100   DIM UNEUR$(350),TYPE$(350)
110   PRINT "PLEASE WAIT ..."
120   PRINT "READING ";UNIQ$
130   OPEN "I",#1,UNIQ$
140   IF EOF(1) THEN 300
150   SEQNO$ = INPUT$(4,#1)
160   SEQNO = VAL(SEQNO$)
170   UNEUR$(SEQNO) = INPUT$(7,#1)
180   TYPE$(SEQNO) = INPUT$(4,#1)
190   INPUT#1,CARRET$
200   GOTO 140
300   CLOSE #1
310   PRINT "READING ";REDATA$
320   PRINT "AND WRITING ";PAIRS$
330   OPEN "O",#3,PAIRS$
340   OPEN "I",#1,REDATA$
350   IF EOF(1) THEN 600
360   PRE$ = INPUT$(7,#1)
370   POST$ = INPUT$(7,#1)
380   INPUT#1,DEN
390   FOR I = 1 TO 306
400   IF PRE$ = UNEUR$(I) THEN PRE$ = STR$(I) : GOTO 450
410   NEXT I
420   PRINT "NEURON MISSING IN PRE COLUMN. PLEASE FIX DATA.":
      END
450   PRINT#3,USING "###";VAL(PRE$);
460   FOR J = 1 TO 306
470   IF POST$ = UNEUR$(J) THEN POST$ = STR$(J) : GOTO 500
```

```
480  NEXT J
490  PRINT "NEURON MISSING IN POST COLUMN. PLEASE FIX DATA.":
     END
500  PRINT#3,USING "###";VAL(POST$);
510  PRINT#3,USING "####";DEN
520  GOTO 350
600  CLOSE #1 : CLOSE #3
700  PRINT "WRITING ";LABEL$
710  OPEN "O",#3,LABEL$
720  FOR K = 1 TO 306
730  PRINT#3,USING "\      \";UNEUR$(K);
740  PRINT#3,USING "\  \";TYPE$(K)
750  NEXT K
760  CLOSE #3 : CLEAR
770  PRINT "DONE." : END
```

--- End of Program ---

**B.3.** *ELEGRAPH.BAS*

**B.3.a. Description**

This program displays the adjacency matrix of *C. elegans*. The colors represent synaptic densities up to 15, using the number to color transformation key of your computer. A keyboard-controlled cursor appears below and to the right of the screen along with a display of the text name and array location of the line (cell) to which the cursor points.

A short command line appears to the right at the bottom. The commands are:

CR = clear and redisplay
R  = sum the densities of efferent connections (rows)
C  = sum the densities of afferent connections (columns)
D  = sum the densities along diagonals of the matrix where 2*MAXPOS-1
     diagonals are possible. If there are MAXPOS entries, e.g., 306, then
     there are 611 diagonals.

The main diagonal is numbered 306. Diagonals beginning on column 2 count in descending order 305,304,...to 1 which is the right upper corner of the matrix. Diagonals beginning in row 2 count 307,308,...611 such that 611 is the lower left corner of the matrix.

A  = sum the densities in a specified rectangular area
S  = swap entries in a row-column pair such that the diagonal entries are preserved
     (i.e., symmetrically)
B  = rearrange a block of cells in ascending or descending order

F = save matrix and labels order to diskette
Q = quit program.

The program as listed is directly usable on an 80386-based, 640 Kb RAM machine, with minor modifications necessary to display the matrix on an 8088 or 8086 level IBM™ or compatible PC. The SCREEN command should be 9 for color display on an 80386-based machine. SCREEN 4 can be used on an IBM XT™-level machine. As interpreted BASIC, however, the program without compilation is uncomfortably slow at the XT level. The program may also be run under HBASIC on a Hercules monochrome monitor using the SCREEN 2 command. At least, EGA display capability is required.

Four arrays contain the data retrieved from diskette from files formatted in LABELG.DAT (A7,A4) and PAIRSG.DAT (I3,I3,I4) formats. These data are stored in GRAPH(3,2475) with densities restricted to the range 1-15; in CX$(310) and CY$(310) which contain text strings for cell name and type. In 8088 processor machines, the cell type may be truncated and cell name kept at 7 characters.

The following outlines the principal program sequence:

| | |
|---|---|
| 10-660 | Main program |
| 690-870 | Subroutine displays matrix and command line |
| 890-940 | Subroutine moves cursor up |
| 960-1010 | Subroutine moves cursor left |
| 1030-1080 | Subroutine moves cursor right |
| 1090-1140 | Subroutine moves cursor down |
| 1160-1480 | Subroutine shifts a set of rows up or down in the table; includes screen control; uses 1440, 1560, 2440, 2610, and 3000 |
| 1440-1480 | Subroutine to sort WORK array in decreasing order; uses 1500 |
| 1500-1540 | Subroutine swaps array location in WORK array |
| 1560-1600 | Subroutine sorts WORK area in increasing order; uses 1500 |
| 1620-1830 | Subroutine sums densities in rectangular area; uses 690 |
| 1850-1900 | Subroutine screen control for swap; uses 1920 |
| 1920-2030 | Subroutine performs symmetrical swap of row/column |
| 2050-2120 | Subroutine screen control for diagonal sums; uses 2520 |
| 2130-2170 | Subroutine screen control for column sums; uses 2360 |
| 2190-2250 | Subroutine screen control for row sums; uses 2440 |
| 2270-2340 | Subroutine to sum diagonal entries |
| 2360-2420 | Subroutine to sum entries in a column |
| 2440-2500 | Subroutine to sum entries in a row |
| 2520-2590 | Subroutine to count number of entries in diagonal |
| 2610-2910 | Subroutine reorders row/columns sets from a WORK template; uses 1920, 3140, and 3200 |
| 2930-2980 | Subroutine for searching tables |
| 3000-3120 | Subroutine (not active in listing) to generate a random number WORK table |
| 3140-3270 | Subroutine to swap entries in a WORK array; uses 1920 and 2930 |
| 3290-3344 | Subroutine to save final table to diskette as files PRSAV.DAT and LAB-SAV.DAT. |

## B.3.b. Directions

Unless changes are necessary in the source code because of the type of computer being used, or because of the type of BASIC interpreter being used, ELEGRAPH.BAS can be loaded and interpreted with adequate speed by a BASICA interpreter in an 80386-based microcomputer.

### B.3.c. Source Code

```
10    ' Program ELEGRAPH.BAS displays adjacency matrix.
20    ' This program also manipulates matrix interactively.
30    ' Written by W. S. Yamamoto, M.D.
40    SCREEN 0
50    CLS : KEY OFF : SCREEN 9
60    DIM GRAPH(3,2475), WORK(3,100)
70    DIM CX$(310) : DIM CY$(310)
80    PRINT "GENERATE DIRECTORY FIRST"
90    INPUT "Input file name of labels - ",L$
100   OPEN "I",1,L$
110   J=0 : LAST=310 : WKND=100
120   IF EOF(1) THEN 160
130   J=J+1 : INPUT#1,CX$(J) : CX$(J)=LEFT$(CX$(J),11)
140   CY$(J) = CX$(J)
150   GOTO 120
160   CLOSE #1
170   MAXPOS=J
180   PRINT "DIRECTORY HAS ";MAXPOS;" NEURONS. NOW MAKE
      GRAPH."
190   INPUT "Enter node-pair file name - ",P$
200   OPEN "I",1,P$
210   N=0 : MD=0 : PRINT "GENERATING NODE-PAIR LIST"
220   IF EOF(1) THEN 310
230   N=N+1 : PRE$ = INPUT$(3,#1) : GRAPH(1,N)=VAL(PRE$)
240   POST$ = INPUT$(3,#1) : GRAPH (2,N)= VAL(POST$)
250   INPUT#1,DS$ : GRAPH(3,N)=VAL(DS$)
260   DENS = VAL (DS$)
270   IF DENS > 15 THEN DENS = 15
280   RES=N-MD*500 : IF RES=1 THEN PRINT "DONE. ";MD*500;
      " ENTRIES."
290   IF RES=1 THEN MD=MD+1
300   GOTO 220
310   CLOSE #1
320   NPAIRS=N
330   PRINT "GRAPH TABLE DONE. " ;NPAIRS; "  PAIRS."
340   INPUT "Strike any key to go on " ,Z$
350   ' Display matrix - Set initial cursor position
360   X = 160 : Y = 160
370   GOSUB 690
```

```
380  ' Shows legend, cells, and cursor
390  MF = 0   ' Moves cursor
400  KEY(11) ON
410  KEY(12) ON
420  KEY(13) ON
430  KEY(14) ON
440  ON KEY(11) GOSUB 900
450  ON KEY(12) GOSUB 970
460  ON KEY(13) GOSUB 1040
470  ON KEY(14) GOSUB 1100
480  G$ = INKEY$       ' Wait for keyboard entry
490  IF LEN(G$) = 0 THEN 480
500  IF ASC(G$) = 13 THEN GOSUB 600    ' 13 is CR
510  IF G$ = "R" THEN GOSUB 2200       ' Row sum
520  IF G$ = "A" THEN GOSUB 1620       ' Area sum
530  IF G$ = "C" THEN GOSUB 2120       ' Column sum
540  IF G$ = "D" THEN GOSUB 2280       ' Diagonal sum
550  IF G$ = "S" THEN GOSUB 1850       ' Swap row/col pair
560  IF G$ = "B" THEN GOSUB 1160       ' Shift block in order
570  IF G$ = "F" THEN GOSUB 3290       ' Save labels and pairs files
580  IF G$ = "Q" THEN 640              ' Quit is 81
590  GOTO 480
600  ' Show cursor, redisplay at old x,y
610  GOSUB 690
620  LOCATE 25,60 : PRINT "Active";
630  RETURN
640  CLS
650  STOP
660  END                 ' End of main. SUBROUTINES follow:
670  '
680  '
690  ' Plot the graph
700  CLS
710  FOR K = 1 TO NPAIRS
720  LINE (2*GRAPH(2,K),GRAPH(1,K)) - (2*GRAPH(2,K) + 1,GRAPH(1,K)),
     GRAPH(3,K)
730  NEXT K
740  LINE (632,Y)-(639,Y),0
750  LINE (2*X,316)-(2*X + 1,320),0,B
760  LOCATE 25,1 : PRINT "POST:";
770  LOCATE 24,1 : PRINT "PRE:";
780  LOCATE 24,50 : PRINT "CHOOSE A,R,C,D,S,B,F,or Q ";
790  GOSUB 810
800  RETURN
810  ' Print cell identity and location
820  LOCATE 24,7 : PRINT CY$(Y);"      ";
830  LOCATE 25,7 : PRINT CX$(X);"      ";
840  LOCATE 24,30 : PRINT Y;
```

```
850  LOCATE 25,30 : PRINT X;
860  LINE (626,Y)-(639,Y) : LINE (2*X,310)-(2*X,320)
870  RETURN
880  '
890  ' Move cursor up
900  IF Y = 1 THEN RETURN
910  LINE (626,Y)-(639,Y),0
920  Y = Y - 1
930  GOSUB 810
940  RETURN
950  '
960  ' Move cursor left
970  IF X = 1 THEN RETURN
980  LINE (2*X,310)-(2*X+1,320),0,B
990  X = X - 1
1000 GOSUB 810
1010 RETURN
1020 '
1030 ' Move cursor right
1040 IF X = MAXPOS THEN RETURN
1050 LINE (2*X,310)-(2*X+1,320),0,B
1060 X = X + 1
1070 GOSUB 810
1080 RETURN
1090 ' Move cursor down
1100 IF Y = MAXPOS THEN RETURN
1110 LINE (620,Y)-(639,Y),0
1120 Y = Y + 1
1130 GOSUB 810
1140 RETURN
1150 '
1160 ' Subroutine to shift set of rows in up or down order
1170 LOCATE 1,1 : INPUT "Rearrange block. Start at row = ";S
1180 IF S<1 OR S>MAXPOS THEN GOTO 1170
1190 INPUT "End block at row = ";F
1200 IF F<1 OR F>MAXPOS THEN 1190
1210 INPUT "Enter A,D,or R for increasing,decreasing or random order ",O$
1220 IF O$ < >"A" AND O$ < > "D" AND O$><"R" THEN GOTO 1210
1230 ' Set up a bubble sort
1240 IF F<S THEN T=F : S=F : F=T
1250 NC=F-S+1 : IF NC>WKND THEN PRINT "BLOCK TOO LARGE.
     MAX = ",WKND
1260 IF O$ = "A" THEN PRINT "Increasing order" : O=1 : GOTO 1310
1270 IF O$="D" THEN PRINT "Decreasing order" : O=2  : GOTO 1310
1280 ' Jump bubble sort and randomize
1290 IF O$="R" THEN PRINT "RANDOMIZE "
1300 GOSUB 3000 : GOTO 1400 ' Go to shuffle
```

```
1310 FOR  K=1 TO NC :  IROW=S+K-1 :  WORK(1,K)=IROW :
     WORK(2,K)=IROW
1320 GOSUB 2440
1330 WORK(3,K)=RSUM : NEXT K
1340 ' Bubble sort converges
1350 CC=NC
1360 WHILE  CC>=1
1370 IF  O=2 THEN  GOSUB 1440
1380 IF  O=1 THEN  GOSUB 1560
1390 CC=CC-1 : WEND
1400 FOR  JJ=1 TO NC :  PRINT WORK(0,JJ);WORK(1,JJ);WORK(2,JJ) :
     NEXT JJ
1410 GOSUB 2610 ' To shuffle rows
1420 INPUT "Shuffle done. Strike <CR> to continue ",Z$
1430  RETURN        ' Return to menu
1440 ' Bubble WORK(2) and WORK(3) in decreasing order
1450 LL=1
1460 IF  WORK(3,LL)<WORK(3,LL+1) THEN GOSUB 1500
1470 LL=LL+1 :  IF LL<=NC-1 THEN GOTO 1460
1480 RETURN
1490 '
1500 ' Subroutine to swap array location in WORK file
1510 T=WORK(3,LL+1) :  E=WORK(2,LL+1)
1520 WORK(3,LL+1)=WORK(3,LL) :  WORK(2,LL+1)=WORK(2,LL)
1530 WORK(3,LL)=T :  WORK (2,LL)=E
1540 RETURN
1550 '
1560 ' Bubble  WORK(2) and WORK(3) in increasing order
1570 LL=1
1580 IF  WORK(3,LL)>WORK(3,LL+1) THEN GOSUB 1500
1590 LL=LL+1 :  IF LL<=NC-1 THEN GOTO 1580
1600 RETURN
1610 '
1620 ' Subroutine to sum a rectangular area
1630 CLS : LOCATE 1,1,1
1640 INPUT "Leftmost col number ",LF
1650 IF LF<1 OR LF>MAXPOS THEN GOTO 1640
1660 INPUT "Rightmost col number ",RT
1670 IF RT<1 OR RT>MAXPOS THEN GOTO 1660
1680 INPUT "Top row number ",TP
1690 IF TP<1 OR TP>MAXPOS THEN GOTO 1680
1700 INPUT "Bottom row number ",BT
1710 IF BT<1 OR BT>MAXPOS GOTO 1700
1720 PRINT "Left=";LF, "Right=";RT, "Top=";TP, "Bottom=";BT
1730 INPUT "Enter E to correct error, C to cancel, <CR> to go on ";E$
1740 IF E$="E" THEN 1640
1750 IF E$="C" THEN 1810
1760 FOR  MM=1 TO NPAIRS
```

```
1770 IF GRAPH(2,MM)<LF OR GRAPH(2,MM) > RT THEN GOTO 1790
1780 IF GRAPH (1,MM)> = TP AND GRAPH(1,MM)< =BT THEN TOT=
     TOT+1
1790 NEXT MM
1800 PRINT "AREA TOTAL IS "; TOT
1810 INPUT "Strike any key to continue ",X$
1820 GOSUB 690
1830 RETURN
1840 '
1850 ' Subroutine to swap a row-column pair
1860 LOCATE 10,1,1 : PRINT "Exchange row/col pair"
1870 INPUT "Row from ",IFROM : INPUT "Row to ",ITO
1880 GOSUB 1920
1890 PRINT "Swap done"
1900 RETURN
1910 '
1920 ' Subroutine to swap a row-column pair
1930 FOR KS=1 TO NPAIRS
1940 RV=GRAPH(1,KS) : CV=GRAPH(2,KS)
1950 IF RV=IFROM THEN GRAPH(1,KS)=ITO
1960 IF CV=IFROM THEN GRAPH(2,KS)=ITO
1970 IF RV=ITO THEN GRAPH(1,KS)=IFROM
1980 IF CV=ITO THEN GRAPH(2,KS)=IFROM
1990 ' Exchange cell name pointers ; Use last cell as scratch area
2000 SWAP CX$(IFROM),CX$(ITO)
2010 SWAP CY$(IFROM),CY$(ITO)
2020 NEXT KS
2030 RETURN
2040 '
2050 ' Subroutine to sum on a diagonal
2060 INPUT "ENTER DIAGONAL TO SUM, -99 TO SKIP ",IDIAG
2070 IF IDIAG=-99 GOTO 2060
2080 RCDIFF=IDIAG-MAXPOS : PRINT "R-C = ";RCDIFF; "MAXPOS =";
     MAXPOS
2090 GOSUB 2520
2100 GOTO 2060
2110 RETURN
2120 LOCATE 1,1
2130 INPUT "ENTER COLUMN TO TOTALIZE,-99 TOSKIP ",ICOL
2140 IF ICOL =-99 GOTO 2170
2150 GOSUB 2360
2160 GOTO 2130
2170 RETURN
2180 '
2190 ' Subroutine to sum rows
2200 LOCATE 1,1,1
2210 INPUT "ENTER ROW TO TOTALIZE, -99 TO SKIP ",IROW
2220 IF IROW =-99 GOTO 2250
```

```
2230 GOSUB 2440
2240 GOTO 2210
2250 RETURN
2260 '
2270 ' Subroutine to sum diagonal
2280 LOCATE 1,1
2290 INPUT "ENTER DIAGONAL TO SUM, -99 TO SKIP ",IDIAG
2300 IF IDIAG=-99 THEN 2340
2310 IF IDIAG < =0 OR IDIAG > = 2*MAXPOS THEN 2290
2320 RCDIFF=IDIAG-MAXPOS : GOSUB 2520
2330 GOTO 2290
2340 RETURN
2350 '
2360 ' Subroutine to calculate a column total
2370 CSUM=0
2380 FOR KC=1 TO NPAIRS
2390 IF GRAPH(2,KC)=ICOL THEN CSUM=CSUM+1
2400 NEXT KC
2410 PRINT "COL TOTAL FOR COL. " ;ICOL; " IS "; CSUM
2420 RETURN
2430 '
2440 ' Subroutine to calculate a row total
2450 RSUM=0
2460 FOR KR=1 TO NPAIRS
2470 IF GRAPH(1,KR)=IROW THEN RSUM=RSUM+1
2480 NEXT KR
2490 PRINT "ROW TOTAL FOR ROW ";IROW ;" IS ";RSUM
2500 RETURN
2510 '
2520 ' Subroutine to count up diagonal
2530 DSUM=0
2540 FOR KD=1 TO NPAIRS
2550 D=GRAPH(1,KD)-GRAPH(2,KD)
2560 IF RCDIFF = D   THEN DSUM=DSUM+1
2570 NEXT KD
2580 PRINT "CONNECTIONS ON DIAGONAL ";IDIAG;" = ";DSUM
2590 RETURN
2600 '
2610 ' Routine to reorder row/col set by WORK array
2620 ' Clear col 0,- and clear completed matches
2630 FOR JC=1 TO WKND
2640    WORK(0,JC)=0
2650    NEXT JC
2660 ' Number of rows to order, Start-Finish+1=NC from 1140
2670 FOR JD=1 TO NC
2680 IF WORK(1,JD)=WORK(2,JD) THEN WORK(0,JD)=WORK(2,JD) :
     GOTO 2750
2690 IF WORK(0,JD)=WORK(2,JD) THEN 2750 ' Data are in place
```

```
2700 IF WORK(0,JD) = 0 THEN GOTO 2730
2710 IF WORK(0,JD) < >0 THEN GOTO 2650
2720 ' Now check for up or down order
2730 IF WORK(2,JD) > WORK(1,JD) THEN GOSUB 3150 : GOTO 2750
2740 IF WORK(2,JD) < WORK(1,JD) THEN GOSUB 3200 : GOTO 2750
2750 NEXT JD  ' Connector
2760 FOR JJ = 1 TO NC + 1 : PRINT WORK(0,JJ);WORK(1,JJ);WORK(2,JJ):
     NEXT JJ
2770 INPUT "Strike any key to continue ",Z$
2780 RETURN
2790 IF WORK(0,JD) = WORK(2,JD) THEN 2750 ' Already matched
2800 IF WORK(2,JD) < WORK(1,JD) THEN GOTO 2870
2810 ITO = WORK(1,JD) : IFROM = WORK(2,JD) : CT = WORK(0,JD) :
     IREF = IFROM-S + 1
2820 IF WORK(0,IREF) > < 0 THEN PRINT "IFROM ZERO FULL
     ERROR" : STOP
2830 GOSUB 1920
2840 WORK(0,JD) = IFROM : WORK(0,IREF) = CT
2850 GOTO 2750
2860 ' Find where 2,jd is
2870 GOAL = WORK(2,JD) : GOSUB 2930
2880 IFROM = WORK(1,GOT) : ITO = WORK(1,JD) : PT = WORK(0,GOT) :
     GOSUB 1920
2890 ' PRINT "3185 GOAL,GOT,PT,IFROM,ITO ",GOAL;GOT;PT;IFROM;ITO
2900 WORK (0,GOT) = WORK(0,JD) : WORK (0,JD) = PT
2910 GOTO 2750
2920 '
2930 ' Subroutine to search for GOAL
2940 GOT = 0
2950 FOR IS = JD TO NC
2960 IF WORK(0,IS) = GOAL THEN GOT = IS
2970 NEXT IS
2980 RETURN
2990 '
3000 ' Subroutine to generate randomized WORK
3010 RANDOMIZE TIMER
3020 MAX = S + NC-1
3030 FOR MM = 1 TO NC
3040 WORK (1,MM) = S + MM-1
3050 L = INT(RND*(NC + 1)) : TEST = 0
3060 IF L < 1 OR L > MAX THEN GOTO 3050
3070 FOR MO = 1 TO MM-1
3080 IF WORK(2,MO) = L + S-1  THEN TEST = 1
3090 NEXT MO
3100 IF TEST = 0 THEN WORK (2,MM) = L-1 + S  ELSE GOTO 3050
3110 NEXT MM
3120 RETURN
3130 '
```

```
3140 ' Subroutine to swap when WORK(0,-) < > 0 and 2 > 1
3150 ' PRINT "WORK(2) > WORK(1) AND JD = ",JD
3160 ITO = WORK(1,JD) : IFROM = WORK(2,JD) : GOSUB 1920
3170 IREF = IFROM-S + 1
3180 WORK(0,IREF) = ITO : WORK(0,JD) = IFROM : RETURN
3190 '
3200 ' Subroutine to swap when 2 < 1
3210 GOAL = WORK(2,JD) : GOSUB 2930
3220 ' PRINT "GOAL = ";GOAL," GOT = ";GOT
3230 IF GOT = 0 THEN PRINT "ERROR" : STOP
3240 FR0M = WORK (1,GOT)      ' Present index of required data
3250 ITO = WORK (1,JD) : GOSUB 1920
3260 WORK(0,GOT) = ITO : WORK (0,JD) = GOAL
3270 RETURN
3280 '
3290 ' Subroutine to save transformed tables
3300 OPEN "O",#2,"LABSAV.DAT"
3310 FOR JO = 1 TO LAST
3320 PRINT# 2,CX$(JO)
3330 NEXT JO
3340 CLOSE #2
3350 ' SAVE NODE PAIRS AND DENSITIES
3360 OPEN "O",#2, "PRSAV.DAT"
3370 FOR JP = 1 TO NPAIRS
3380 PRINT# 2,USING "###"; GRAPH(1,JP);GRAPH(2,JP);
3390 PRINT #2, USING "####";GRAPH(3,JP)
3400 NEXT JP
3410 CLOSE #2
3420 CLS : PRINT "Labels and node pairs saved"
3430 RETURN
3440 END
```

--- End of Program ---

**B.4.** *PLOT100.BAS*

### B.4.a. Description

This program prints an image of the adjacency image of the nerve net, with presynaptic neurons in row position and post synaptic neurons in column position. Normal ordering is that row and column indices are ordered identically by neuron name. The graph matrix is printed as a pattern of dots on 8.5 x 11.0 inch paper using an HP Deskjet® or comparable printer.

The main program uses a series of subroutines to perform the task. The user is notified at various points of what is needed or of the progress achieved. First, Subroutine

1030-1300 requests the user to name the file that lists the names of neurons in sequential order (e.g., UNIQue list and cell type). This will be the reference file created by MAKEDATA.BAS in the format A7,A4. The example provided in this program is LABELG.DAT. It creates a directory in memory. Then, it requests a file containing neural connections in I3,I3,I4 format. This will be the data file created by MAKEDATA.BAS in the format I3,I3,I4. The example provided in this program is PAIRSG.DAT (same as the input data file used by DISPLAY.DAT). The data are mapped into an array, GRAPH (3,2500). Density data are limited to the range 1 to 15 to allow for future color use, but are not used in this version.

In Subroutine 470-600, the numerical data are converted into a bit map with alternate bits empty so that adjacent bits will not blur on printing. A message is printed every 500 entries to notify the user that blank screen is not a malfunction.

Subroutine 630-680 is the main print routine, along with lines 80-110 which set up the printer. The control commands are applicable to the HP Deskjet® printer which is capable of 100 dots per inch (dpi) resolution. Modifying the parameter in statement 110 provides for other density of display, notably 75 dpi. Subroutine 720 puts tic marks spaced every 5 neurons along the top, and Subroutine 870 puts corresponding marks across the bottom of the graph. During printing, line 250-330 print tic marks at the beginning and end of every fifth row starting with 1.

There is an unused listing for Subroutine at 1390-1470 to save the data to diskette as files PRSAV.DAT and LBSAV.DAT. This is used only to save the labels and density tables that the program generates, so that they can be viewed or reviewed as text files in the course of investigation.

The program was written in BASICA because of intrinsic flexibility in control of graphics which this language provides, even if FORTRAN is the idiom for most of the numerical computational programs in this set.

**B.4.b. Directions**

If there is a need to print a bitmapped graph in the examples, PLOT100 program will run in a BASICA interpreter. For differences in hardware and hardware set-up, relevant subroutine numbers have been identified above to facilitate editing of the code.

**B.4.c. Source Code**

```
10    ' Program to plot neuron matrix on DESKJET (R) or equivalent,
20    ' 100dpi version. Written by W. S. Yamamoto, M.D.
30    DIM B$(40),BIT%(4),G%(90)
40    DIM GRAPH(3,2500)
50    DIM CX$(320) : DIM CY$(320)
60    ' Initialize values
70    BIT%(0)=0 : BIT%(1)=64 : BIT%(2)=16 : BIT%(3)=4 : BIT%(4)=1
80    ' Set up printer to print line and blank line
```

242

```
90    WIDTH "LPT1:",255        ' Disable CR-LF
100   OPEN "LPT1:" AS #1
110   PRINT #1,CHR$(27);"*t100R";        ' Set up dots per in.
120   ' Allows 90 byte line with 4 bits set per byte
130   S$ = CHR$(0)
140   ' GOSUB 720        ' Put header tics in
150   GOSUB 1030 ' Get data from diskette into GRAPH(3,2500)
160   IF MAXPOS > 360 THEN MAXPOS = 360 ' Set bounds to plot size
170   FOR ROWNO = 1 TO MAXPOS
180   FOR K = 1 TO 90 : G%(K) = 0 : NEXT K      ' Clear line array
190   FOR PN = 1 TO NPAIRS
200   C% = 0
210   IF ROWNO = GRAPH(1,PN) THEN C% = GRAPH(2,PN) ' Relevant bit
220   GOSUB 470     ' Locate and mark the bit
230   NEXT PN
240   ' Convert array to string
250   D$ = ""
260   TNO = ROWNO MOD 5
270   IF TNO = 1# THEN  D$ = D$ + CHR$(252)       ' Left tic mark
280   IF TNO < > 1 THEN D$ = D$ + CHR$(0)        ' No tic
290   FOR KL = 1 TO 88 : D$ = D$ + CHR$(G%(KL)) : NEXT KL
300   ' Add tic mark every 5 lines
310   TNO = ROWNO MOD 5
320   IF TNO = 1# THEN  D$ = D$ + CHR$(63)        ' Tic mark
330   IF TNO < > 1 THEN D$ = D$ + CHR$(0)        ' No tic
340   GOSUB 640       ' Print one row of matrix in graph
350   PRINT "DONE PRINT ";ROWNO;" TIC NO = ";TNO
360   NEXT ROWNO
370   ' Print tics at bottom of figure
380   GOSUB 870
390   CLOSE (1)
400   LPRINT CHR$(13),CHR$(13);"Names of files are: " L$;" AND ";P$
410   INPUT "COMMENTS: ",LS$
420   LPRINT CHR$(13);"Comments: ";LS$
430   STOP
440   END
450   '
460   '
470   ' Subroutine to compute byte and bit location
480   IF C% > MAXPOS THEN C% = MAXPOS      ' Protect from overflow
490   IF C% = 0 THEN GOTO 590
500   BT% = C% MOD 4
510   BY% = C%\4      ' Integer divide truncates
520   IF BT% < > 0 THEN BY% = BY% + 1
530   IF BT% = 0 THEN BT% = 4
540   ' Is bit already set ?
550   TEMP% = G%(BY%) : NUBIT% = BIT%(BT%)
560   DUP% = TEMP% AND NUBIT%
```

```
570  IF DUP%=0 THEN G%(BY%)=G%(BY%) OR NUBIT%   ' Enter bit
580  IF DUP% < > 0 THEN PRINT "Error. Duplicate bit in byte ",BY%
590  DUP%=0
600  RETURN
610  '
620  '
630  ' Subroutine to print line and blank line, this is the main print routine
640  PRINT #1,CHR$(27);"*r1A"
650  PRINT #1,CHR$(27);"*b90W";D$;
660  PRINT #1,CHR$(27);"*b1W";S$;
670  PRINT #1,CHR$(27);"*rB";
680  RETURN
690  '
700  '
710  ' Subroutine to setup and print vertical tic marks spaced 5
720  FOR J=1 TO 86 STEP 5 : G%(J)=128 : G%(J+1)=32   ' 21H
730  G%(J+2)=8 : G%(J+3)= 2 : G%(J+4)=0    ' 10H
740  NEXT J
750  D$=CHR$(0)
760  FOR JT=1 TO 89 : D$=D$+CHR$(G%(JT)) : NEXT JT
770  PRINT #1,CHR$(27);"*r1A"
780  FOR LN=1 TO 6
790  PRINT #1,CHR$(27);"*b90W";D$;
800  NEXT LN
810  PRINT #1,CHR$(27);"*b1W";S$;
820  PRINT #1,CHR$(27);"*b1W";S$;
830  PRINT #1,CHR$(27);"*rB";
840  RETURN
850  '
860  '
870  ' Subroutine to print tic marks at bottom of graph
880  FOR J=1 TO 86 STEP 5 : G%(J)=128 : G%(J+1)=32   ' 21H
890  G%(J+2)=8 : G%(J+3)= 2 : G%(J+4)=0    ' 10H
900  NEXT J
910  D$=CHR$(0)
920  FOR JT=1 TO 89 : D$=D$+CHR$(G%(JT)) : NEXT JT
930  PRINT #1,CHR$(27);"*r1A"
940  PRINT #1,CHR$(27);"*b1W";S$;
950  PRINT #1,CHR$(27);"*b1W";S$;
960  FOR LN=1 TO 6
970  PRINT #1,CHR$(27);"*b90W";D$;
980  NEXT LN
990  PRINT #1,CHR$(27);"*rB";
1000 RETURN
1010 '
1020 '
1030 ' Subroutine to get nerve net data from diskette
1040 PRINT "GENERATING DIRECTORY"
```

```
1050 INPUT "Input file name of labels - ",L$
1060 OPEN "I",2,L$
1070 J = 0
1080 IF EOF(2) THEN 1120
1090 J = J + 1 : INPUT#2,CX$(J)
1100 CY$(J) = CX$(J)
1110 GOTO 1080
1120 CLOSE #2
1130 MAXPOS = J          ' Number of neurons in name list
1140 PRINT "DIRECTORY HAS ";MAXPOS;" NEURONS. START PAIRS
     LIST"
1150 INPUT "Enter node-pair file name - ",P$
1160 OPEN "I",2,P$
1170 N = 0 : MD = 0 : PRINT "GENERATING NODE-PAIR LIST"
1180 IF EOF(2) THEN 1270
1190 N = N + 1 : PRE$ = INPUT$(3,#2) : GRAPH(1,N) = VAL(PRE$)
1200 POST$ = INPUT$(3,#2) : GRAPH (2,N) = VAL(POST$)
1210 INPUT#2,DS$ : GRAPH(3,N) = VAL(DS$)
1220 DENS = VAL (DS$)
1230 IF DENS > 15 THEN DENS = 15
1240 RES = N-MD*500 : IF RES = 1 THEN PRINT "DONE ";MD*500;
     " ENTRIES"
1250 IF RES = 1 THEN MD = MD + 1
1260 GOTO 1180
1270 CLOSE #2
1280 NPAIRS = N
1290 PRINT "GRAPH TABLE DONE. " ;NPAIRS; " PAIRS"
1300 RETURN
1310 '
1320 '
1330 ' Subroutine to save transformed tables
1340 OPEN "O",#3,"LABSAV.DAT"
1350 FOR JO = 1 TO LAST
1360 PRINT# 3,CX$(JO)
1370 NEXT JO
1380 CLOSE #3
1390 ' Save node pairs and labels to diskette
1400 OPEN "O",#4, "PRSAV.DAT"
1410 FOR JP = 1 TO NPAIRS
1420 PRINT# 4,USING "###"; GRAPH(1,JP);GRAPH(2,JP);
1430 PRINT #4, USING "####";GRAPH(3,JP)
1440 NEXT JP
1450 CLOSE #4
1460 PRINT "Labels and node pairs saved"
1470 RETURN
1480 END
```

--- End of Program ---

## C. The FORTRAN Programs

### C.1. *MAKELIST.FOR*

#### C.1.a. Description

MAKELIST.FOR is a FORTRAN program which works directly with a modified version of the PREPOSGO.PRN file named PREPOSGO.LIS. Its purpose and logic is the same as that of MAKEDATA.BAS, but produces only two final products: the data files PAIRSG.DAT and LABELG.DAT, which are identical to those produced by MAKEDATA.BAS (see note below).

The sample data file PREPOSGO.LIS was created using an electronic spreadsheet. The PREPOSGO.PRN file from Disk 1 was imported into a worksheet, and the file was broken into its respective columns by using the *Parse* function of the spreadsheet. Column headings were added, and column widths and entry justifications were adjusted as specified in Chapter III. The Type column was readjusted to format A4, instead of A7. The second data block, from Entry Numbers 2368 to 2463 were deleted. Except Type, Pre, Post, and Den columns, all other columns were deleted. Cell types "G" (Ventral Cord Neuron Group) and "P" (Pharyngeal Neuron Group) were moved, to become the second to the last and last data blocks, respectively. Each ganglion was then sorted alphabetically using the *Sort* function of the spreadsheet. The Uniq column was recreated as the last (fifth) column in A7 format by using the *Uniq* function of the spreadsheet, the input data block being the Pre column, the criterion block being a Pre column heading and an empty cell below it, and the output block being a new Pre column (which becomes the new Uniq column). Entries WE, WI, WM, and WN were then appended to the new Uniq column. These five columns, Type, Pre, Post, Den, and Uniq were then saved (without column headings) as the ASCII file PREPOSGO.LIS in A4,A7,A7,I4,A7 format. Note that rows with blank Den column entries were not deleted. An editor was used to verify that PREPOSGO.LIS was saved in proper format, modifying as necessary to conform to the format.

MAKELIST.FOR expects that PREPOSGO.LIS be on floppy disk loaded in floppy drive A. The program reads the records sequentially, and writes two output files that are coordinated through the record numbers in sequential order; i.e., the record numbers represent data, but are not explicitly written since they can be retrieved through indexing of loops or arrays. The normal termination for the first part is a message stating the number of unique records read and the FLAG set at 4.

As mentioned earlier, the output files are PAIRSG.DAT and LABELG.DAT:

(1) PAIRSG.DAT is written on the hard disk, as Pre, Post, Den(sity) in format I3,I3,I4 for numerical processing of the data, particularly in matrix form. This format and its use was discussed in program MAKEDATA.BAS.

(2) LABELG.DAT is also written on the hard disk, as Uniq(ue) Cell, Type pairs in A7, A4 form. The sequential position of the cell names correspond to the record number in the PREPOSGO.LIS file, and inversion of the numeric form to cell

names is linked through the record number. Thus, there are no more then 302 neurons in the label list, plus the four hypothetical "cells" representing the external and internal environments of the nervous system. The order of cells in this LABELG.DAT list is the order of appearance in the presynaptic cell list made in ganglionic order. Use of files like LABELG.DAT have been shown earlier in the BASIC programs.

NOTE: The PAIRSG.DAT files produced by MAKEDATA.BAS and MAKELIST.FOR differ only in the sequence of the lines of data in the numerically translated files. MAKEDATA.BAS derives its line sequence from PREPOSGO.REF, and MAKELIST.FOR from PREPOSGO.LIS. When rearranged numerically, however, the PAIRSG.DAT files produced by either program are identical.

### C.1.b. Directions

MAKELIST.FOR may be immediately compiled to MAKELIST.EXE by using a suitable FORTRAN compiler, and executed, although its output files PAIRSG.DAT and LABELG.DAT are already provided in Disk 2. Changes in code will require an editor and program recompilation.

### C.1.c. Source Code

```
      PROGRAM MAKELIST
C Written by W. S. Yamamoto, M.D.
C Modified by T. B. Achacoso, M.D.
      CHARACTER*7 PRE,POST,TEMP,SPC,UNIQ
      CHARACTER*4 T
      INTEGER DENS,FLAG
      INTEGER MAXPOS
      COMMON MAXPOS
      CHARACTER*7 CELL(350)
      COMMON CELL
      CHARACTER*4 TYPE(350)
      COMMON TYPE
C Setup ordered neuron list from unique list
      OPEN (4, FILE = 'A:PREPOSGO.LIS')
      FLAG = 0
      TEMP = '        '
      SPC = TEMP
      J = 0
100   READ (4,150,END = 110) T,PRE,POST,DENS,UNIQ
C Option to see what is being read from disk on screen
C     WRITE (*,170) T,PRE,POST,DENS,UNIQ
  170 FORMAT (1X,A4,A7,A7,I4,A7)
      IF (UNIQ .EQ. TEMP) GOTO 100
      IF (UNIQ .EQ. SPC ) FLAG = FLAG + 1
      IF (FLAG .GT. 3 ) GOTO 110
      IF (UNIQ .EQ. SPC ) GOTO 100
```

```
        J = J + 1
        TEMP = UNIQ
        CELL(J) = TEMP
        TYPE(J) = SPC
        GOTO 100
150     FORMAT (A4,A7,A7,I4,A7)
 110    MAXPOS = J
        WRITE (*,998) J,FLAG
 998    FORMAT (1H ,' RECORDS READ ',I4, '  FLAG= ',I4)
160     REWIND (4)
C Create numeric pairs list in PAIRSG.DAT
        OPEN (5, FILE = 'PAIRSG.DAT', STATUS = 'NEW')
        NUM=0
400     READ (4,425,END=430) T,PRE,POST,DENS
C Option to see what is being read from disk on screen
C       WRITE (*,171) T,PRE,POST,DENS
 171    FORMAT (1X,A4,A7,A7,I4)
        CALL FINDPOS(PRE,J,T)
        CALL FINDPOS(POST,K,SPC)
C Omit entries with zero efferent connections
        IF (DENS .EQ. 0) GOTO 400
        WRITE (5,428) J,K,DENS
        NUM=NUM+1
        GOTO 400
425     FORMAT (A4,A7,A7,I4)
428     FORMAT (I3,I3,I4)
430     CLOSE (4)
        WRITE (*,997) NUM
 997    FORMAT ( '  PAIRS IN LIST = ',I6)
        ENDFILE (5)
        CLOSE (5)
        WRITE (*,450) MAXPOS
450     FORMAT (' Matrix size = ', I4)
C Make disk file LABELG.DAT
        OPEN (5, FILE = 'LABELG.DAT', STATUS = 'NEW')
        DO 500 I = 1, MAXPOS
        IF ( CELL(I) .EQ. 'WE     ') TYPE(I)='WE  '
        IF ( CELL(I) .EQ. 'WI     ') TYPE(I)='WI  '
        IF ( CELL(I) .EQ. 'WM     ') TYPE(I)='WM  '
        IF ( CELL(I) .EQ. 'WN     ') TYPE(I)='WN  '
        WRITE (5,550) CELL(I),TYPE(I)
500     CONTINUE
550     FORMAT (A7,A4)
        ENDFILE (5)
        CLOSE (5)
        END
        SUBROUTINE FINDPOS(TC,POS,TY)
        CHARACTER*7 TC,CELL(350)
```

```
      INTEGER  POS,X,MAXPOS
      COMMON  MAXPOS,CELL,TYPE
      CHARACTER*4  TYPE(350),TY,SP4
      SP4 = '    '
      X  =  1
100   IF (TC .EQ. CELL(X)) GOTO 400
      X  =  X  +  1
      IF (X .LE. MAXPOS) GOTO 100
400   POS  =  X
C Enter cell type in array if presynaptic
      IF ( TY .NE. SP4 ) TYPE(X) = TY
      RETURN
      END
```

--- End of Program ---

## C.2. *CONNECT.FOR*

### C.2.a. Description

This program uses transitive closure to compute the connectivity of a neural network listed as in PAIRSG.DAT and LABELG.DAT (product files of MAKEDATA.BAS or MAKELIST.FOR above); i.e., in preganglionic, postganglionic order along with the synaptic densities. The list is converted to a full numerical matrix including all zeroes. Various subroutines allow conversion to Boolean form, computation of the transpose, conversion to matrix of the corresponding undirected graph, and generating powers of the matrix $(1 + M)$.

The program is written as a family of subroutines, some of which are not used in the program if executed as shown. These are included in the source code to provide facility to perform other computation, and to allow saving or retrieving of results in compact form. As listed, the program requires memory space for at least five 320x320 integer arrays. It is too large for manipulation on a personal computer with a 640 Kb RAM limitation. The program has been executed on DEC-VAX 780 and CRAY Y-MP/832 machines, where computation times for closure on a 306x306 network were 33 minutes and 12 seconds, respectively.

The Subroutines are as follows:

Subroutine MXMF (A,NRA,B,C) forms product AxB in C. All arrays are dimen-
      sioned in COMMON rather than with the PARAMETER construction.
Subroutine BOOLAD (A,B,AD,NRA) forms sum A + B in AD.
Subroutine BOOLEQ (A,B,ANS,NZ,NRA,NCA) if A = B ANS = 0 counts zero
      entries in B.
Subroutine MATCOPY (A,B,NRA,NCA) copies A to B.
Subroutine MATRANS (A,B,NRA,NCA) copies transpose of A to B.
Subroutine SUMOUT (A,NRA,NCA,ITYPE,NAME) unused by MAIN as listed
      here forms row and column sums and prints results.

Subroutine ARROUT (A,B,NRA,LTH) forms row sums and column sums and puts total in adjacent columns by index LTH in array B. Index is usually exponent in M**LTH.

Subroutine PRTARY (B,NRA,LTH,ANS) prints B in sets of 10 columns with proper labels for cells using row index as label number.

Subroutine SORT (B,IX,NRT) bubble sorts the last three columns of array B: column 318 in descending order, truncates terminal ones, and then sorts column 319 in ascending order by synaptic densities. This tends to leave the array sorted into Sensory (high output-low output), Interneuron, and Motor (low output-high input) order.

Subroutine NAMES (B,NRA) by referencing LABELG.DAT converts a cell numerical name to text name and stores text in array B(311-317).

Subroutine STOREBIT (AT,LOUT,NRA) compresses expanded array of dimension NRA into array LOUT(320,20) as a bitmap 16 bits numbered left to right in the word. Only NRA rows are converted but no adjustment is made for excess columnar space. This subroutine prepares sparse data for file storage.

Subroutine GETBIT (AT,LIN,NRA) expands bit map into Boolean array, one bit per array location, thus reversing action of STOREBIT.

## C.2.b. Directions

CONNECT.FOR can be compiled and executed in a computer of appropriate memory and processing speed. The subroutines themselves, however, may be used or modified for smaller computer programs as needed.

## C.2.c. Source Code

```
      PROGRAM  CONNECT
C Compute connectivity using closure,
C version reads C. elegans files. Written by W. S. Yamamoto, M.D.
C Compute closure and output matrix sums
      INTEGER A,B,C,AD,ANS,AT
      DIMENSION  A(320,320),B(320,320),C(320,320),AD(320,320)
     1 ,AT(320,320)
      COMMON  MRA,MCA
      NRA = 0
      NCA = 0
      NZER = 0
      LTH = 1
      MRA = 320
      MCA = 320
      NSTOP = 20
C Open input and output files
      OPEN  (9,FILE = 'netlis.bol',STATUS = 'new')
C Clear arrays
      DO  120  K = 1,MCA
      DO  110  I = 1,MRA
```

```
            A(I,K) = 0
            B(I,K) = 0
            C(I,K) = 0
            AD(I,K) = 0
            AT(I,K) = 0
110     CONTINUE
120     CONTINUE
C Read in data in sparse list from appropriate file
C Lower case name necessary for UNICOS use
        OPEN (17,FILE = 'pairsg.dat')
C Count rows and columns occupied
        NRT = 0
        NCT = 0
  70    READ (17,950,END = 80) IROW,ICOL,IVAL
 950    FORMAT (I3,I3,I4)
C Check for valid row and column number
        IF (ICOL .GE. 303) GOTO 70
        IF (IROW .GE. 303) GOTO 70
        IF ( ICOL .GT. NCT) NCT = ICOL
        IF (IROW .GT. NRT) NRT = IROW
        IF (ICOL .GT. MCA) GOTO 20
        IF (IROW .GT. MRA) GOTO 20
        A(IROW,ICOL) = IVAL
        IF ( IVAL .LE. 0) WRITE (*,1010) IROW,ICOL,IVAL
1010    FORMAT (' Blank density at ',I4,'Col = ',I4,' Val ',I4)
        GOTO 70
  80    CONTINUE
        NRA = NRT
        IF (NRT .LT. NCT) NRA = NCT
        NCA = NRA
C Convert A matrix to Boolean
        ICNT = 0
        DO 10 J = 1,NRA
        DO 10 K = 1,NCA
        IF (A(J,K) .NE. 0) THEN
          ICNT = ICNT + 1
          A(J,K) = 1
        ENDIF
  10    CONTINUE
C Print description
        WRITE (*,960) NRT,NCT,NRA, ICNT
 960    FORMAT (' INPUT DIMENSION ROWS = ',I4,' COL ',I4,
     1  'USED = ',I4 , ' NON-ZERO = ',I8)
        CLOSE (17)
C Adding transpose of data will make graph undirected
C       CALL MATRANS(A,B,NRA,NCA)
C       CALL BOOLAD(A,B,A,NRA)
C COMMENT OUT LINE 40-41 TO RUN AS DIRECTED GRAPH
```

```
C For closure calculation, set diagonal = 1
C Set diagonal =0 otherwise
      DO 50 I=1,NRA
   50 A(I,I)=0
C Use BOOLEQ to count zeroes
      CALL BOOLEQ (A,A,ANS,NZER,NRA,NCA)
C Now copy it into old AD position
      CALL MATCOPY (A,AD,NRA,NCA)
      WRITE (*,910) LTH
  910 FORMAT ( ' PATHS OF LENGTH ', I5)
      WRITE (*,930) LTH,ANS,NZER
C When using ARROUT array B should not be used
      CALL ARROUT (A,B,NRA,LTH)
C Compute A*A
      CALL MXMF(A,NRA,A,C)
      LTH=LTH+1
      WRITE (*,910)LTH
  900 FORMAT (1H ,7I8)
C Copy Boolean A*A to AT NEW MATRIX
      DO 90 I=1,NRA
      DO 90 J=1,NCA
C Clear array as you go
      AT(I,J)=0
      IF ( C(I,J) .NE. 0) AT(I,J)=1
   90 CONTINUE
C Check old AD against AT
      CALL BOOLEQ (AD,AT,ANS,NZER,NRA,NCA)
      WRITE (*,930) LTH,ANS,NZER
      IF (ANS .EQ. 0) GOTO 150

C Unequal copy new into old position

      CALL MATCOPY(AT,AD,NRA,NCA)
      CALL ARROUT (AT,B,NRA,LTH)
C Cycle higher path lengths
C MULTIPLY A X OLD
  100 LTH=LTH+1
      CALL MXMF(A,NRA,AD,AT)
      WRITE (*,910) LTH
      CALL BOOLEQ(AD,AT,ANS,NZER,NRA,NCA)
C Output sums of AT
      CALL ARROUT (AT,B,NRA,LTH)
C Computes zeroes of AT is the new matrix
      WRITE(*,930) LTH,ANS,NZER
      IF ( ANS .EQ. 0) GOTO 160
C Unequal - copy new to old
      CALL MATCOPY (AT,AD,NRA,NCA)
      IF (LTH .LT. NSTOP) GOTO 100
```

```
C All paths computed
C Note BOOLEQ counts zeroes ITYPE = 0 when matrix is dense
  150 WRITE (*,940) LTH,NRA,NZER,ANS
  940 FORMAT (' No closure NSTOP exceeded at ',I3,' Cells = ',I4,
     1 ' zeroes = ',I5, 'ANS is ',I2)
  190 IX = 2*LTH-1
      CALL SORT(B,IX,NRA)
      CALL NAMES (B,NRA)
      CALL PRTARY (B,NRA,LTH,ANS)
C Count cycles by adding up diagonal entries
      ISUM = 0
      DO 170 K = 1,NRA
  170 ISUM = ISUM + AT(K,K)
      WRITE (*,980) NCA,ISUM
  980 FORMAT (' Cells in cycles = ',I4,' Sum of diagonal entries =
     1 ',I4)
C Output bitmap to NETLIS.BOL
C
C THIS COMMENTED SECTION MAKES A TEST PATTERN FOR
    DEBUGGING
C      DO 220 M = 1,20
C      DO 220 N = 1,320
C      AT(M,N) = 0
C  220 CONTINUE
C      DO 210 N = 1,16
C      IC = (N-1)*16 + N
C  210 AT(1,IC) = 1
      CALL STOREBIT (AT,C,NRA)
C
C THIS COMMENTED SECTION TESTS RESTORATION OF BIT
    MATRIX
C      DO 200 L = 1,10
C  200 WRITE (9,990) (C(L,M),M = 1,20)
C  990 FORMAT (10I6)
C Get C back into A and test print
      CALL GETBIT (A,C,NRA)
C      WRITE (*,206) (A(1,lm),lm = 1,200)
C 206 FORMAT (1H ,16i3)
C
      CLOSE (9)
      CALL EXIT
C
C Exit when closure has been computed
  160 WRITE (*,920) LTH, NRA,NZER
  920 FORMAT (' Closure at length ',I3,' for cells ',I6,
     1 ' zeroes = ',I5)
      CALL ARROUT (AT,B,NRA,LTH)
  930 FORMAT (' At loop ',I4,' equality is ',I3,' zeroes ',I6)
```

```
        GOTO 190
C
C Error exit for bad read
   20 WRITE (*,970) IROW,ICOL,MRA
  970 FORMAT (' Error input dimension ROW= ',I4,' col ',
     1 I4,' MRA MAX= ',I4)
        GOTO 190
        END
C
C
C Subroutine to Multiply A and B
        SUBROUTINE MXMF(A,NRA,B,C)
        COMMON MRA,MCA
        DIMENSION A(MRA,MCA),B(MRA,MCA),C(MRA,MCA)
        INTEGER A,B,C
C Multiply A and B
        DO 230 K=1,NCA
         DO 220 I=1,NRA
          DO 210 J=1,NRA
          C(I,K)=C(I,K)+A(I,J)*B(J,K)
210       CONTINUE
220      CONTINUE
230     CONTINUE
C Convert product to Boolean
        DO 250 i=1,nra
        DO 250 J=1,NCA
  250 IF ( C(I,J) .NE. 0) C(I,J)=1
        RETURN
        END
C
C Boolean addition
        SUBROUTINE BOOLAD (A,B,AD,NRA)
        INTEGER A,B,AD
        COMMON MRA,MCA
        DIMENSION A(MRA,MCA),B(MRA,MCA),AD(MRA,MCA)
        DO 10 I=1,NRA
        DO 10 J=1,NCA
        AD(I,J)=A(I,J)+B(I,J)
        IF ( AD(I,J) .NE. 0) AD(I,J)=1
   10 CONTINUE
        RETURN
        END
C
C Check two matrices for equality count zeroes
        SUBROUTINE BOOLEQ (A,B,ANS,NZ,NRA,NCA)
        INTEGER A,B,ANS
        COMMON MRA,MCA
        DIMENSION A(MRA,MCA),B(MRA,MCA)
```

254

```
      ANS = 0
      NZ = 0
      DO 10 I = 1,NRA
      DO 10 J = 1,NCA
      AVAL = 0
       IF (B(I,J) .LE. 0) THEN
       NZ = NZ + 1
       BVAL = O
       ENDIF
      IF (B(I,J) .GT. 0) BVAL = 1
   20 IF ( A(I,J) .NE. 0) AVAL = 1
      IF (AVAL .NE. BVAL) ANS = 1
   10 CONTINUE
      RETURN
      END
C
C Matrix copy A to B
      SUBROUTINE MATCOPY(A,B,NRA,NCA)
      INTEGER A,B
      COMMON MRA,MCA
      DIMENSION A(MRA,MCA),B(MRA,MCA)
      DO 10 I = 1,NRA
      DO 10 J = 1,NCA
        B(I,J) = A(I,J)
   10 CONTINUE
      RETURN
      END
C
C Matrix transpose, Transpose is in B
      SUBROUTINE MATRANS(A,B,NRA,NCA)
      INTEGER A,B
      COMMON MRA,MCA
      DIMENSION A(MRA,MCA),B(MRA,MCA)
      DO 10 I = 1,NRA
      DO 10 J = 1,NCA
        B(J,I) = A(I,J)
   10 CONTINUE
      RETURN
      END
C
C Subroutine to output sparse matrix to file
C List either entries or zeroes depending on sparsity
      SUBROUTINE SUMOUT(A,NRA,NCA,ITYPE,NAME)
      INTEGER A,RSUM,CSUM
      COMMON MRA,MCA
      DIMENSION A(MRA,MCA)
      WRITE (*,900) NAME,NRA
      WRITE (9,900) NAME,NRA
```

```
  900 FORMAT (' Net is sparse named ',I6, ' Size =',I4)
C Output only row and column sums ignore ITYPE
      RSUM = 0
      CSUM = 0
      DO 10 I = 1,NCA
      DO 20 J = 1,NRA
       RSUM = RSUM + A(I,J)
       CSUM = CSUM + A(J,I)
   20 CONTINUE
      WRITE (*,910) I,RSUM,CSUM
  910 FORMAT ( ' Row no. ',I4,' Rowsum = ',I6,' Colsum = ',I6)
      RSUM = O
      CSUM = 0
   10 CONTINUE
   50 RETURN
      END
C
C Cumulate row and col sums in array B for output
      SUBROUTINE ARROUT(A,B,NRA,LTH)
      INTEGER A,B,RSUM,CSUM
      COMMON MRA,MCA
      DIMENSION A(MRA,MCA),B(MRA,MCA)
      IROUT = 2*LTH-1
      ICOUT = IROUT + 1
      NCA = NRA
C Compute row and col sums
      RSUM = 0
      CSUM = 0
      DO 10 I = 1,NCA
      DO 20 J = 1,NRA
       RSUM = RSUM + A(I,J)
       CSUM = CSUM + A(J,I)
   20 CONTINUE
C Insert sums in B
      B(I,IROUT) = RSUM
      B(I,ICOUT) = CSUM
      RSUM = 0
      CSUM = 0
   10 CONTINUE
      RETURN
      END
C
C Print B array in sets of five rxc pairs
      SUBROUTINE PRTARY (B,NRA,LTH,ANS)
      INTEGER A,B,ANS
      COMMON MRA,MCA
      DIMENSION B(MRA,MCA)
C Print only NRA sets up to LTH
```

```
      ICYCLE = (LTH-1)/5 + 1
      DO 10  K = 1,ICYCLE
      JA = 10*K-9
      JB = 10*K
      JC = JB/2
      WRITE (*,910)JA,JB
  910 FORMAT (1H1,' Printing RSUM for ',I4,' THRU ',I4,/)
      DO 20 L = 1,NRA
C Decide how to to print labels -K < ICYCLE print short else long
      IF (K .GE. ICYCLE) GOTO 30
      WRITE(*,900) L,(B(L,J),J = JA,JB),(B(L,M),M = 304,310)
      GOTO 20
C Print labelled sequence
   30 WRITE (*,920) L,(B(L,J),J = JA,JB),(B(L,M),M = 304,320)
   20 CONTINUE
  900 FORMAT (1H ,I4,2X,10I5,3X,7A1)
  920 FORMAT (1H ,I4,2X,10I5,3X,7A1,4X,7A1,2X,3I5)
   10 CONTINUE
      RETURN
      END
C
C
C Symmetrical sort on sums of R/C in array B where ARROUT
C is stored. Cell name in 320, row sum in 318, col sum in 319.
C Copy column start at IX to 318-320
      SUBROUTINE SORT(B,IX,NRT)
      INTEGER B
      COMMON MRA,MCA
      DIMENSION B(MRA,MCA)
C Move data to last 3 cols
      DO 200 J = 1,NRT
      B(J,320)  = J
      B(J,318) = B(J,IX)
  200 B(J,319) = B(J,(IX + 1))
C Bubble sort row descending size
      ILAST = NRT
  210 DO 220 M = 1,ILAST-1
      IF ( B(M,318) .LT. B(M + 1,318)) THEN
      NX = M + 1
      ITR  = B(M,318)
      ITC  = B(M,319)
      ITN  = B(M,320)
      B(M,318) = B(NX,318)
      B(M,319) = B(NX,319)
      B(M,320) = B(NX,320)
      B(NX,318) =  ITR
      B(NX,319) =  ITC
      B(NX,320) =  ITN
```

```fortran
      ENDIF
  220 CONTINUE
      ILAST = ILAST -1
      IF ( ILAST .GE. 2) GOTO 210
C Rows sorted into descending order, exclude ones
      DO 230 KL=1,NRT
      ILAST = NRT-KL+1
      IF ( B(ILAST,318) .GT. 1) GOTO 240
C No isolated rows
  230 CONTINUE
C Mark last neuron with row sum=1
  240 IEND=ILAST
C Check for empty list ILAST=1
      IF (IEND .LE. 1) GOTO 270
C Sort shortened list by col sums ascending
  250 DO 260 M=1,IEND-1
      ILAST=IEND-M+1
      IF (B(M,319) .GT. B(M+1,319) ) THEN
      NX=M+1
      ITR=B(M,318)
      ITC=B(M,319)
      ITN=B(M,320)
      B(M,320)=B(NX,320)
      B(M,319)=B(NX,319)
      B(M,318)=B(NX,318)
      B(NX,318)=ITR
      B(NX,319)=ITC
      B(NX,320)=ITN
      ENDIF
  260 CONTINUE
C Shorten bubble loop
      IEND=IEND-1
      IF ( IEND .GE. 2) GOTO 250
  270 RETURN
      END
C
C Subroutine to read LABELG.DAT and enter names
      SUBROUTINE NAMES (B,NRA)
      INTEGER B
      COMMON MRA,MCA
      DIMENSION B(MRA,MCA),LABEL(7)
C Read labels into B(N,311) THRU B(N,317) A1 format
      OPEN (18,FILE='labelg.dat')
      DO 10 J=1,NRA
      READ (18,1010,END =30) NO,(LABEL(K),K=1,7)
 1010 FORMAT (1X,I3,7A1)
C Transfer to B array
C Find row location for sorted name
```

258

```
      DO 40 NF = 1,NRA
      NZ = 0
      IF (NO .EQ. B(NF,320 )) GOTO 60
   40 CONTINUE
C Check for mismatch
      IF (NZ .EQ. 0) GOTO 50
   60 NZ = NF
      DO 20 JJ = 311,317
      NN = JJ-310
      B(NO,JJ-7) = LABEL(NN)
   20 B(NZ,JJ) = LABEL (NN)
   10 CONTINUE
   30 CLOSE (18)
      RETURN
C Error exit name doesn't match
   50 WRITE (*,900) NO,NZ
  900 FORMAT (' Error in cell name ',I4, ' assigned to ',I4)
      GOTO 30
      END
C
C Subroutine to store bit pattern
      SUBROUTINE STOREBIT (AT,LOUT,NRA)
      COMMON MRA,MCA
      INTEGER AT
      DIMENSION AT(MRA,MCA),LOUT (MRA,MCA)
C  Do up to 320 rows in order
      DO 10 I = 1,NRA
C Check 20 sets of 16 columns each
      DO 20 J = 1,20
C Calculate start of each word
      K = 16*(J-1)
      MAPBIT =  0
C Compute location and store bits - left to right 1-16 natural order
      DO 30 L = 1,16
      IF ( AT(I,K + L) .NE. 0) THEN
      MAPEXP = 16-MOD(K + L,16)
         IF (MAPEXP .EQ. 16) THEN
            MAPBIT = MAPBIT + 1
            GOTO 30
         ENDIF
      MAPBIT  =  MAPBIT + 2**MAPEXP
      ENDIF
   30 CONTINUE
C Save the data
      LOUT (I,J) =  MAPBIT
   20 CONTINUE
   10 CONTINUE
      RETURN
```

```
      END
C
C
C Subroutine to read bit pattern stored as integers into array
      SUBROUTINE GETBIT (AT,LIN)
C To fill 320x320 array from 320x20 words
      COMMON MRA,MCA
      INTEGER AT
      DIMENSION AT(MRA,MCA),LIN(MRA,MCA)
C Fill rows in order. Clear output array
      DO 5 I=1,MRA
      DO 5 J=1,MCA
    5 AT(I,J)=0
C Expand word by word
      DO 10 I=1,MRA
      DO 20 J=1,20
        IWORK=LIN(I,J)
      DO 30 K=1,16
        IF ( IWORK .EQ. 0) GOTO 40
        ILOC= (J-1)*16+K
        IVAL= IWORK-2**(16-K)
        IF ( IVAL.GE. 0) THEN
           AT(I,ILOC)=1
           IWORK=IVAL
           GOTO 30
        ENDIF
   30 CONTINUE
C Handle empty word
   40 CONTINUE
   20 CONTINUE
   10 CONTINUE
      RETURN
      END
```

--- End of Program ---

## C.3. *EIGCRAY.FOR*

### C.3.a. Description

The purposes of this program are to exemplify the code of some matrix manipulations, to present the parameters required to compute for the eigenvalues of a real general nonsymmetric matrix, and to provide the code that calls the pertinent computing subroutine from an EISPACK* library. This is one of the programs used to compute for over 10,000 eigenvalues of the *C. elegans* synaptic connectivity matrix and its experimental matrix variations using the Cray Y-MP/832 [49]. Results of these computations are discussed with the issue of network stability in Chapter II.

*EISPACK is a systematized collection of subroutines which compute the eigenvalues and/or eigenvectors of 6 classes of matrices. It is a product of the National Activity to Test Software (NATS) Project. A published guide for use of these subroutines is: Smith, B. T., Boyle, J. M., Dongarra, J. J., Garbow, B. S., Ikebe, Y., Klema, V. C., and Moler, C. B. Matrix Eigensystem Routines - EISPACK Guide. *In* "Lecture Notes in Computer Science (Volume 6)" (G. Goos and J. Hartmanis, Eds.), pp. 1-551, Springer-Verlag, New York, ed. 2, 1976.

### C.3.b. Directions

After setting the desired parameters within the FORTRAN code and/or modifying the code to produce an experimental matrix, a batch file is made and is used to submit this program as a job for the Cray, both for compilation and execution. The input matrix is Pre, Post, Den in I3,I3,I4 format (e.g., PAIRSG.DAT), and like EIGCRAY.FOR, is "fetched" by Cray through the batch file during job execution.

### C.3.c. Source Code

```
C  EIGCRAY.FOR
C
C  Code revised by Theodore B. Achacoso, M.D. from EISPACK,
C  used to compute eigenvalues of a real general matrix.
C
      PROGRAM EIGCRAY
C
      WRITE(*,*)' '
      WRITE(*,*)'EIGCRAY Program: used to compute eigenvalues'
      WRITE(*,*)'of a real general nonsymmetric matrix.'
      WRITE(*,*)'Randomly signs negative (-)'
      WRITE(*,*)'50% of about 90,000 matrix elements'
C
C NM is the reserved matrix dimension;
C NM minimum is 306 for PAIRSG.DAT.
C N is the actual matrix dimension used in computation;
C N is 302 for 302 neurons.
C DIAG is the matrix diagonal.
C APERCENT is % of negative matrix elements; signed randomly.
C
C *CHANGE NM, N, DIAGONAL, & APERCENT PARAMETERS HERE*
      NM = 310
      N = 302
      DIAG = 0.0
      APERCENT = 0.5
C  ********************************************************************
C
      CALL TEST02(NM,N,DIAG,APERCENT)
C
      STOP
      END
C
```

```
        SUBROUTINE TEST02(NM,N,DIAG,APERCENT)
C
C  Matrix A is nonsymmetric:
C
C
        DIMENSION A(NM,NM)
        DIMENSION FV1(NM)
        DIMENSION IV1(NM)
        DIMENSION WI(NM)
        DIMENSION WR(NM)
        DIMENSION Z(NM,NM)
C
C  Set all matrix values to zero initially:
C
        DO 7 J = 1,N
        DO 5 K = 1,N
        A(J,K) = 0.0
5       CONTINUE
7       CONTINUE
C
C  Set values of the matrix:
C
        I = 1
9       READ(5,11,END = 13)J,K,NENS
11      FORMAT(I3,I3,I4)
        A(J,K) = NENS
        I = I + 1
        GOTO 9
13      NFILL = I-1
C
        WRITE(*,*)' '
        WRITE(*,'(I6)')NFILL
        WRITE(*,*)' '
C
C  Randomly sign a percentage of matrix elements:
C
        DO 16 J = 1,N
        DO 14 K = 1,N
        IF(RANF().LE.APERCENT) THEN
        A(J,K) = -A(J,K)
        ENDIF
14      CONTINUE
16      CONTINUE
C
C  Set matrix diagonal to DIAG:
C
        DO 18 L = 1,N
        A(L,L) = DIAG
```

```fortran
18      CONTINUE
C
C   Find row sums:
        ADIAG = 0.0
        SDIAG = 0.0
        DO 21 J = 1,N
        DO 20 K = 1,N
        ADIAG = ADIAG + A(J,K)
        IF(K.EQ.N) THEN
        SDIAG = SDIAG + ADIAG
        WRITE(*,'(1F8.1)')ADIAG
        ENDIF
20      CONTINUE
        ADIAG = 0.0
21      CONTINUE
        WRITE(*,*)' '
        WRITE(*,'(1F8.1)')SDIAG
        WRITE(*,*)' '
C
C   Get eigenvalues only. This is the subroutine call to EISPACK:
C
        MATZ = 0
        CALL RG(NM,N,A,WR,WI,MATZ,Z,IV1,FV1,IERR)
C
C   Print:
C
        WRITE(*,*)' '
        WRITE(*,*)'EIGCRAY, which computes the eigenvalues'
        WRITE(*,*)'of a real nonsymmetric matrix, 50% negative.'
        WRITE(*,*)' '
        WRITE(*,*)'Error flag = ',IERR
        WRITE(*,*)' '
        WRITE(*,*)'Eigenvalues:'
        WRITE(*,*)'     Real part          Imaginary part'
        WRITE(*,*)' '
        NPOSEIG = 0
        DO 25 I = 1,N
          WRITE(*,'(I6,1X,2G14.6)')I, WR(I), WI(I)
C   Count the number of eigenvalues with positive real parts
        IF(WR(I).GT.0.0) NPOSEIG = NPOSEIG + 1
25      CONTINUE
        WRITE(*,*)' '
        WRITE(*,'(I6)')NPOSEIG
        WRITE(*,*)' '
        RETURN
        END
```

--- End of Program ---

Chapter V

## Bibliography and Annotations

*And so these men of Indostan*
*Disputed loud and long,*
*Each in his own opinion*
*Exceeding stiff and strong,*
*Though each was partly in the right,*
*And all were in the wrong!*
-The Blind Men and the Elephant, J.G. Saxe

### A. Books We Used and Why

Poe writes "While I pondered weak and weary, Over many a quaint and curious Volume of forgotten lore;...," thus reporting upon a more or less solitary activity in scholasticism. Our effort is not science in the sense of learning new facts about an objective reality, but an exercise in automated pondering. If the emphasis is upon pondering over volumes, the task is both vastly simpler and impossibly more complex than in Poe's century. Automated indexing and teleprocessed bibliography make bibliographies easy to make in order to fulfill the evidential form for scholarship. Reading them and collating the attendant ideas and discoveries into a coherent thesis, however, is virtually impossible, and makes us amateur scholars. A contemporary automated bibliography for the ideas in this volume will easily exceed the volume, and would likely be unusable. In all our interests, we encounter the physical impossibility of finding, citing, and crediting all references that deal with the multiple fields of study that converge to make a consideration of *C. elegans'* "wormness" an exciting adventure. In reporting only those sources over which we pored, we do injustice, no doubt, to many whose work is germane, essential, or even contrary to the grist of premises we use. To them we sincerely apologize, but without a sense of guilt.

What we have tried to do is to provide a start for anyone who also happens to stumble upon the idea that computer-assisted speculation forms an enterprise not unlike logic-assisted theology that defined the period of human intellectual evolution commonly known as the Scholastic Period. New knowledge floods us again, and for our Irish scholar, we find the digital computer and the algorithm to look at the oldest of medical sciences, anatomy.

We hoped to contribute "un petit, d'un petit" to the flood as a scheme for reassembly of analyzed concepts, to give a sense of objective reality to subjectively comprehended phenomena like individuality, or in our case, any awareness that accompanies wormhood. The computer, as others have now said in more concise or trenchant form, provides an intellectual watershed dealing with the revealed truths of the laboratory and the understanding of those truths by conceptual reassembly on a scale the unassisted nervous system seems unable to absorb. That computers, in turn, may be flooded is symptomatically revealed in the tendency for automated bibliographic search to set a first horizon at about 5 years, with the attendant prejudice that older findings and ideas are all incorrect or outmoded. It is not, therefore, entirely with perversity that most of the references cited here are old.

The references we cite are tools which we found necessary to seed the speculation. Speculation, it seems, is all the better when ignorance of detail removes barriers to imagination. Different tools will yield different perspectives and behavior. In the enterprise of this book, one seeks only to create by analogical intellectual process in others the pleasure that comes from the personal, private experience we derive from "understanding" something. Unlike the Scholastic scholar, we thus seek not the truth but a temporary satiety of our curiosity about being. The references listed are the distorted lens through which we have looked.

## B. References to *C. elegans* Anatomy

Of course, the first set of references must be the factual information about the nematode. Since none of our ventures has taken us into the laboratory, we depend entirely upon the reported results in literature. Again, only a small summary list of sources in included.

1. Albertson, D. G., and Thomson, J. N. The Pharynx of *Caenorhabditis elegans*. *Phil. Trans. R. Soc. Lond.* **275**(B), 299-325 (1976).

2. Kenyon, C. The Nematode *Caenorhabditis elegans*. *Science* **240**, 1448-1453 (1988).

3. Ward, S., Thomson, J. N., White, J. G., and Brenner, S. Electron Microscopical Reconstruction of the Anterior Sensory Anatomy of the Nematode *Caenorhabditis elegans*. *J. Comp. Neur.* **160**, 313-338 (1978).

4. Ware, R. W., Clark, D., Crossland, K., and Russel, R. L. The Nerve Ring of the Nematode *Caenorhabditis elegans*. *J. Comp. Neur.* **162**, 71-110 (1975).

5. White, J.G. Neuronal Connectivity in *C. elegans*. *Trends Neurosci.* **June**, 277-283 (1985).

6. White, J. G., Southgate, E., Thomson, J. N., and Brenner, S. The Structure of the Ventral Nerve Cord of *Caenorhabditis elegans*. *Phil. Trans. R. Soc. Lond.* **275**(B), 28-348 (1976).

Mostly earlier work, entries [1] to [6] provide details on the anatomy and neuronal connectivity of *C. elegans*. They complete or augment information provided in pertinent sections of the main references, [11] and [12].

7. Roberts, L. The Worm Project. *Science* **248**, 1310-1313 (1990).

Part of the history of *C. elegans* as an organism for research in Chapter I was summarized from this "research news" article.

8.  Sulston J. E., and Horvitz, H. R. Post-embryonic cell lineages of the nematode *Caenorhabditis elegans*. *Dev. Biol.* **56**, 110-156 (1977).

    Source of the *C. elegans* line drawings in Figure 1.1.

9.  Sulston, J. E., Schierenberg, E., White, J. G., and Thomson, J. N. The embryonic cell lineage of the nematode *Caenorhabditis elegans*. *Dev. Biol.* **100**, 64-119 (1983).

    Source of Figure 1.4, depicting the cell lineage of embryonic key blast cell "D".

10. Warwick, N. "The Biology of Free Living Nematodes." Clarendon Press, Oxford, 1975.

    Contains effective detail for behavioral modeling, particularly locomotion, and practical discussion of culture methods. Surveys taxonomy and characterizes species.

11. White, J. G., Southgate, E., Thomson, J. N., and Brenner, S. The Structure of the Nervous System of the Nematode *Caenorhabditis elegans*. *Phil. Trans. R. Soc. Lond.* **314(B 1165)**, 1-340 (1986).

    A main reference, especially for neuronal connections (synapses and gap junctions), and for neuron functional type and laterality.

12. Wood, W. B., *et al*. *In* "The Nematode *Caenorhabditis elegans* (Monograph 17)" (W. B. Wood and Community of *C. elegans* Researchers, Eds.), pp. 1-489. Cold Spring Harbor Laboratory Press, Cold Spring Harbor, NY, 1988.

    A main reference, especially for neuron classes, parts list and embryology, and general motor circuitry. The anatomy and part of the history of *C. elegans* as an organism for research in Chapter I were summarized from this book.

13. Zuckerman, B. M. *et al*. *In* "Nematodes as Biological Models (Volume 1. 'Behavior and Developmental Models' and Volume 2. 'Aging and Other Model Systems')" (B. M. Zuckerman, Ed.), vol. 1 pp. 1-312 & vol. 2 pp. 1-306. Academic Press, New York, 1980.

    Allow useful comparison of the *C. elegans* neuromotor circuit with that of *Ascaris*.

## C. Mathematical and Computational Tools

14. Achacoso, T. B., Fernandez, V., Nguyen, D. C., and Yamamoto, W. S. Computer Representation of the Synaptic Connectivity of *Caenorhabditis elegans*. *In*

"Proceedings of the Thirteenth Annual Symposium on Computer Applications in Medical Care" (L. C. Kingsland III, Ed.), pp. 330-334. IEEE Computer Society Press, Washington, DC, 1989.

15. Aleksander, I. *et al. In* "Neural Computing Architectures" (I. Aleksander, Ed.), pp. 1-401, M.I.T. Press, Cambridge, MA, 1989.

Uneven review of neural networks from a more or less European perspective. Contains chapter precis of Rumelhart and McClelland on PDP machines.

16. An der Heiden, U. Structures of Excitation and Inhibition. *In* "Lecture Notes in Biomathematics (Volume 21. 'Theoretical Approaches to Complex Systems')" (R. Heim and G. Palm, Eds.), pp. 75-88, Springer Verlag, New York, 1978.

This discussion may not be unique, but has the virtue of terseness and generality. The latter part of the paper dealing with existence and uniqueness are not essential to our conceptualization of a framework for the modeling problem.

17. Baccala, L. A., Nicolelis, M. A. L., Yu, C., and Oshiro, M. Structural Analysis of Neural Circuits Using the Theory of Directed Graphs. *Comput. Biomed. Res.* **24**, 7-28 (1991).

Demonstration of the use of directed graph algorithms and matrices to visualize properties of complicated (i.e., many parts) connected systems in particular neural systems. Contrasts real systems with similar ones generated by random process.

18. Box, G., and Jenkins, G. M. "Time Series Analysis: Forecasting and Control." Holden Day, San Francisco, 1970.

Fundamental text for us on discretized time series analysis. Develops a unified insight into integrated auto-regressive moving-average (ARIMA) processes. Provides time-based perspective for the time-active view of neurons as against the "logical" view. Special emphasis on invertibility of forms is exceptionally useful.

19. Bracewell, R. N. Numerical Transforms. *Science* **248**, 697-704, (1990).

A brief and effective introduction into the subject of convolution transforms, summarizing a list of the most commonly used ones. It serves as a suitable quick entree into more detailed books like Elliott and Rao [21], and the rather forbiddingly large arena of publications in the engineering and mathematical literature.

20. Chen, W. K. "Theory of Nets." John Wiley and Sons, New York, 1990.

Relatively new treatment of nets from perspective of undirected graphs. Discussion of triple operations come from here.

21. Elliott, D. F., and Rao, K. R. "Fast Transforms: Algorithms, Analyses, Applications." Academic Press, New York, 1982.

A comprehensive discussion of discrete transforms and their implementation. Covers transforms unfamiliar in neuroanatomy. Also pursues a discussion of generalized transforms and families of discrete transforms which may be relevant to the discretized view of neuronal activity.

22a. Goldberg, S. "Introduction to Difference Equations." John Wiley and Sons, New York, 1961. [reprinted unabridged and slightly corrected by Dover Publications, Inc., New York, 1986]

22b. Mickens, R. "Difference Equations." Van Nostrand Reinhold Co., New York, 1987.

Difference equations form a structure naturally adapted to represent systems characterized by discrete time or spatial variables. In digital computation, continuous mathematical forms are translated into discretized methods. For neural systems, it is as well to start with the discretized formulation. These texts, one older and the other more recent, form an introduction into the extent and variety of tools.

23a. Hastings, H. M. The May-Wigner Stability Theorem. *J. theor. Biol.* **97**, 155-166 (1982).

23b. Hastings, H. M. The May-Wigner Stability Theorem for Connected Matrices. *Bull. (New Ser.) Am. Math. Soc.* **7**, 387-388 (1982).

23c. Hastings, H. M. Stability of Large Systems. *BioSystems* **17**, 171-177 (1984).

23d. McMurtrie, R. E. Determinants of Stability of Large Randomly Connected Systems. *J. theor. Biol.* **50**, 1-11 (1975).

The preceding four references and reference [27] sample model considerations on stability and connectivity, which form a theme in Chapter II. While these references are quantitatively minimal, these inclusions provide the theorems on connectivity and the motivation for stability arguments. McMurtrie also addresses closely the problem of submatrix organization and coupling coefficients.

24. Kaufmann, A. "Graphs, Dynamic Programming, and Finite Games." Academic Press, New York, 1967.

A basic, if dated, text on the mathematical treatment of graphs. Algorithms on transitive closure and simple paths were taken from here. Difficult format to read, but useful after one becomes familiar with it.

25. MacGregor, Ronald J. "Neural and Brain Modeling." Academic Press, San Diego, CA, 1987.

In addition to a comprehensive review of the subject of neuronal and neuronal network models, the approach is biased in the direction of physiological and biochemical verisimilitude rather than on theoretical elegance. Both program and perspective are thus in sympathy with the orientation pursued in this text: to study an existing nervous system rather than to produce behavior of a predetermined type by using algorithms on networks. The codes supplied are implicitly difference equations, particularly in the case of the "point" process neuron. Although large networks may be simulated using code directly adopted from this text, major problems will be found in deciding upon a proper display of output and inputs, and most specifically, in determining what behavior has been simulated.

26. Marmarelis, P. Z., and Marmarelis, V. Z. "Analysis of Physiological Systems, The White Noise Approach." Plenum Press, New York, 1978.

While the principal focus of this work appears to be the provision of a rigorous method for system analysis on experimental data, particularly those occurring in the nervous system, its use of the integral equation description of systems by convolution kernels makes it complementary to the Heiden [16] approach. The methods depend conceptually upon white noise signals, which provide the necessary simplifications for deriving methods for defining mechanism in firing patterns in networks. Starting with the Volterra expansion, the conceptual development to the Wiener-Lee orthogonalization, and finally the Lee-Schetzen demonstration of relationships between correlation properties of signals and the convolution kernels is made. Analyses on laboratory data are presented.

27. May, R. M. "Stability and Complexity in Model Ecosystems." Princeton Univ. Press, Princeton, NJ, ed. 2, 1974.

Extensive and readable review of mathematical problems in the representation of ecosystems. This and related sources form the basis for considerations of connectivity and stability as applied to *C. elegans*. The best summary of concepts from ecology that seems useful in stability-connectivity problems.

28. Milgram, M., and Atlan, H. Probabilistic Automata as a Model for Epigenesis of Cellular Networks. *J. theor Biol.* **103**, 523-547 (1983).

29. Panter, Philip F. "Modulation, Noise, and Spectral Analysis." McGraw-Hill Book Co., New York, 1963.

In 30 years, the literature in this subject area has burgeoned incredibly, particularly in the growth of pulse-modulated signals. Many newer books can provide discussion of relevant, simpler, and more exact ideas in the digitization of angle-modulated signals. This reference is cited only because it is the one with which we have greatest familiarity, found to be useful at a nonexpert level, with painstaking attention to the expansion of some formulations.

30. Stein, R. B., Leung, K. V., Mangeron, D., Oguztoreli, M. N. Improved Neuronal Models for Studying Neural Networks. *Kybernetik* **15**, pp. 1-9, 1974.

Modifies the J.D. Cowan representation for neuron model. Companion paper to the small network discussion [31] which included daisy petal and 3-neuron cycles. Like the Cowan model, output focuses upon firing rate averages and loses the progression in time feature of neural processing.

31. Stein, R. B., Leung, K. V., Mangeron, D., Oguztoreli, M. N. Properties of Small Neural Networks. *Kybernetik* **14**, pp. 223-230, 1974

    Shows model solutions for neurons connected in daisy petal and 3-cycles. Primary concerns are stability and oscillations.

32. Wilkinson, J. H. "The Algebraic Eigenvalue Problem." Clarendon Press, Oxford, 1965.

    Comprehensive mathematical treatise on eigenvalues. Source of Gerschgorin Theorem used in Chapter II.

## D. Mathematical Models

33. Caianello, E. R., DeLica, A., and Ricciardi, L. M. Reverberations and Control of Neural Networks. *Kybernetik* **4**, pp. 10-18, 1967.

34. Feldman, J. L., and Cowan, J. D. Large Scale Activity in Neural Nets: I and II. *Biol. Cybernetics* **17**, pp. 29-51, 1975.

    The models emphasize the representation of inhibitory neurons in network models, along with the concepts leading to hysteresis (i.e., multivalued functions). They deal extensively with probabilistic firing patterns in neuron systems. They are cited here largely because of their relationship to the Heiden [16] perspective.

35. Hartline, H. K., and Ratliffe, F. Spatial Summation of Inhibitory Influences in the Eye of Limulus and the Mutual Interaction of Receptor Units. *J. Gen Physiol.* **41**, pp. 1049-1066, 1958. See also: *J. Gen Physiol.* **40**, p. 357, 1957.

    Describes a linear equation model for summation of inhibition on neurons in pairs resembling the daisy petals and three-way nets. Discussed by An der Heiden [16, Equation 4, p. 1065] is the generalization for the model, with arithmetic addition and no time lags as the principal features.

36. McCulloch, W. S., and Pitts, W. F. A Logical Calculus of the Ideas Immanent in Nervous Activity. *Bull. Math. Biophys.* **5**, pp. 115-137, 1943.

37. Shannon, C. E. "Automata Studies." Princeton Univ. Press, Princeton, NJ, 1956.

270

In particular, see chapter by Kleene on logical majority organs (Kleene, S.C. Representation of Events in Nerve Nets and Finite Automata, pp. 3-41), and discussion by Von Neuman proposing neural nets as transition probability machines, i.e., finite state determinstic machines.

38. Wilson, H. R. and Cowan, J. D. Excitatory and Inhibitory Interactions in Localized Populations of Model Neurons. *Biophysical J.* **12**, pp. 1-24, 1972.


## E. Perspectives

The following references stimulate and review perspectives on the functioning of nervous systems and neurons:

39. Achacoso, T. B., and Yamamoto, W. S. Artificial Ethology and Computational Neuroethology: A Scientific Discipline and Its Subset by Sharpening and Extending the Definition of Artificial Intelligence. *Pers. Biol. Med.* **33**, 379-389 (1990).

40. Adrian, E. D. "The 1946 Waynflete Lecture: The Physical Background of Perception." Oxford Univ. Press, 1947.

41. Bishop, G. Natural History of the Nerve Impulse. *Physiol. Rev.* **36**, pp. 376-399, 1956.

42. Cherniak, C. The Bounded Brain: Toward Quantitative Neuroanatomy. *J. Cognitive Neurosci.* **2**, 58-68 (1990).

   In retrospect, a haunting prevision of the paradigm shift in neurophysiology, from electrical impulse networks to computation by chemical soups.

43. Cherniak, C. Undebuggability and Cognitive Science. *Communications of the ACM* **31**, 402-412 (1988).

44. Eccles, J. C. "The Physiology of Nerve Cells." Johns Hopkins Univ. Press, Baltimore, MD, 1957.

   Perhaps one should no longer cite this, since many and more modern paradigm monographs on the physiology of the nerve cell are available. Furthermore, this one precedes the contemporary neurochemical perspective on neurons, and probably talks too much about electrical characterization of excitation and the membrane. It provides, however, a historical perspective.

45. Householder, A. S. Neural Structure in Perception and Response. *Psychological Reviews* **54**, pp. 169-176, 1947.

Early discussion of time-dependent subnetworks with discrete sequential signalling for computation of discrimination, transposition, constancy and other psychological processes. Leads to formulations with difference equations. Referenced by Goldberg [22a]. See also: Householder, A. S. and Landahl, H. D. "Mathematical Biophysics of the Central Nervous System." The Principia Press, Bloomington, IN, 1945.

46. Purves, D. "Body and Brain." Harvard Univ. Press, Cambridge, MA, 1988.

Develops the trophic theory of nervous system development, and argues for this as a postembryonic mechanism that allows for the individuation of nervous systems despite being within the same species. It references *C. elegans* as an example of one organism without such mechanism.

47. Walter, W. G. "The Living Brain." W. W. Norton, Inc., London, ed. 2, 1963.

Written from the perspective of the technology of the mid-century EEG and CW electronic dominance. From an electroencephalographer's perspective, this book emphasizes speculation on the oscillatory nature of electrical activity, and the coherent activity in the brain. In particular, the perspective is from a frequency-time and control paradigm for conceptual and physical models with behavior.

48. Young, D. "Nerve Cells and Animal Behavior." Cambridge Univ. Press, Cambridge, 1989.

Reviews neuronal properties, relates circuits to elementary intrinsic responses, and has a good collection of small nets and what they do behaviorally. Strong on the auditory system.

49. Computations in the Cray Y-MP/832 were supported in part by grant number DMB90001P from the Pittsburgh Supercomputing Center, Pittsburgh, PA, USA.

# INDEX

## THE BLIND MEN AND THE ELEPHANT

### (A HINDOO FABLE)

It was six men of Indostan
  To learning much inclined,
Who went to see the Elephant
  (Though all of them were blind),
That each by observation
  Might satisfy his mind.


The *First* approached the Elephant,
  And happening to fall
Against his broad and sturdy side,
  At once began to bawl:
'God bless me! but the Elephant
  Is very like a wall!'


The *Second*, feeling at the tusk,
  Cried, 'Ho! what have we here
So very round and smooth and sharp?
  To me 'tis mighty clear
This wonder of an Elephant
  Is very like a spear!'


The *Third* approached the animal,
  And happening to take
The squirming trunk within his hands,
  Thus boldly up and spake:
'I see,' quoth he, 'the Elephant
  Is very like a snake!'


The *Fourth* reached out an eager hand,
  And felt about the knee.
'What most this wondrous beast is like
  Is mighty plain,' quoth he;
''Tis clear enough the elephant
  Is very like a tree!'

The *Fifth* who chanced to touch the ear,
   Said: 'E'en the blindest man
Can tell what this resembles most;
   Deny the fact who can,
This marvel of an Elephant
   Is very like a fan!'


The *Sixth* no sooner had begun
   About the beast to grope,
Than, seizing on the swinging tail
   That fell within his scope,
'I see,' quoth he, 'The Elephant
   Is very like a rope!'


And so these men of Indostan
   Disputed loud and long,
Each in his own opinion
   Exceeding stiff and strong,
Though each was partly in the right,
   And all were in the wrong!


*Moral*


So oft in theologic wars,
   The disputants, I ween,
Rail on in utter ignorance
   Of what the others mean,
*And prate about an Elephant*
   *Not one of them has seen!*


- John Godfrey Saxe